Biodiversity of Aquatic Resources

Dr. Mamta Rawat is currently working as a Scientist and Head of Water and Health Programs of Ecology and Rural Development Society, Jodhpur, Rajasthan, with a theme of "Water Solution to Thar Desert Ecosystem via Community Participation". She is actively engaged in research activities on various health and environmental issues of Thar Desert of Rajasthan since 1999. Apart from this she has keen interest in the area of groundwater contaminations, eco-toxicology, water pollution, ecological engineering and wastewater treatment. She was awarded with Young Scientist Fellowship in year 2005 with a project from Department of Science and Technology, Govt. of India. She has successfully completed other research projects and research outcome are recognized at national and international level. Dr. Rawat is associated with various professional and scientific bodies. Presently she is a member of editorial board in the international journal named *"Indian Journal of Environment and Fisheries"* and also referees for few Academic Press journals related with the same issues. She has 35 research papers in peer reviewed reputed national and international journals, conferences, seminars and symposiums. Recently Dr. Rawat has published a book entitled "Microbial Community and Water borne Diseases in Thar Desert, India: A health and Sanitation Survey with respect to water quality of Thar Desert, Rajasthan, India" from LAP-Lambert Academic Publishing Saarbrüken, Deutschland, Germany.

Dr. Sumit Dookia, is working in the Thar Desert on endangered fauna since 1999. His doctoral work was on the Ecology of Indian Gazelle/Chinkara (*Gazella bennettii*) from this area, and awarded University Research Scholarship from J. N. V. University, Jodhpur, Rajasthan, for the same. He also worked with Desert Regional Station, Zoological Survey of India, Jodhpur on Faunal diversity of Thar Desert. During his doctoral work, he comes close to the local communities and starts working with them for the conservation of desert wildlife. He also worked in Satpura-Maikal landscpae, particularly in Kanha, Pench and Satpura Tiger Reserve in Madhya Pradesh with Project Tiger Directorate, New Delhi and Wildlife Institute of India, Dehradun and with Bombay Natural History Society, Mumbai, for Bird-Hazard studies for Indain Air Force. Also worked as a Scientist at Ecology and Rural Development Society with a theme of "Conservation through Community Participation" and supported by The Ruffords Small Grants Foundation, UK. He authored more then 30 reserch papers, book chapters, symposia and conference papers of national and international repute. Currently Dr. Dookia is involved in teaching and research on conservation issues of biodiversity and natural resource management.

Biodiversity of Aquatic Resources

— Editors —

Mamta Rawat
Scientist-in-Charge,
Ecology & Rural Dev. Society
Jodhpur – 342 005
Rajasthan

Sumit Dookia
Assistant Professor
University School of Env. Mgt.
GGS Indraprastha University
New Delhi – 110 075

2012
DAYA PUBLISHING HOUSE®
New Delhi - 110 002

© 2012 EDITORS
ISBN 9788170359739

Published by : **Daya Publishing House®**
A Division of
Astral International Pvt. Ltd.
– ISO 9001:2008 Certified Company
4760-61/23, Ansari Road, Darya Ganj,
New Delhi - 110 002
Phone: 23245578, 23244987
Fax: (011) 23260116
e-mail : dayabooks@vsnl.com
website : www.dayabooks.com

Laser Typesetting : **Classic Computer Services**
Delhi - 110 035

Printed at : **Chawla Offset Printers**
Delhi - 110 052

PRINTED IN INDIA

Preface

This book "*Biodiversity of Aquatic Resources*" is an anthology of recent original research in the field of aquatic ecology and the applied aspects across the country. It is an assemblage of up-to-date information of rapid advances and developments taking place in the field of aquatic ecology. The book is a sole omnibus of 20 chapters which discusses studies on different aspects like fisheries, diatoms, vermicomposting, bacteriology, macrophytes, rotifers, issues related to water like monitoring, water resources and stress, molassess, mangrove vegetation, avifauna and other aspects respectively contributed by eminent scientist and academicians active in the respective areas.

Broadly the book is divided into five sections; including various issues of aquatic regime like, aquatic ecology, conservation and management approaches, pollution, and water issues of water resources and miscellaneous. The publication also throws light on the challenges aquatic regime is facing in overall country, with the hope that it will not only be helpful to students and researchers but also to policy makers, environmental lawyers and rest. It may form the basis for initiating conservation measures for securing the long term future of aquatic biodiversity. We hope that the government will rise to the challenge and use this document to further make progress to conserve our biodiversity in the state.

With a view to compile the information available in the country on the various undiscussed facts of aquatic ecosystem, the formulated peer reviewed book thus serves as a reference tool for all concerned college students, researchers, scientists, policy makers, environmental lawyers and all concerned for the noble cause of aquatic biodiversity conservation who wish to gain a better understanding of our biological wealth, the issues and problem to it. This work would not have materialized if not been supported wholeheartedly by the contributors of this book from various parts of country. My special thanks and appreciation go to the scientists whose contribution has enriched this volume. I owe my thanks to Dr. Anita Shindey, Dr. D. K. Sinha, Dr. Jyoti Verma, Mamata Pandey, K. Khelchandra Singh, Dr. S.V.A. Chandrasekhar, Dr. S. Singh, Dr. Pawar Vijaykumar Bhikusing, Poonam Nomulwar, Dr. M. A. Khan, Dr. Pankaj Gupta, G. M. Narasimha Rao, Dr. Shalu Mittal, Dr. C. Sivaperuman, Dr. N. Shiddamallayya, Dr. Sarita Mehra, Dr. S. Srikantaswamy, Kh. Usha and Dr. P.C. Verma. I sincerely acknowledge their support and encouragement. Also I would like to thank Mr. Anil Mittal of Daya Publishing House, for bringing out this volume in time. Suggestions for further improvement of the book are most welcome.

Last but not least, we will always remain a debtor to my entire well wisher for their blessings without which this volume would not have come into existence.

Mamta Rawat
rawatscorner@gmail.com

Sumit Dookia
sumitdookia@gmail.com

Contents

Part II–Pollution: Monitoring and Management

Part III–Water Resources and Other Issues

Part IV–Miscellaneous Aspects

Sketch of Contributors

Chapter No.	Name	Official Address/E-mail
1	Anita Shindey	Department of Zoology, Osmania University College for Women, Koti, Hyderabad, A.P.
	N. Sree Ram Kumar	Department of Zoology, Nizam College, Basheerbagh, Hyderabad, A.P.
2	D. K. Sinha	Reader and Head, Department of Chemistry, K.G.K. (P.G.) College, Moradabad – 244 001, U.P. dkskgk@rediffmail.com
	Navneet Kumar	Assistant Professor, Department of Applied Science and Humanities, College of Engineering, Teerthankar Mahaveer University, Moradabad – 244 001, U.P. navkchem@gmail.com
3	Jyoti Verma	Department of Zoology, H.N.B. Garhwal University, Srinagar – 246 174, Uttarakhand diatombuster@gmail.com
	Prakash Nautiyal	
4	Mamata Pandey	School of Life Science, Sambalpur University, Jyotivihar – 768 019, Orissa pndy_mmt @ rediffmail.com
	T V Rao	tvrao1951@gmail.com
	S. K. Satapathy	RTRS, Baripada, Mayurbhanj, Orissa drsubratsatapathy@gmail.com

Chapter No.	Name	Official Address/E-mail
5	K. Khelchandra Singh	Ecology Laboratory, Department of Life Sciences, Manipur University, Canchipur, Imphal – 795 003, Manipur k_khel@yahoo.com
	B. Manihar Sharma	
	Khuraijam Usha	
6	S.V.A. Chandrasekhar	Freshwater Biological Research Station , Zoological Survey of India, Plot 366/1, Attapur (V), Hyderguda (P.O.), Hyderabad – 500 048, A.P.
7	S. Singh	Department of Fisheries, Indira Gandhi Agricultural University, Raipur – 492 006, Chhattisgarh sattupoonia@yahoo.com
	M.S. Chari	
	Om Prakash	
	H.K. Vardia	
8	Pawar Vijaykumar Bhikusing	Dnyanopasak College, Parbhani – 431 401, M.S. vbpawar08@rediffmail.com
9	Poonam Nomulwar	Depatment of Zoology and Fishery Science, N.E.S. Science College, Nanded, Maharashtra
	P.M. Patil	
10	M.A. Khan	Department of Geography, Govt. Lohia P. G. College, Churu – 331 001, Rajasthan
	Hemant Mangal	
11	Pankaj Gupta	Department of Chemistry, Basic Environmental Engineering and Disaster Management, Alwar Institute of Engineering and Technology, Alwar, Rajasthan pgupta1975@yahoo.com
12	G.M. Narasimha Rao	Department of Botany, Andhra University, Visakhapatnam – 530 003, Andhra Pradesh gmnrao_algae@hotmail.com
13	Shalu Mittal	Department of Zoology, B.S.A. (P.G.) College, Mathura, U.P. shalumittalgarg@gmail.com
	Sangeet Prabha	
	P.K. Agarwal	dragarwalpk@gmail.com

Chapter No.	Name	Official Address/E-mail
14	C. Sivaperuman	Zoological Survey of India, Andaman and Nicobar Regional Centre, Hodda, Port Blair – 744 102, A&N c_sivaperuman@yahoo.co.in
	E.A. Jayson	Division of Forest Ecology and Biodiversity Conservation, Kerala Forest Research Institute, Peechi, Thrissur – 680 653, Kerala
15	N. Shiddamallayya	National Ayurveda Dietetics Research Institute, Jayanagar, Bangalore, Karnataka snmathapati@gmail.com
	M. Pratima	Department of P.G. Studies and Research in Botany, Gulbarga University, Gulbarga, Karnataka
16	Sarita Mehra	Rajputana Society of Natural History, Kesar Bhawan, 16/747, P.No.90, B/d Saraswati Hospital, Ganeshnagar, Pahada, Udaipur – 313 001 , Rajasthan indiagreenmunia@yahoo.co.in spmehra@yahoo.com
	Satya Prakash Mehra	Laboratory of Biodiversity and Molecular Development, Department of Zoology, Maharshi Dayanand Saraswati University, Ajmer – 305 009 Rajasthan
	Krishan Kumar Sharma	kksmds@gmail.com
17	S. Srikantaswamy	Department of Studies in Environmental Science, University of Mysore, Manasagangotri, Mysore – 570 006, Karnataka
	B.K. Harish Kumara	
18	Kh. Usha	Ecology Laboratory, Department of Life Sciences, Manipur University, Canchipur, Imphal – 795 003, Manipur khuraijamusha36@gmail.com
	Manihar Sharma	
	K. Khelchandra Singh	
19	P.C. Verma	Scientific Officer (G), Environmental Studies Section Health Physics Division, BARC, Mumbai, Mh. pcverma@barc.gov.in
	A.G. Hegde	
	L.L. Sharma	College of Fisheries, Maharana Pratap University of Agriculture and Technology, Udaipur, Rajasthan
20	Mamta Rawat	Scientist in Charge, Ecology and Rural Development Society, Jodhpur – 342 005, Rajasthan rawatscorner@gmail.com

Part I

Ecology of Lentic and Lotic Environment

Biodiversity of Aquatic Reseources (2012)

Editors: **Mamta Rawat & Sumit Dookia**

Published by: **DAYA PUBLISHING HOUSE, NEW DELHI**

Pages **3-19**

Chapter 1

Ecological Studies of Holy Pond Papnesh, Bidar, Karnataka

N. Shiddamallayya[1]* *and M. Pratima*[2]

[1]*National Ayurveda Dietetics Research Institute, Jayanagar, Bangalore, Karnataka*
[2]*Depatment of P.G. Studies and Research in Botany, Gulbarga University, Gulbarga, Karnataka*

ABSTRACT

Ecological studies of holy pond Papnash with reference to chemistry and quality variations by domestic activities of pilgrims of Papnash temple and influx of run off from the surroundings has been studied over a period of two years (February-2002 to January-2004). In this period, dissolved oxygen, carbon dioxide, alkalinity, calcium, chloride, nitrite, phosphate, silicate, sulphate, total solids, organic matter and COD were within the stipulated range, where as the pH, total hardness, and BOD of water were exceeding the limits of ISI and WHO standards indicate the impact of domestic activities performed by the pilgrims. The statistical cluster analysis indicated pH, hardness, Ca, Cl, and alkalinity are the dominant factors and phosphate and COD are the key factors in regulation of aquatic environment in the holy pond.

Keywords: Ecology, Holy pond, Sewage water, Pollution.

---

* Corresponding Author E-mail: snmathapati@gmail.com

Introduction

The standard life style of human being is creating lot of problems by contributing pollutants to the water body. Freshwater bodies are in the verge of loss of potability due to improper management, misuse and lack of knowledge in community health. Some time blind faith also leads to many undesirable activities like taking bath in holy water bodies wash their sins and open up the gates of heaven to the soul to rest peacefully after death. Such beliefs are adding pollutants and lead to enrich micro flora in the water bodies. It affects on the quality of life in and around the water body.

Historical resume of literature indicates efforts to consolidate information pertaining to the interaction of various parameters of water bodies, from 19th century itself in different parts of the world. In recent years, the studies on standing water bodies have been carried out with different orientation for specific problems. Limonlogical investigations are aimed at the assessment of the physico-chemical and biological parameters of the water body. Of all the chemical parameters, nutrient elements play a significant role in determining the tropic status of the water body. It is known that these nutrients act as carriers of essential substances. A deficiency in the concentration of nutrients leads to the reduction of plankton production in natural waters.

In India, a considerable amount of information is available on limnology. Investigations on different limnological aspects of water bodies such as tanks, ponds and reservoirs were initiated about seven decades back. Many of the researchers have contributed towards specific aspects of freshwater studies. Hosmani and Bharati (1975) have worked on hydrobiology of freshwater ponds and lakes of Dharawad. Goel and Trivedy (1984) have studied the disposal practices in major cities of Indian freshwater bodies and resultant effects. Vyas and Nama (1991) have assessed some physico-chemical and biotic factors in Akharaj-ji-ka talab a pollution prone pond of Jodhpur. Ahmed and Singh (1993) have recorded the diurnal changes in the physico-chemical properties of water in a freshwater tank of Dholi, Bihar. Rao *et al.* (1993) have made an assessment on seasonal dynamics of physico-chemical factors in a tropical high altitude Ooty lake. Kumar (1994) have worked on the limnological aspects of freshwaters of tropical wetland of Santhal Praganas, Bihar. Mukhopadhyay (1996) has made the limnological investigations in lotic and lentic freshwater bodies in and around Darjeeling, West Bengal, Swarnalatha and Rao (1997) have showed the inter relationships of certain physico-chemical factors in Osmania University pond, Hyderabad. Katiyar and Belsare (1997) have investigated the rapid organic pollution indicators of three urban lakes for monitoring purposes in Bhopal. Patel and Sinha (1998) have documented the pollution load in the ponds of Burla area, near Hirakud Dam Orissa. Naganandini and Hosmani (1998) have reported the ecology of certain inland water body of Mysore district. Swarnalatha and Rao (1998) have studied the Banjara Lake with reference to physico-chemical characteristics and water pollution

Raghavendra and Hosmani (2002) have conducted the hydrobiological study of Mandakally Lake, a polluted water body at Mysore. Harsha and Malammanavar (2004) have assessed the environmental variables in Gopal Samy pond, Chitradurga.

Sachidanandamurthy and Yajurvedi (2004) have assessed monthly variations in physico-chemical factors of a Yennehole lake in Mysore city. Garode and Agarkar (2004) have evaluated the drinking water quality of Chandai reservoir in Chikhli, Buldana, Maharastra.

However, such reports are very limited on the water bodies of Bidar except Angadi *et al.* (2005) and Shiddamallayya and Pratima (2006, 2007a and 2007b). The selected work for the study is to note the changes in the water by blind belief, unhygienic activities, improper management and others are causing the change in the ecological conditions of the freshwater body of holy pond of Papnash Pond, Bidar.

Materials and Methods

The surface water samples collected once in a month from two sites by using wide mouthed clean Iodine treated Poly Vinyl Chloride container during 8.30 am to 10.00 am through out the study period (February–2002 to January–2004). Standard methods were employed for the estimation of various parameters such as atmospheric temperature, water temperature, pH, dissolved oxygen, free carbon dioxide (CO_2), alkalinity, hardness, calcium (Ca), magnesium (Mg), chloride (Cl), nitrite, phosphate, silicate, sulphate (SO_4), total solids, organic matter, biological oxygen demand (BOD) and chemical oxygen demand (BOD) (Trivedy and Goel, 1986; APHA, 1995; Aneja, 1996 and Gupta, 2001).

Pearson's correlation matrix and hierarchical cluster analysis of physico-chemical characters was carried out by using SPSS version 10.3 software to correlate relationship between physico-chemical factors and dominant and key factors of the water body.

Study Area

Bidar is the head quarter of the district lies at 17° 55′ North latitudes and 77° 32′ East longitudes, located at 551 meters above mean sea level. It presents semiarid climatic conditions with temperature varying from 9°C to 44°C. It has several places of interest for tourism. Such as Papnash temple, Narasinha zara, Gurunanaka zara and fort. The perennial holy pond situated near the Papanash temple at the western side of the city and covers an area of 0.25 square km the mean depth of pond is 2.1 m with a maximum depth of 4 m during monsoon and minimum being 0.75 m during summer season. The pond is under the cover of large population of *Ipomea fistula* Mart (Figure 1.1).

Results and Discussion

The physico-chemical factors varied with the seasons. During the first year of study in summer, pH, alkalinity, total solids and chemical oxygen demand, in monsoon, free carbon di-oxide and silicate and in winter, nitrite and BOD were recorded very high. The maximum concentration of magnesium, organic matter in summer; hardness, calcium, chloride, phosphate and sulphate in monsoon and water temperature and dissolved oxygen during winter season of the successive year (Table 1.1).

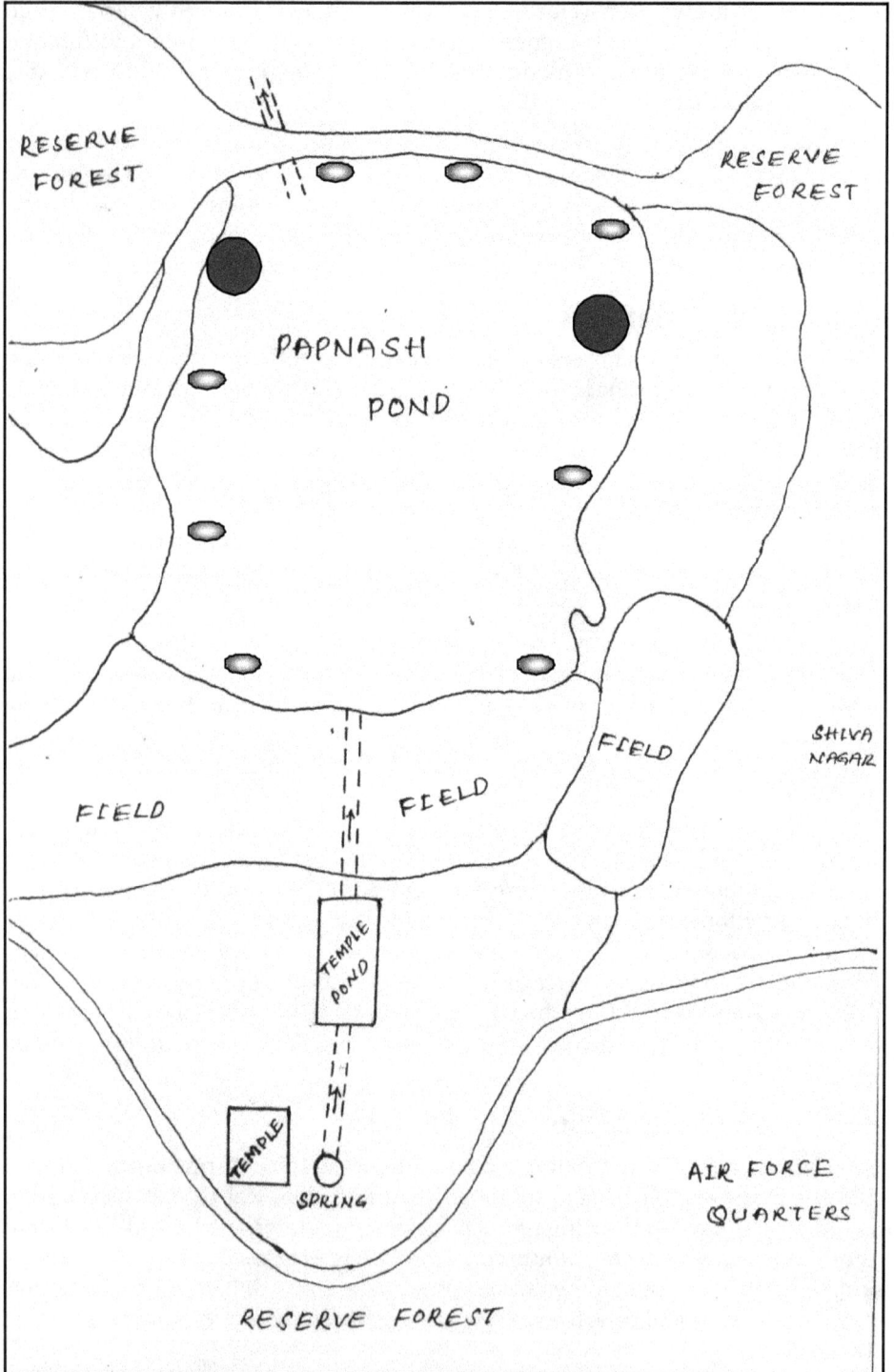

Figure 1.1: The Map of Papnash Pond

Table 1.1: Physico-chemical Factors of Papnash Pond, Bidar (February 2002 to January 2004)

Parameters	Summer (Feb. 2002 to May 2002)	Monsoon (Jun. 2002 to Sep. 2002)	Winter (Oct. 2002 to Jan. 2003)	Summer (Feb. 2003 to May 2003)	Monsoon (Jun. 2003 to Sep. 2003)	Winter (Oct. 2003 to Jan. 2004)
Atmospheric temperature	29–35	24.75–29.5	24–25	28–38	25.5–29.5	26–31.5
Water temperature	21–25.25	24.5–25.25	18.5–23.75	23.5–28	25.75–26.25	20.5–28.5
pH	7.75–8.55	7.2–7.95	7–7.85	7.6–8.2	7.05–7.45	7.05–7.43
Dissolved oxygen	4.13–5.925	1.825–3.645	1.215–2.435	1.22–3.645	4.255–6.28	4.86–6.28
Carbon di oxide	16.4–28.78	4.4–31.9	8.8–26.4	23.1–29.1	7.72–24.2	12.1–28.6
Alkalinity	155–232.5	110–210	115–212.5	197.5–225	130–125	112.5–125
Hardness	171–203	164–197	139–170	179–208	126–323	129–169
Calcium	42.1–46.5	49.3–56.1	43.3–57.7	28.1–48.9	37.3–87.8	34.5–55.7
Magnesium	14.1–21.25	9.5–18.05	6.35–15.15	14.85–26.55	8.05–25.35	6.35–13.4
Chloride	19.9–30.45	24.15–36.95	24.1–31.95	27.7–39.05	19.2–39.75	16.35–31.95
Nitrite	0.012–0.5035	0.0045–0.1405	0.0135–0.8125	0.0175–0.1425	0.002–0.2645	0.0295–0.375
Phosphate	0.0085–0.03555	0.047–0.0715	0.010–0.025	0.0385–0.102	0.255–0.465	0.14–0.22
Silicon	14.9–22.5	16.15–22.8	12.25–18.1	12.95–20	8.95–21	6.75–15.7
Sulphate	1.15–11.3	2.65–16.5	1.1–5.8	2.9–8.95	5.525–20.335	3.316–11.375
Total solids	54–208	67.5–172	80–147.5	100.5–143	43–172	50.2–85
Organic matter	0.14–2.285	0.865–3.735	0.755–2.59	1.07–6.7275	2–5.705	0.425–2.545
B O D	0.61–3.24	3.245–6.075	3.04–6.28	1.62–3.65	2.025–3.44	1.62–4.865
C O D	9.6–21	00–3.8	3.2–13.3	2.4–12.8	3.8–11	0.8–8.6

All the physico-chemical values are expressed in mgl^{-1} except pH, atmospheric temperature, and water temperature.

The highest pH in summer was due to low water levels and increased photosynthetic activity (Mali and Gajaria, 2004). Similarly dissolved oxygen during winter was attributed to low temperature (Kumar, 1994 and Harsha and Mallammanawar, 2004). The maximum contents of free CO_2 during monsoon was due to influx of water to the water body and increase in number of overgrazing micro organisms. Similar observation was also made by Shastri and Pendse (2001) in Dahikuta reservoir, Nasik. The highest content of calcium recorded in monsoon indicates that, the rain water carries calcium to the water body. Das (2002) have also recorded increase of calcium content during monsoon in Dahikuta reservoir, Nasik and reservoirs of Andhra Pradesh respectively.

The highest concentration of nitrite was recorded in winter (Bailey-Watts, 1986; Kirsten and Arnold, 1993; Rao *et al.*, 1993 and Sachidanandamurthy and Yajurvedi, 2004). Similarly phosphate during monsoon was due to runoff from surrounding agricultural fields (Prakasam and Johnson, 1992; Raghavendra and Hosmani, 2002; Sachidanandamurthy and Yajurvedi, 2004 and Mali and Gajaria, 2004). The highest silicate concentration during monsoon was due to the death and decomposition of diatoms in the water body. Similarly by Govindsamy *et al.* (2000) in coastal water biotopes of Coromandel coast. The maximum total solids concentration during summer was due to evaporation (Sachidanandamurthy and Yajurvedi, 2004) and BOD in winter (Bhat and Pathak, 1992).

Statistical Analysis

Pearson's Correlation Matrix

Every ecosystem has two important components as abiotic and biotic. Each has its own relationship with another. Pearson's correlation matrix (two tailed) used to record significant correlation and presented as year wise and both years together of the water body is presented in Tables 1.2–1.4.

In the present work it was noted that, the significant correlation of atmospheric temperature correlation with dissolved oxygen, pH and phosphate are due to availability of maximum carbonates and bicarbonates lead to increase in pH with an increase of temperature. Ahmed and Singh (1993) have also recorded a similar positive correlation between atmospheric temperature and dissolved oxygen in freshwater tank at Dholi (Bihar). The increases in phosphate, silica and COD content have shown positive relation with the increase of water temperature.

pH has shown a very close relationship with magnesium and atmospheric temperature, dissolved oxygen, hardness, phosphate and total solids. Similar relation of pH with magnesium was observed by Shastri and Pendse (2001) in Dahikhuta reservoir, Nasik; with dissolved oxygen by Ahmed and Singh (1993) in freshwater tank at Dholi (Bihar). The significant positive relation of dissolved oxygen with pH was recorded as observed by Simon (2002). Free carbon di-oxide has shown negative relationship with sulphate and nitrite. Similarly Vyas and Nama (1991) and Mukhopadhyay (1996) have recorded lowest carbon di-oxide uses with the increase of sulphate.

A close positive relation of alkalinity with hardness, magnesium, chloride, phosphate, silicate, total solids and organic matter was observed. The decomposition of organic matter enhances the alkalinity by liberating the carbon di-oxide, which on reacting with water forms bicarbonates. Decomposition of organic matter also increases the concentration of magnesium, hardness, chloride, phosphate silicate and total solids. Similar, relation between alkalinity and organic matter was observed by Sachidanandamurthy and Yajurvedi (2004) in Yennehole lake and alkalinity and hardness by Vijay Kumar and Ramesha (2002) in Jagat tank, Gulbarga.

The hardness of water has shown a close positive relation with pH, alkalinity, magnesium, calcium, chloride, total solids and organic matter. It is obviously evident that the increase of calcium, magnesium, total solids, and organic matter increases the hardness of water. The increase in chloride is due to washing of clothes and inflow from surrounding agricultural fields (Varma and Thakur, 1998; Kumar, 1994 and Patel and Sinha, 1998). Calcium has shown positive relation with hardness and chloride and negative relation with COD. The calcium was abundant in surface water along with magnesium are responsible for the hardness (Prasad and Manjula, 1980 and Das, 2002).

Hydrogen ion-concentration, hardness, alkalinity, phosphate, total solids and organic matter have shown positive relation with magnesium and BOD has shown negative correlation (Kumar *et al.*, 1994). Chloride concentration varied with the fluctuation of alkalinity, hardness, calcium, phosphate, total solids and organic matter. These findings are similar as observed by Trivedy *et al.* (1988) have noticed that the increase of alkalinity, hardness, and total solids with the increase of chloride concentration.

The nitrite content of water exhibited a positive correlation with the concentration of sulphate and BOD and a negative relation with free CO_2. Katiyar and Belsare (1997) have reported a similar positive relation of nitrite with phosphate. A significant positive relation of phosphate with temperature, pH, alkalinity, magnesium, chloride, silicate, total solids and organic matter were recorded; Katiyar and Blesare (1997) have also found a positive correlation between phosphate and nitrite in Mansarover lake, Bhopal.

Silica content has shown close relation with water temperature, alkalinity, phosphate, sulphate, total solids and organic matter. Swarnalatha and Rao (1997, 1998) have also noticed a positive relation of silica with alkalinity. Sulphate has shown a positive relation with nitrite and silicate and negative relation with free CO_2.

Total solids content has shown a very close relation with alkalinity, organic matter, hardness, magnesium, chloride, phosphate and silicate. Swarnalatha and Rao (1997, 1998) have also observed that, the total solids having positive relation with alkalinity. Similarly with organic matter by Trivedy *et al.* (1988) and Mishra *et al.* (1992).

Organic matter was increased with the increase of alkalinity, hardness, magnesium, chloride, phosphate, silicate and total solids. The increase of BOD with the increase of nitrite was noticed (Rao *et al.*, 1993). Negative relation of BOD with

Table 1.2: Pearson's Correlation Coefficients for Physico-chemical Characteristics of Papnash Pond, Bidar during February 2002 to January 2003

	A	B	C	D	E	F	G	H	I	J	K	L	M	N	O	P	Q	R
A	+1.000																	
B	+0.523	+1.000																
C	+0.473	+0.473	+1.000															
D	+0.750**	+0.235	+0.624*	+1.000														
E	+0.008	-0.018	+0.170	+0.205	+1.000													
F	+0.423	+0.358	+0.510	+0.321	-0.503	+1.000												
G	+0.306	+0.377	+0.753**	+0.355	+0.239	+0.216	+1.000											
H	-0.278	-0.278	-0.180	-0.462	+0.013	+0.076	+0.244	+1.000										
I	+0.189	+0.339	+0.657*	+0.361	-0.120	+0.426	+0.586*	-0.199	+1.000									
J	-0.069	+0.258	-0.172	-0.263	+0.266	+0.159	+0.036	+0.175	+0.173	+1.000								
K	-0.024	-0.264	-0.244	-0.246	-0.480	+0.005	-0.362	-0.168	-0.130	+0.670	+1.000							
L	+0.126	+0.718**	+0.007	-0.216	+0.271	-0.030	+0.506	+0.429	+0.071	+0.357	-0.384	+1.000						
M	+0.061	+0.600*	+0.274	+0.089	-0.242	+0.203	+0.520	-0.034	+0.487	-0.100	-0.114	+0.500	+1.000					
N	+0.029	+0.442	+0.089	-0.147	-0.608*	+0.361	+0.301	+0.208	+0.421	-0.040	-0.076	+0.356	+0.608*	+1.000				
O	+0.102	+0.295	+0.551	+0.080	+0.199	+0.255	+0.524	-0.093	+0.568	+0.562	+0.135	+0.216	+0.119	-0.012	+1.000			
P	-0.041	+0.256	-0.260	-0.544	-0.267	-0.144	+0.055	+0.253	+0.115	+0.372	+0.322	+0.441	+0.089	+0.513	+0.219	+1.000		
Q	-0.089	+0.092	-0.522	-0.464	+0.127	-0.266	-0.198	+0.401	-0.619*	+0.264	-0.221	+0.438	-0.273	+0.170	-0.301	+0.402	+1.000	
R	-0.497	-0.057	+0.383	+0.463	-0.305	+0.233	-0.196	-0.556*	+0.158	-0.409	+0.407	-0.572	-0.219	-0.288	+0.071	-0.172	-0.539	+1.000

*: is significant at the 0.05 level (2-tailed); **: is significant at the 0.01 level (2-tailed).

A: Atmospheric temperature; B: Water temperature; C: pH; D: Dissolvec oxygen; E: Carbon dioxide; F: Alkalinity; G: Hardness; H: Calcium; I: Magnesium; J: Chloride; K: Nitrite; L: Phosphate; M: Silicon; N: Sulphate; O: Total solids; P: Organic matter; Q: Biological oxygen demand; R: Chemical oxygen demand.

Table 1.3: Pearson's Correlation Coefficients for Physico-chemical Characteristics of Papnash Pond, Bidar during February 2003 to January 2004

	A	B	C	D	E	F	G	H	I	J	K	L	M	N	O	P	Q	R
A	+1.000																	
B	-0.007	+1.000																
C	+0.709**	-0.053	+1.000															
D	-0328	+0.104	-0.539	+1.000														
E	+0. 472	-0.165	+0.489	-0.420	+1.000													
F	+0.271	+0.259	+0.468	-0.374	+0.489	+1.000												
G	+0.004	+0.119	+0.039	+0.182	-0.011	+0.563	+1.000											
H	-0.273	+0.069	-0.346	+0.382	-0.190	+0.339	+0.898**	+1.000										
I	+0.484	+0.056	.+0694*	-0.258	+0.384	+0.581*	+0.591*	+0.200	+1.000									
J	+0.355	-0.254	+0.436	-0.200	+0.058	+0.580*	+0.660*	+0.503	+0.504	+1.000								
K	-0.117	-0.071	-0.027	+0.020	-0.585*	-0.108	+0.114	+0.124	-0.130	+0.300	+1.000							
L	+0.823**	+0.009	+0.732**	-0.546	+0.388	+0.699*	+0.305	+0.026	+0.617*	+0.613*	-0.055	+1.000						
M	+0.318	+0.255	+0.403	-0.418	+0.053	+0.595*	+0.295	+0.057	+0.469	+0.302	+0.275	+0.595*	+1.000					
N	-0.497	-0.114	-0.348	+0.211	-0.684*	+0.114	+0.443	+0.503	-0.043	+0.269	+0.679*	-0.161	+0.311	+1.000				
O	+0.412	+0.097	+0.543	-0.207	+0.346	+0.735**	+0.776**	+0.502	+0.769**	+0.774**	-0.073	+0.630*	+0.277	-0.057	+1.000			
P	+0.541	-0.086	+0.575	-0.120	+0.146	.+0690*	+0.571	+0.327	+0.761**	.+0601*	-0.086	-0.780**	+0.466	+0.159	+0643*	+1.000		
Q	-0.325	-0.109	-0.191	+0.072	-0.289	-0.057	+0.232	+0.337	-0.301	-0.301	+0.664*	-0.247	-0.084	+0.343	+0.155	-0.402	+1.000	
R	-0.117	+0.719**	-0.038	+0.066	-0.425	+0.365	+0.091	+0.149	-0.176	+0.079	+0.268	+0.091	+0.375	+0.254	+0.008	-0.002	+0.331	+1.000

*: is significant at the 0.05 level (2-tailed); **: is significant at the 0.01 level (2-tailed).

A: Atmospheric temperature; B: Water temperature; C: pH; D: Dissolved oxygen; E: Carbon dioxide; F: Alkalinity; G: Hardness; H: Calcium; I: Magnesium; J: Chloride; K: Nitrite; L: Phosphate; M: Silicon; N: Sulphate; O: Total solids; P: Organic matter; Q: Biological oxygen demand; R: Chemical oxygen demand.

Table 1.4: Pearson's Correlation Coefficients for Physico-chemical Characteristics of Papnash Pond, Bidar during February 2002 to January 2004

	A	B	C	D	E	F	G	H	I	J	K	L	M	N	O	P	Q	R
A	+1.000																	
B	+0.273	+1.000																
C	+0.509*	+0.098	+1.000															
D	+0.266	+0.184	+0.086	+1.000														
E	+0.229	+0.080	+0.207	-0.044	+1.000													
F	+0.345	+0.312	+0.378	+0.030	+0.232	+1.000												
G	+0.063	+0.156	+0.155	+0.186	+0.045	+0.416*	+1.000											
H	-0.243	+0.015	-0.225	+0.122	-0.118	+0.239	+0.847**	+1.000										
I	+0.185	+0.259	+0.544**	+0.248	-0.044	+0.352	+0.260	-0.017	+1.000									
J	+0.191	-0.029	+0.169	-0.214	+0.140	+0.398	+0.538**	+0.433*	+0.162	+1.000								
K	-0.066	-0.187	-0.124	-0.146	-0.523**	-0.044	-0.013	+0.039	-0.112	+0.185	+1.000							
L	+0.484*	+0.394	+0.321	-0.355	+0.325	+0.326	+0.310	+0.100	+0.122	+0.501*	-0.236	+1.000						
M	+0.221	+0.370	+0.295	-0.184	-0.060	+0.428*	+0.320	+0.043	+0.303	+0.194	+0.105	+0.541**	+1.000					
N	-0.169	+0.254	-0.065	-0.034	-0.625**	+0.264	+0.294	+0.326	+0.345	+0.136	+0.168	+0.156	+0.397	+1.000				
O	+0.269	+0.198	+0.470*	-0.050	+0.269	+0.499*	+0.633**	+0.342	+0.467**	+0.687**	+0.037	+0.434*	+0.219	-0.029	+1.000			
P	+0.281	+0.085	+0.141	-0.334	-0.057	+0.299	+0.427*	+0.284	+0.161	+0.515**	+0.122	+0.622**	+0.337	+0.346	+0.458*	+1.000		
Q	-0.205	-0.099	-0.310	-0.243	-0.085	-0.168	+0.108	+0.300	-0.477*	+0.276	+0.152	+0.109	-0.149	+0.227	-0.071	-0.015	+1.000	
R	+0.251	+0.195	+0.185	+0.338	-0.340	+0.272	-0.005	-0.082	+0.117	-0.156	+0.359	-0.315	-0.072	-0.146	+0.044	-0.096	-0.229	+1.000

*: is significant at the 0.05 level (2-tailed); **: is significant at the 0.01 level (2-tailed).

A: Atmospheric temperature; B: Water temperature; C: pH; D: Dissolved oxygen; E: Carbon dioxide; F: Alkalinity; G: Hardness; H: Calcium; I: Magnesium; J: Chloride; K: Nitrite; L: Phosphate; M: Silicon; N: Sulphate; O: Total solids; P: Organic matter; Q: Biological oxygen demand; R: Chemical oxygen demand.

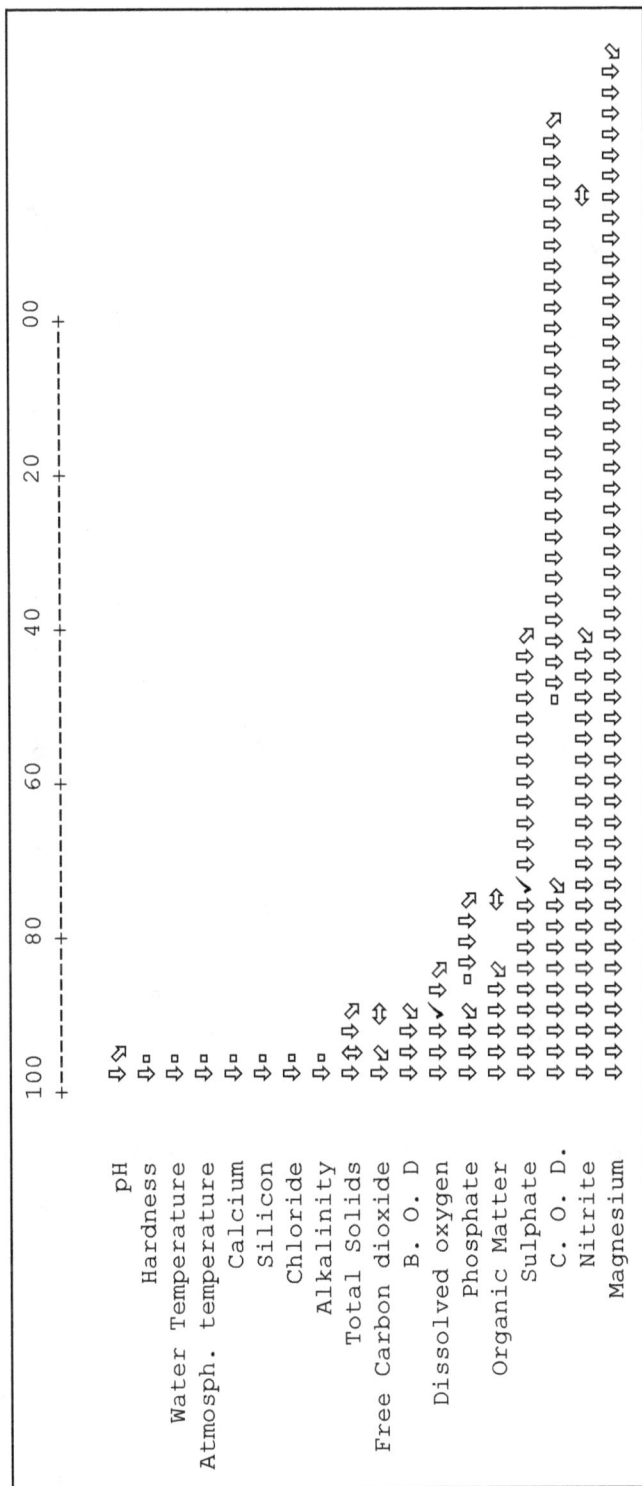

Figure 1.2: The Dendrogram of Percentage Similarity between Physico-chemical Factors and Primary Productivity of Papnash Pond, Bidar (February–2002 to January–2003)

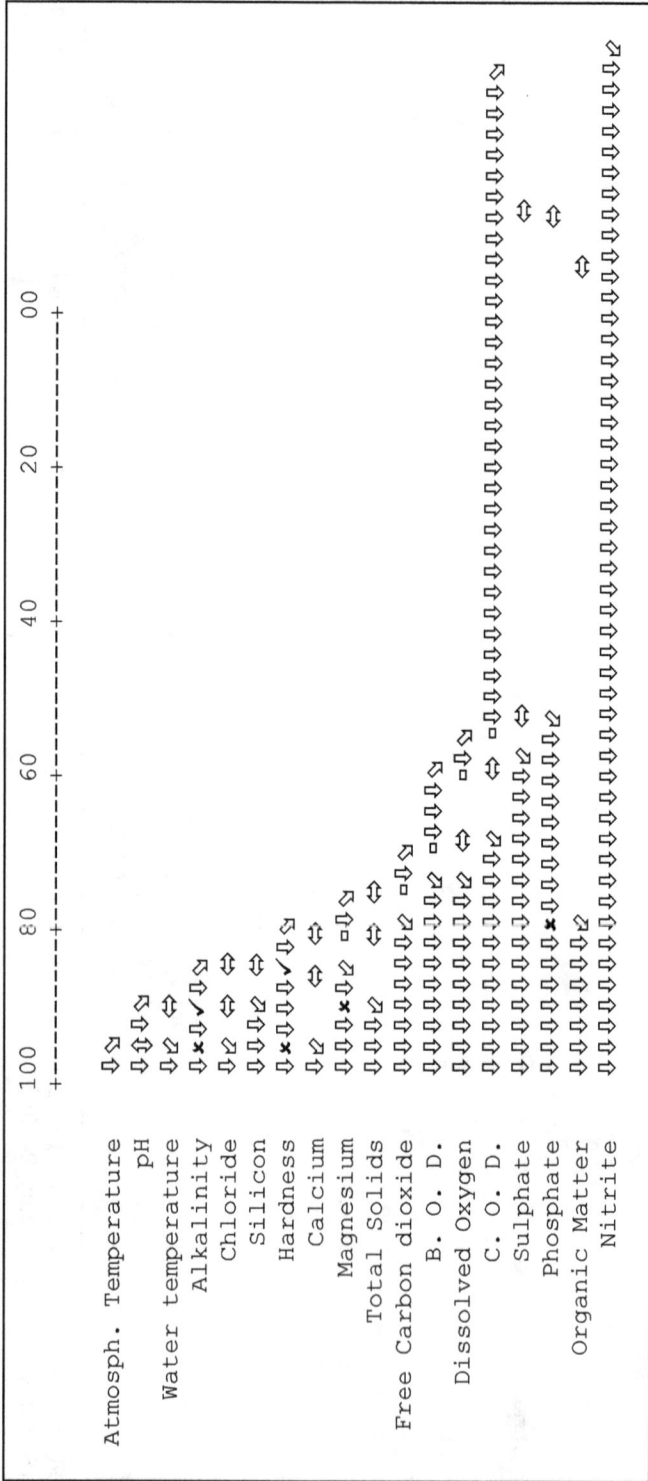

Figure 1.3: The Dendrogram of Percentage Similarity Between Physico-chemical Factors and Primary Productivity of Papnash Pond, Bidar (February–2003 to January–2004)

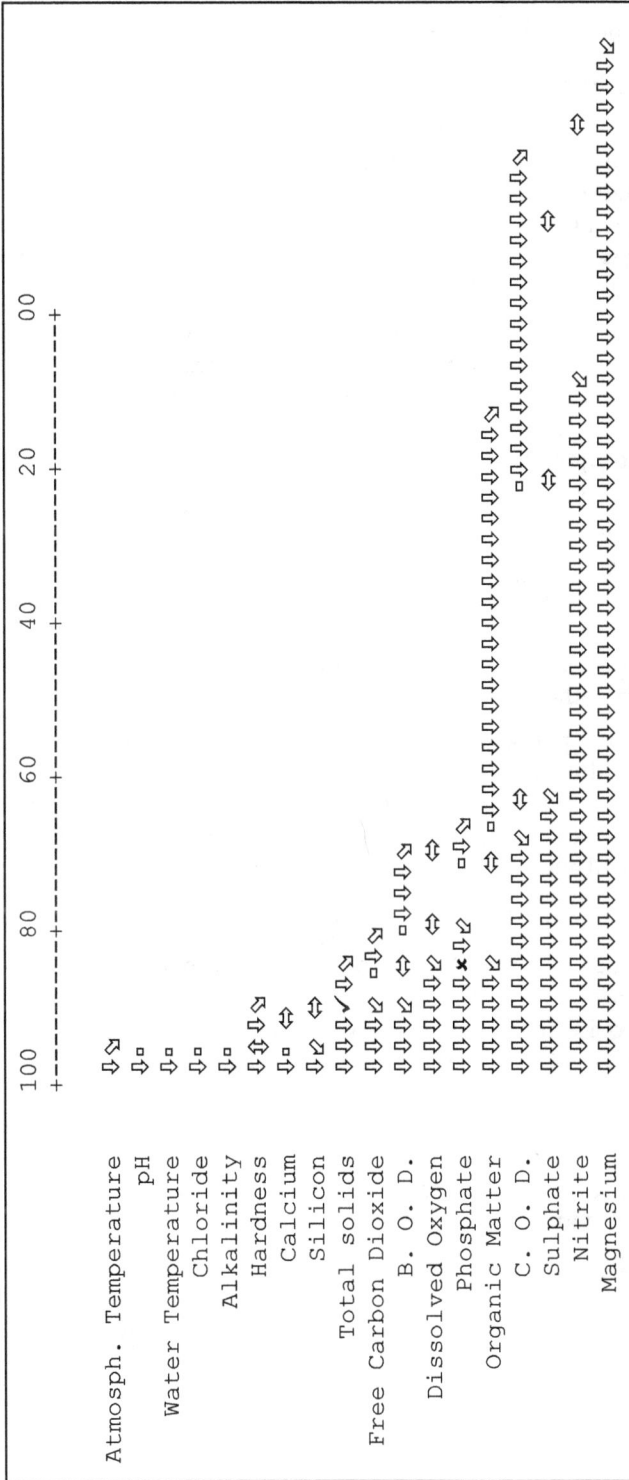

Figure 1.4: The Dendrogram of Percentage Similarity between Physico-chemical Factors and Primary Productivity of Papnash Pond, Bidar (February–2002 to January–2004)

magnesium was evident. COD has shown a positive relation with water temperature and negative relation with calcium.

Hierarchical Cluster Analysis

Hierarchical cluster analysis is applied to identify relatively homogenous group of factors and to measure the distance or similarity in the present work on aquatic ecosystem. The similarity is calculated by comparison of characters of each group parameters. High similarity factors are grouped together in a separate cluster. Simple matching coefficients are arranged to form a similarity matrix and presented as dendrograms (Figures 1.2 to 1.4).

The hierarchical cluster over a period of February 2002 to January 2003 is presented in Figure 1.1, which indicates that pH, hardness, water temperature, atmospheric temperature, calcium, silicate, chloride, alkalinity, total solids and free CO_2 are closely related factors and are dominating factors on other factors. The total solids, dissolved oxygen, phosphate, sulphate and COD are inter linked with others and are regulating factors of the pond.

The cluster over a period of February 2003 to January 2004 is presented in the form of dendrogram in Figure 1.2. The seven factors are dominant factors, such as atmospheric temperature, pH, water temperature, alkalinity, chlorine, hardness and calcium. The factors like pH, alkalinity, hardness, magnesium, free CO_2, BOD, dissolved oxygen, COD, phosphate are interrelated with other factors and are monitoring factors of the water body.

The cluster over a period of February 2002 to January 2004 is presented in Figure 1.3,which shows atmospheric temperature, pH, water temperature, chloride, alkalinity, hardness, calcium, silicate are the dominant factors of the water body. The factors such as hardness, total solids, free CO_2, BOD, phosphate, organic matter and COD are inter linked and controlling factors of the water body.

Recommendations

Physico-chemical factors of Papnash pond varied with season. The concentration of pH, total hardness, and BOD of the studied sample showed more values when compared to the guidelines for drinking water quality standards of World Health Organzation (WHO, 2004). This indicates that the quality of water is deteriorating due to entry of run off, sewage and activities of pilgrims and surrounding. There is an urgent need to restore the quality of potable water of the Papnash Pond to overcome the problem of acute shortage of potable water to Bidar city and to recharge the groundwater table for irrigation. The better management of water body is potential to use for aquaculture and for other benefits of the mankind.

As Papnash Pond is a perennial source of water to the Bidar city which comes under temperate zone, the pond is under threat of getting completely contaminated. There is an emergency to educate the people who visit the temple nearby the water source to not misuse the potable water in the name of holy bath and blind belief.

Acknowledgement

The authors are indebted to late Dr. S.B. Angadi, Reader, Dept. of Botany, for his valuable guidance. We also thank the Head of the Dept. of Botany, Gulbarga University, Gulbarga; Assistant Director i/c, NADRI, Bangalore and Director, CCRAS, New Delhi for providing the facilities and encouragement.

References

1. Ahmed, S.H. and Singh, A.K. (1993). Correlation between physico-chemical factors and zooplanktons during diurnal variations in a freshwater tank at Dholi (Bihar), India, *J. Environ. Biol.*, 14: 95–105.

2. American Public Health Association (APHA), American Wastewater Association (AWWA) and Water Pollution Control Federation (WPCF) (1995). *Standard Methods for the Examination of Water and Wastewater*, 19th Edition. Washington D.C., p. 1220.

3. Aneja, K.R. (1996). *Experiments, in Microbiology, Plant Pathology, Tissue Culture and Mushroom Cultivation*, 2nd Edition. Wishwa Prakashan, New Delhi, p. 451.

4. Angadi, S.B., Shiddamallayya, N. Shiddamallayya and Patil, P.C. (2005). Limnological studies of Papnash Pond, Bidar (Karnataka). *J. Environ. Biol.*, 26: 213–216.

5. Bailey-Watts, A.E. (1986). Seasonal variation in size spectra of phytoplankton assemblages in Loch Leven, Scotland. *Hydrobiologia*, 138: 25–42.

6. Bhatt, S.D. and Pathak, J.K. (1992). Assessment of water quality and aspects of pollution in a stretch of river Gomti (Kumaun: Lesser Himalaya). *J. Environ. Biol.*, 13(2): 113–126.

7. Das, A.K. (2002). Phytoplankton primary production in some selected reservoirs of Andhra Pradesh. *Geobios*, 29(9): 52–57.

8. Garode, A.M. and Agarkar, S.V. (2004). Evalution of drinking water quality of Chandai reservoir water. *Indian Hydrobiol.*, 7: 173–176.

9. Goel, P.K. and Trivedy, R.K.(1984). Some considerations on sewage disposal to freshwaters and resultant effects. *Poll. Res.*, 3: 7–12.

10. Govindasamy, C., Kannan, L. and Azariah, J. (2000). Seasonal variation in physico-chemical properties and primary production in the coastal water biotopes of Coromandel coast, India. *J. Environ. Biol.*, 21(1): 1–7.

11. Gupta, P.K. (2001). *Methods in Environmental Analysis Water, Soil and Air*. Agrobios (India), p. 424.

12. Harsha, T.S. and Mallammanavar, S.G. (2004)., Assessment of phytoplankton density in relation environmental variables in Gopalswamy pond at Chitradurga, Karnataka. *J. Environ. Biol.*, 25: 113–116.

13. Hosmani, S.P. and Bharati, S.G. (1975). Hydrobiological studies in ponds and lakes of Dharwar. III. Occurrence of two euglenoid blooms. *The Karnataka Uni. J. Sci.*, 20: 151–156.

14. Katiyar, S.K. and Belsare, D.K. (1997). Limnological studies on Bhopal lakes: Freshwater protozoan communities as indicators of organic pollution. *J. Environ. Biol.*, 18: 271–282.

15. Kirsten, O. and Arnold, N. (1993). Stress and disturbance in the phytoplankton community of a shallow, hypertrophic lake. *Hydrobiologia*, 249: 15–24.

16. Kumar, A. (1994). Periodicity and abundance of rotifers in relation to certain physico-chemical characteristics of two ecologically different ponds of Santhal. Paraganas (Bihar). *Indian. J. Ecol.*, 21(1): 54–59.

17. Kumar, A., Tiwari, M.G. and Kumar, A. (1994). An assessment of drinking water quality of Santhal Praganas (Bihar). *J. Mendel.*, 11: 119–120.

18. Mali, K.N. and Gajaria, S.C. (2004). Assessment of primary productivity and hydrobiological characterization of a fish culture pond, Gujarat. *Indian Hydrobiol.*, 7: 113–119.

19. Mishra, S.R., Sharma, S. and Yadav, R.K. (1992). Phytoplanktonic communities in relation to environmental conditions of lentic waters at Gwalior (M.P.). *J. Environ. Biol.*, 13: 291–296.

20. Mukhopadhyay, S.K. (1996). Limnological investigations in lotic and lentic freshwater bodies in and around Darjeeling, West Bengal, India. *Geobios*, 23: 101–106.

21. Naganandini, M.N. and Hosmani, S.P. (1998). Ecology of certain inland waters of Mysore district occurrence of cyanophycean bloom at Hosakere lake. *Poll. Res.*, 17(2): 123–125.

22. Patel, N.K. and Sinha, B.K. (1998). Study of the pollution load in the ponds of Burla area near Hirakud dam of Orissa. *J. Environ. and Poll.*, 5: 157–160.

23. Prakasam, V.R. and Johnson, P. (1992). Ecology of Quilon canal (T–S canal) with reference to physico-chemical characteristics. *J. Environ. Biol.*, 13: 221–225.

24. Prasad, B.N. and Manjula, S. (1980). Ecological study of blue-green algae in river Gomati. Indian, *J. Environ. Health*, 22: 151–168.

25. Raghavendra and Hosmani, S.P. (2002). Hydrobioloical study of Mandakally lake: A polluted waterbody at Mysore. *Nature Environ. and Poll. Tech.*, 1: 291–293.

26. Rao, V.N.R., Mohan, R., Hariprasad, V. and Ramasubramanian, R. (1993). Seasonal dynamics of physico-chemical factors in a tropical high altitude lake: an assessment in relation to phytoplankton. *J. Environ. Biol.*, 14: 63–75.

27. Sachidanandamurthy, K.L. and Yajurvedi, H.N. (2004). Monthly variations in water quality parameters (physico-chemical) of a perennial lake in Mysore city. *Indian Hydrobiol.*, 7: 217–228.

28. Shastri, Y. and Pendse, D.C. (2001). Hydrobiological study of Dahikhuta reservoir. *J. Environ. Biol.*, 22(1): 67–70.

29. Shiddamallayya, N. and Pratima, M. (2006). Influence of abiotic factors and polluting sources in the structure of cyanophycean community in the freshwater Tank Bhalki, Karnataka. *J. of Indian Hydrobiol.*, 9: 45–50.

30. Shiddamallayya, N. and Pratima, M. (2007a). Studies on phytoplankton community with special reference to physico-chemical factors in Bhalki tank, Bhalki, Karnataka, India. *J. of Ecol. Environ. and Conser.*, 13(2): 361–365.

31. Shiddamallayya, N. and Pratima, M. (2007b). A study on phytoplankton community of Papnash pond, Bidar, Karnataka, India. *J. of Ecol. Environ. and Conser.*, 13(2): 367–370.

32. Simon, A.T. (2002). Seasonal evaporative concentration of an extremely turbid water-body in the semiarid tropics of Australia. *Lakes and Reserv. Res. and Manage.*, 7: 103–107.

33. Swarnalatha, P. and Rao, A.N. (1997). Interrelationships of physico-chemical factors of a pond. *J. Environ. Biol.*, 18(1): 67–72.

34. Swarnalatha, P. and Rao, A.N. (1998). Ecological studies of Banjara lake with reference to water pollution. *J. Environ. Biol.*, 19: 179–186.

35. Trivedy, R.K. and Goel, P.K. (1986). *Chemical and Biological Methods for Water Pollution Studies.* Environmental Publication, Karad, India, p. 215.

36. Trivedy, R.K., Goel, P.K., Shrotri, A.C., Ghadge, M.R. and Khatavkar, S.D. (1998). Quality of lentic water resources in south-western Maharastra, India. *Perspectives in Aqua. Biol. (National Seminar),* p. 215–235

37. Varma, M.C. and Thakur, P.K. (1998). Assessment of drinking water quality of an industrial township of south Bihar. *J. Environ. and Poll.*, 5: 17–21.

38. Vijay Kumar, K. and Ramesha, I. (2002). Insect abundance in relation to physico-chemical characteristics of pond water at Gulbarga, Karnataka. *J. Curr. Sci.*, 2: 185–188.

39. Vyas, N. and Nama, H.S. (1991). Pollution ecology of freshwater reservoir at Jodhpur, with special reference to microorganisms. *Geobios*, 18: 33–37.

40. WHO (2004). *Guidelines for Drinking Water Quality*, 3rd Edition, Vol. 1. World Health Organization, Geneva, p. 515.

Biodiversity of Aquatic Reseources (2012)
Editors: Mamta Rawat & Sumit Dookia
Published by: DAYA PUBLISHING HOUSE, NEW DELHI

Pages 20-28

Chapter 2

Diatoms Flora: An Important Bioresource of the Central Highland Rivers, India

Jyoti Verma and Prakash Nautiyal*

Department of Zoology, H.N.B. Garhwal University, Srinagar – 246 174, Uttarakhand

ABSTRACT

Investigations were undertaken on freshwater epilithic diatoms of three Vindhyan Rivers, the Ken, Tons and Paisuni (Central Highland, India). Thirty three epilithic samples were obtained between November 2003 to April 2004, by scraping 3 x 3 cm surface area of cobbles at the 11 sampling stations on three rivers situated between latitude 23°30' to 26°N and longitude 78°30' to 82°30' E. In all 293 diatom taxa (species, varieties and forms) belonging to 49 genera were identified, along with some unidentified forms. The diatom flora belonged to three sub orders one centric, two pennate and nine families, two centric (Thalassiosiraceae, Meloseriaceae) and seven pennate (Fragilariaceae, Eunotiaceae, Achnanthaceae, Naviculaceae, Epithemiaceae, Bacillariaceae, Surirellaceae).Only three Centrale taxa were recorded. The Araphidineae was represented by Fragilariaceae having five genera and 28 taxa. The Raphidineae was represented by six families; Eunotiaeae, Achnantheacae, Naviculaceae, Epithemiaceae, Bacillariaceae and Surirelliaceae in the Vindhya. Eunotiaceae was represented by six taxa of *Eunotia*. Achnantheacae comprised 26 taxa of four genera (*Achnanthidium, Achnanthes, Planothidium, Cocconeis*) in the flora. Naviculaceae

* Corresponding Author E-mail: diatombuster@gmail.com

was represented by 183 taxa of 30 genera, constituting the bulk of flora. 21 genera were similar among the Vindhya rivers, however nine were different; *Gomphocymbelopsis* in the Ken and Paisuni, *Anomoeneis* in the Paisuni and Tons and *Hippodonta* in the Ken and Tons, *Frustulia, Mastogloia* and *Stauroneis* in the Ken only while *Amphipleura, Aneumastus* and *Scoliopleura* in the Paisuni only. Within the Vindhya region, Ken and Paisuni with 26 genera were relatively richer than Tons having 23 genera. Epithemiaceae comprised two genera; *Epithemia* and *Rhopalodia* with one and two species, respectively. Bacillariaceae consisted of 4 genera (*Bacillaria, Denticula, Hantzschia, Nitzschia*) and thirty five taxa. Surirellaceae was represented by *Surirella* (9 taxa). The raphids (mono and biraphids) accounted for 89.4 per cent of the Vindhya flora. The species-rich genera were *Navicula* sensu stricto (32), *Cymbella* sensu stricto (31), *Nitzschia* (22), *Synedra* (19), *Gomphonema* (15), *Achnanthidium* (14), *Cymbopleura* (14) and *Amphora* (12).

Keywords: Centrale, Central highland diatom, Indian Subcontinent, Pennate.

Introduction

Diatom flora has been studied scantily in India, mostly from southern parts of the Peninsular India, its islands Andaman and Nicobar and some from Himalaya (Gandhi, 1998; Dickie, 1882; Carter, 1926; Gosh and Gaur, 1991; Rout and Gaur, 1994; Nautiyal and Nautiyal, 1999 a, b; Khan, 2002; Nautiyal *et al.*, 2004 a, b). Hence, a study was made to generate information on floristic composition of diatoms in the Vindhya region for the first time. The present study also assumes significance in light of the river linking programmes like the Ken–Betwa link, especially in view of the climate change threat. The knowledge on their floristic composition and distribution will help to understand their use as environmental indicators.

Study Area

The rivers selected in Vindhya region were located between 23°30' to 26° N, 78°30' to 82°30' E (Table 2.1). The Vindhya Rivers Ken, Paisuni and Tons flow north from low (north of Narmada, around Tropic of Cancer) to high latitude along southern fringe of the Indo-Gangetic Plains. The Ken and Paisuni are right-bank tributaries of the Yamuna River, while Tons that of the Ganga. Ken (340 km) and Tons (305 km) were relatively larger drainage compared to the Paisuni (100 km) having an average gradient of 0.91 m/km, 2.0 m/km and 1.02 m/km, respectively. The land is primarily used for agriculture purposes along the banks of the Ken and the Tons River. Patches of forest (*Shorea robusta*) occur in case of the Ken River. In case of the Paisuni River the headwaters is covered by forest and only lower stretch of the river is used for agriculture practices. Cement and stone quarrying industry exist in the Vindhya region though none were located in the immediate vicinity of the sampling stations. Except for the mouth zone these rivers vary topographically. The Physico-chemical characteristics were also studied from these rivers (Tables 2.1 and 2.2).

Materials and Methods

The conditions for intensive sampling are available only during stable flow (November to May) was deemed suitable for present study as the rivers are in floods

Table 2.1: Geographical Coordinates of the Sampling Stations in Different Rivers of the Central Highland Rivers

River System	Rivers/Stations with Acronym	Latitude (N)	Longitude (E)	Altitude (m asl)	Distance from Source (km.)
Yamuna	Ken River				
(lower stretch)	Shahnagar K1	23° 59' 00"	80° 15' 45"	400	ca.10
	Panna K2	24°43'18"	80° 11' 25"	200	142.5
	Banda K3	25°29'24"	80° 19' 16"	140	267.5
	Chilla K4	25° 46' 54	80° 32' 08"	100	340
	Paisuni River				
	Anusuya P1	25° 08' 14	80° 51' 01"	160	10
	Chitrakut P2	25° 13' 54	80° 54' 36"	140	26
	Purwa P3	25° 16' 17	80° 52' 28"	100	42
Ganga	Tons River				
(middle stretch)	Amdara T1	24° 00' 00"	80° 26' 21"	400	20
	Maihar T2	24° 16' 35	80° 48' 53"	400	56
	Satna T3	24° 34' 07	80° 54' 36"	323	98
	Chakghat T4	25° 03' 28	81° 42 '21"	100	232.5

Table 2.2: Physical and Chemical Characteristics (Minimum and Maximum) at Different Stations of the Central Highlands Region

Stream/ River	A T (°C)	WT (°C)	CV (cm s^{-1})	pH	C (µmho cm^{-1})	DO (mg l^{-1})
Ken	11–32	15–31	0-42	7.0-7.5	165-420	8.6-10.7
Paisuni	10–40	16–30	2.8-40	7.0-7.7	170-440	8-11.5
Tons	17–33	18–33	1.5-35	7.0-7.8	160-420	8.2-10.5

AT: Air temperature; WT: Water temperature; CV: Current velocity; C: Conductivity; DO: Dissolved oxygen.

during monsoon; mid June to mid September. Benthic diatoms were obtained from eleven stations on 3 rivers (from source to mouth) in the central highland region. Diatom samples were collected by scraping the cobble surface with a brush from an area of 3 x 3 cm^2 and preserved in 4 per cent formaldehyde on site itself. Samples were treated with Hydrochloric acid-peroxide, washed repeatedly and mounted in Naphrax. Each slide was examined under bright field by PLANAPO x 100 oil immersion objective to record the flora. Identifications were made according to standard literature (Schmidt 1874-1959, Hustedt 1931-1959, Krammer and Lange-Bertalot 1991a, b, 1999, 2004, Lange-Bertalot 2001, Metzeltin and Lange-Bertalot 2002, Krammer 2003, Lange-Bertalot *et al.*, 2003, Werum and Lange-Bertalot 2004, Metzeltin *et al.*, 2005). Sarode and Kamat (1984) and Gandhi (1998) were also consulted for further identification. The permanent mounts have been adequately stored at the Aquatic Biodiversity Unit, Department of Zoology, H. N. B. Garhwal University, Srinagar (Uttarakhand) where preparation of the slides and their microscopic examination was undertaken.

Table 2.3: The Number of Species Occurring in Various Genera Recorded from the Vindhyan Rivers

Sl.No.	Genera	K	P	T
	THALASSIOSIRACEAE			
1.	*Aulacoseira*	1	1	
2.	*Cyclotella*	1	1	2
	Genera/Species	2/2	2/2	1/2
	FRAGILARIACEAE			
3.	*Diatoma*	2	1	4
4.	*Fragilaria*	1	1	2
5.	*Staurosira*	1	2	1
6.	*Synedra*	13	14	15
7.	*Tabellaria*	1		1
	Genera/Species	5/28	4/18	5/23
	EUNOTIACEAE			
8.	*Eunotia*	3	4	4
	Genera/Species	1/3	1/4	1/4
	ACHNANTHACEAE			
9.	*Achnanthes*	1	1	
10.	*Achnanthidium*	8	11	10
11.	*Planothidium*	3	5	3
12.	*Cocconeis*	6	3	3
	Genera/Species	4/18	4/20	4/16
	NAVICULACEAE			
13.	*Amphipleura*		1	
14.	*Amphora*	10	8	8
15.	*Anomoeoneis*		1	1
16.	*Brachysira*	1	2	2
17.	*Caloneis*	5	3	5
18.	*Cymbella*	21	22	20
19.	*Cymbopleura*	12	6	10
20.	*Encyonema*	4	4	4
21.	*Diploneis*	5	5	4

Contd...

Table 2.3–Contd...

Sl.No.	Genera	K	P	T
22.	*Frustulia*	1		
23.	*Gomphocymbelopsis*	1	1	
24.	*Gyrosigma*	2	2	3
25.	*Gomphonema*	11	10	12
26.	*Mastogloia*	1		
27.	*Navicula*	27	28	29
28.	*Navicula sensu lato*	3	4	3
29.	*Adlafia*	1	1	2
30.	*Aneumastus*		2	
31.	*Craticula*	4	3	3
32.	*Diadesmis*	1	2	1
33.	*Fallacia*	1	1	2
34.	*Geissleria*	1	1	1
35.	*Hippodonta*	1		2
36.	*Luticola*	6	3	7
37.	*Placoneis*	2	1	2
38.	*Sellaphora*	5	5	5
39.	*Neidium*	2	4	2
40.	*Pinnularia*	3	7	3
41.	*Scoliopleura*		1	
42.	*Stauroneis*	2		
	Genera/Species	26/133	26/128	23/131
	EPITHEMIACEAE			
43.	*Epithemia*			1
44.	*Rhopalodia*		1	2
	Genera/Species	0/0	1/1	2/3
	BACILLARIACEAE			
45.	*Bacillaria*			1
46.	*Denticula*	1	1	1
47.	*Hantzschia*	1	1	
48.	*Nitzschia*	21	20	25
	Genera/Species	3/23	3/22	3/26
	SURIRELLACEAE			
49.	*Surirella*	8	7	5
	Genera/Species	1/8	1/7	1/5
	Total Genera/Species	42/205	42/202	39/211

Results and Discussion

The diatom flora of Vindhya Rivers consisted of 293 species and 49 genera. The Vindhyan Rivers examined during present study consisted largely of Pennale elements. Centrale species were few and belonged to the suborder Coscinodiscineae, Family Thalassiosiraceae (*Aulacoseira* 1 species; *Cyclotella* 2 species) and Melosiraceae (*Melosira* 1 species). Rest 46 genera and 289 taxa were Pennales. Numerically, suborder Araphidineae comprising family Fragilariaceae (6 genera, 38 taxa) and Raphidineae consisting family Achnanthaceae (4 genera, taxa), Naviculaceae (30 genera, 183 taxa) and Bacillariaceae (4 genera, 35 taxa) accounted for a large proportion (Table 2.3). The remaining few taxa belonged to Eunotiaceae, Epithemiaceae and Surirellaceae. The study on floral composition has thrown up some broad regional and local patterns of distribution.

The centric diatoms comprise just 1 per cent of the total flora (3 taxa). The araphids accounted for 10.5 per cent (38 spp.) and the raphids (mono and biraphids) accounted for 42 genera with 258 taxa (80 per cent) however monoraphids accounted for 8 per cent (26 spp.). The major share in floristic composition was family Naviculaceae (62 per cent), Bacillariaceae (12 per cent) and Fragilariaceae (9.5 per cent) from Central Highland. *Navicula* (32), *Cymbella* (31), *Gomphonema* (15) and *Cymbopleura* (14) were species rich genera in family Naviculaceae, *Synedra* (19) in family Fragilariaceae and *Nitzschia* (32) in family Bacillariaceae. The Vindhya was richer in *Navicula* sensu stricto, *Nitzschia* and *Cymbella* sensu stricto. *Navicula* sensu lato had highest number of species in the flora reported from Bombay and Salsette (24 spp. Gonzalves and Gandhi 1952-1954), Mysore State (11 spp. reported by Gandhi, 1958) and South India (20 spp. reported Krishnamurthy, 1953) compared to only few (9 spp.) from Uttar Pradesh (U.P.) in the Gangetic Plains. However, flora was quite different in these regions as Bombay and Salsette in the west had more species of *Neidium* and *Stauroneis* in case of Naviculaceae, while *Achnanthes* in case of Achnanthaceae and *Eunotia* in case of Eunotiaceae, the dominant families. In South India including the Mysore State and U. P. most of the flora belonged to Naviculaceae. Nitzschiaceae was conspicuously absent from Bombay and Salsette Islands. Investigations in the Central Highland revealed 293 diatom taxa and 49 genera from 11 sites. Naviculaceae (62 per cent), Bacillariaceae (12 per cent) and Fragilariaceae (9.5 per cent) dominate the flora, respective genera being *Navicula* sensu stricto, *Nitzschia* and *Synedra*. The floral composition of this region was quite different from the rest of India.

Recommendations

Diatoms are excellent indicators of the ecological condition of aquatic system, and have been used for water quality assessment for many decades, and can be used for same in this region too. The importance of diatoms in river systems owing to their role as major primary producers and established as indicator value has been neglected in India. The study will help to classify streams of India on the basis of stress, useful for deciding their best possible use. This type of study will generate knowledge of diatom flora from different parts of India and the extent of diversity. The presence of diverse flora can also serve as an argument for the protection of their habitats. So this type of study should be initiated from the other parts of Indian subcontinent.

Acknowledgement

The academic support by the Head, Department of Zoology, H.N.B. Garhwal University is acknowledged.

References

1. Carter, N. (1926). Freshwater algae from India. *Records of Botanical Survey of India* 9: 263–302.

2. Dickie, G. (1882). Notes on algae from the Himalayas. *Journal of Linnean Botanical Society*, 19: 230–232.

3. Gandhi, H.P. (1958). The freshwater diatoms flora of the Hirebhasgar Dam area, Mysore State. *Journal of Indian Botanical Society*, 37: 249–265.

4. Gandhi, H.P. (1998). *Freshwater Diatoms of Central Gujarat* (with a review and some others). Bishen Pal Singh, Mahendra Pal Singh, Dehradun, India, pp. 324.

5. Gonzalves, E.A. and Gandhi, H.P. (1952). A systematic account of the diatoms of Bombay and Salsette–I. *Journal of Indian Botanical Society*, 31: 117–115.

6. Gonzalves, E.A. and Gandhi, H.P. (1953). A systematic account of the diatoms of Bombay and Salsette–II. *Journal of Indian Botanical Society*, 32: 239–263.

7. Gonzalves, E.A. and Gandhi, H.P. (1954). A systematic account of the diatoms of Bombay and Salsette–III. *Journal of Indian Botanical Society*, 33: 338–350.

8. Ghosh, M. and Gaur, J.P. (1991). Structure and interrelation of epilithic algal communities in two deforested streams at Shillong, India. *Hydrobiologia*, 122: 105–116.

9. Hustedt, F. (1931–1959). *Die Kieselalgen Deutschlands Oesterrichs und der Schweiz Bd. 7, Teil 2*, Translated by N. G. Jensen as The Pennate Diatoms 1985. pp. 918. Koenigstein: Koeltz Scientific Books.

10. Khan, M.A. (2002). Phycological studies in Kashmir. Algal biodiversity, p. 69–93. In: *Ethology of Aquatic Biota*, (Ed.) A. Kumar. APHA Publ. Corp., New Delhi.

11. Krammer, K. (2002). Diatoms of europe. Diatoms of the European Inland waters and comparable habitats. (ed. Lange–Bertalot) Vol. 3. Cymbella. 584 p., 194 pl. A.R.G. Gantner Verlag K.G., FL 94191 Ruggell. Distributed by Koeltz Scientific Books, Konigstein.

12. Krammer, K. (2003). Diatoms of Europe: Diatoms of European Inland Waters and Comparable Habitats. *Cymbopleura, Delicata, Navicymbula, Gomphocymbellopsis* and *Afrocymbella*. pp. 530. Ruggell: A. R. G. Gantner, K. G. Verlag.

13. Krammer, K. and Lange-Bertalot, H. (1991a). Suβwasserflora von Mitteleuropa, Bacillariophyceae, Band 2/3, 3. Teil: Centrales, Fragilariaceae, Eunotiaceae. pp. 576. Stuttgart: Gustav Fischer Verlag.

14. Krammer, K. and Lange-Bertalot, H. (1991b). Suβwasserflora von Mitteleuropa, Bacillariophyceae, Band 2/4, 4. Teil: Achnanthaceae, Kritische Erg.nzungen zu Navicula (Lineolatae) und Gomphonema Gesamtliteraturverzeichnis. pp. 437. Stuttgart: Gustav Fischer Verlag.

15. Krammer, K. and Lange-Bertalot, H. (1999). Suβwasserflora von Mitteleuropa. Bacillariophyceae, Band 2/2, 2. Teil: Bacillariaceae, Epithemiaceae, Surirellaceae. p. 1-610. Berlin: Spectrum Academischer Verlag, Heidelberg.

16. Krammer, K. and Lange-Bertalot, H. (2004). Süsswasser Flora von Mitteleuropa. Bacillariophyceae Band 2/4, 4. Teil : Achnanthaceae. P. 68. Berlin: Spektrum Akademischer Verlag, Heidelberg.

17. Krishnamurthy, V. (1954). A contribution to the diatom flora of South India. *Journal of Indian Botanical Society*, 33: 354–381.

18. Lange-Bertalot, H. (2001). Diatoms of Europe: Diatoms of European inland waters and comparable habitats. *Navicula* sensu stricto 10 genera separated from *Navicula* sensu lato, *Frustulia*. P. 526. Ruggell: A.R.G. Gantner and K.G. Verlag.

19. Lange-Bertalot, H., Cavacini, P., Tagliaventi, N. and Alfinito, S. (2003). Diatoms of Sardina. Biogeography–Ecology–Taxonomy, Iconographia Diatomologica. P. 438. Ruggell: A.R.G. Gantner and K.G. Verlag.

20. Metzeltin, D. and Lange-Bertalot, H. (2002). Diatoms from the Island Continent Madagascar. Annotated Diatom Micrographs Iconographia, Diatomologica 11, pp. 726. Ruggell: A.R.G. Gantner and K.G. Verlag.

21. Metzeltin, D., Lange-Bertalot, H. and Garcia-Rodriguez, F. (2005). Diatoms of Uruguay. Taxonomy–Biogeography–Diversity. Annotated Diatom Micrographs Iconographia, Diatomologica. P. 726. Ruggell: A.R.G. Gantner and K.G. Verlag.

22. Nautiyal, R. and Nautiyal, P. (1999a). Altitudinal variations in the pennate diatom flora of the Alaknanda–Ganga river system in the Himalayan stretch of Garhwal region. In: *Proceedings of Fourteenth International Diatom Symposium*, (Eds.) S. Mayama, M. Idei and I. Koizumi. Koeltz Scientific Books, Koenigstein, pp. 85–100.

23. Nautiyal, R. and Nautiyal, P. (1999b). Spatial distribution of diatom flora in Damodar river system of Chhota Nagpur. In: *The Fourth Indian Fisheries Forum*, (Ed.) M. Joseph, Kochi, pp. 17–22.

24. Nautiyal, P., Kala, K. and Nautiyal, R. (2004a). A preliminary study of the diversity of diatoms in streams of the Mandakini basin Garhwal Himalaya. In: *Proceedings of 17th International Diatom Symposium*, (Ed.) M. Poulin. Ottawa, Canada, 2002 Biopress, Bristol, pp. 235–269.

25. Nautiyal, P., Nautiyal, R., Kala, K. and Verma, J. (2004b). Taxonomic richness in the diatom flora of Himalayan streams (Garhwal, India). *Diatom*, 20: 123–132.

26. Rout, J. and Gaur, J.P. (1994). Composition of dynamics of epilithic algae in a forest stream at Shillong (India). *Hydrobiologia*, 291: 61–74.

27. Sarode, P.T. and Kamat, N.D. (1984). *Freshwater Diatoms of Maharashtra*. Saikripa Prakashan, Aurangabad, pp. 338.

28. Schmidt, A. *et al.* (1874–1959). *Atlas der Diatomaceen–kunde*. Leipzig: Aschersleben.

29. Werum, M. and Lange-Bertalot, H. (2004). Diatoms in springs from Central Europe and elsewhere under the influence of hydrogeology and anthropogenic impacts. Ecology–Hydrogeology–Taxonomy Iconographia Diatomologica. p.p. 480. Ruggell: A.R.G. Gantner and K.G. Verlag.

Biodiversity of Aquatic Reseources (2012)
Editors: *Mamta Rawat & Sumit Dookia*
Published by: DAYA PUBLISHING HOUSE, NEW DELHI

Pages **29-49**

Chapter 3

Distribution Pattern and Present Scenario of Mangroves and Associated Flora of Andhra Pradesh

*G.M. Narasimha Rao**

*Department of Botany, Andhra University,
Visakhapatnam – 530 003, Andhra Pradesh*

ABSTRACT

Mangroves are the most productive and bio-diverse wet lands on earth. Indian mangrove forests accounts for about 5 per cent of the total mangrove ecosystems of the Globe. In India mangrove forest spread out over an area of more than 4300 square kilometers along the estuarine water of the aquatic ecosystem. Mangrove forests of Andhra Pradesh stand for the second largest mangrove ecosystems in India. Andhra Pradesh mangroves spreads in the estuarine regions such as Godavari, Krishna, Vamsadhara, Sarada and Varaha estuarine complex and Meghadrigedda rain fed drain of Visakhapatnam.

Mangroves and associated flora occurring in different estuarine ecosystems of Andhra Pradesh were studied using transect with quadrant (16 m²) and quadrant samples were analyzed. In the present study 12 mangrove species, 7 mangrove associated species, 6 species of halophytes, 5 estuarine macro algae and 2 sea grasses were reported. The abundance of mangroves varied an all these estuaries.

* E-mail: gmnrao_algae@hotmail.com

Maximum mangrove cover was reported in Godavari estuary followed by Krishna estuary. In Vamsadhara, Sarada and Varaha estuarine complex and Meghadrigedda rain fed drain vegetation is poor and scanty. Halophytes and associated plants with only 2 to 5 typical mangroves species were reported. Transect studies in the Godavari and Krishna estuaries revealed that mangroves and halophytes are reported up to 50 to 100 meters from the water front, where as in remaining estuaries the mangrove forest extends up to 20 to 35 meters only from water front to barren zone. Aquaculture in recent times play critical role in the existence of the mangroves. Most of the mangrove forest, halophytic and barren zones are converted into either aqua forms or for manufacturing of salt. In Visakhapatnam (Meghadrigedda rainfed drain) this ecosystem gradually converted into various industrial activities. On the whole mangrove cover of the Andhra Pradesh reducing in an alarming rate. If the conservation and management programmes will not be taken, the beautiful ecosystem may be vanishing especially in tiny estuaries like Sarada and Varaha estuarine complex and Visakhapatnam.

Keywords: Distribution, Mangroves, Halophytes, Algae, Estuaries of Andhra Pradesh.

Introduction

Mangrove is the combination of Portuguese term "Manque" for an individual tree and grooves the English word for a group or strand of trees. Mangroves are coastal tropical formations found along the broader of the sea and lagoons reaching up to edges of the river to the point where the water is saline, growing in swampy soils and covered by the sea during high tides. These forests act as natural coast guard during cyclones from the high tides of the sea. Mangroves have been a source of astonishment for the laymen and of interest for the scientist, particularly because of strange morphological adaptations such as slit roots for mechanical support, pnuematophores for respiration, and vivipary for proper germination and special physiology for removal of excess salts through leaves. Some authors consider mangroves as possible transitional species in the evolution from aquatic to terrestrial plant life. Godavari mangroves of Andhra Pradesh stand for the second largest mangrove ecosystems in India.

Mangrove populations of Andhra Pradesh have been studied by several authors. Among all these studies, mangroves ecosystems of the Godavari estuary received much attention. (Mathuda,1959; Rao,1959; Sidhu,1963; Raju,1968; Blasco,1975; Umamaheswara Rao and Narasimha Rao, 1988; Narasimha Rao,1989, 1995; Bhaskara Rao *et al.*, 1992; Narasimha Rao *et al.*, 2000; Narasimha Rao and Dora, 2009. Similarly significant work has been done on the mangroves of Krishna estuary (Venkanna, 1991; Venkanna and Narasimha Rao, 1993; Krishna Rao and Narasimha Rao, 1994; Venkateswara Rao and Narasimha Rao, 2008). Work on mangroves and associated flora of Sarada and Varaha estuarine complex was studied by Narasimha Rao and Venkanna (1996). Narasimha Rao and Vanilla Kumari (1997). Mangroves of Meghadrigedda of Visakhapatnam were carried out by Venkateswarlu *et al.* (1972), Venkanna *et al.* (1989) and Narasimha Rao (2008). Mangrove populations of Vamsadhara estuary was studied by Narasimha Rao and Murthy (2010).

Mangroves ecosystems of the globe were extensively studied by several authors. Mangroves ecosystems of the globe divided into two categories, viz, the old world mangroves and the new world mangroves. Chapman (1976) reported the existence of 68 species of mangroves in world wide. Saenger *et al.* (1983), observed the 6 distinct mangrove regions in the world mangrove ecosystem. The total cover of the mangroves estimated by several authors (Bunt, 1992; Twilley, 1998; Spalding *et al.*, 1997; Valiela *et al.*, 2001; IUCN, 2006) with varying figures 10 million ha to 24 million ha). According to FAO (2007), the most extensive mangrove area is found in Asia, followed by Africa and North and Central America. Duke (1992) has reported the 68 true mangrove plant species in the world mangrove species. In India several authors (Banerjee *et al.*, 1989; Banerjee and Rao, 1990) have studied the mangrove ecosystems and reported the number of true mangroves species varying from 37 to 47.

The present study deals the quantitative data on mangrove populations of Vamsadhara estuary near Naupada, Meghadrigedda rain fed drain of Visakhapatnam, Sarada and Varaha estuarine complex of Rambilli, Coringa, Gaderu, Balusutippa, Masanitippa, and Pandi of Godavari estuary, Yeta chettu dibbapalem and Sorlagondi reserve forest of Krishna estuary (Plates 3.1 and 3.2) using the line transect. Simultaneously data collected on the hydrographical parameters of the different study sites of the Andhra Pradesh. Information was also collected on the distribution and occurrence of estuarine algae and sea grasses in the estuaries of Andhra Pradesh.

Study Sites

The mangrove forests of Godavari estuary spreads from Chollangi near Kakinada to Pandi region near Amalapuram. The Gautami branch of Godavari is a typical estuary and forest situated between 82° 12' and 82° 21' E and 16° 31' and 16° 54' N (Figure 3.1). The entire forest divided into 5 stations for collection of data, namely Coringa, Gaderu, Balusutippa, Masanitippa and Pandi. Mangrove forest of Krishna estuary occurring between Yetachettu Dibba palem to Sorlagondi and vegetation distributed between 15° 43' 16° 00' N 80° 45' 81° 10' E (Figure 3.1). The Sarada and Varaha are two small rivers in the east coast of India that flow into the Bay of Bengal near Vatada in Andhra Pradesh (17° 22' 30½ N and 82° 47' 30½ E) Near the confluence, the muddy and swampy regions exist, harboring the mangrove vegetation which occupy nearly 8 to 10 square kilometers area of the estuarine system. Mangrove ecosystem of Vishakhapatnam lies between latitudes 17° 14' 30½ to 17° 45' 00½ and 83° 16' 25½ to 83° 21' 30½ E. Near the confluence, muddy and swampy regions exist, harboring the mangrove vegetation which occupy nearly 6 to 8 square kilometers of the estuarine system. Vamsadhara estuary located between 18 °32' 73 N½ to 84° 19' 80½ E on the East Coast of India, near Naupada and in between Visakhapatnam and Chilaka lakes. Vamsadhara is the fourth largest river in Andhra Pradesh originated in the Eastern Ghats of Orissa and bifurcated into major and minor branches; major branch merged into Bay of Bengal at Kalingapatnam and minor branch flows through various parts and finally merges into Bay of Bengal at Bhavanapadu. Mud flats and some islands like structures are formed with mangrove plants. Mangrove forest occurs from Meghavaram to Bhavanapadu. The mouth region which is Bhavanapadu to

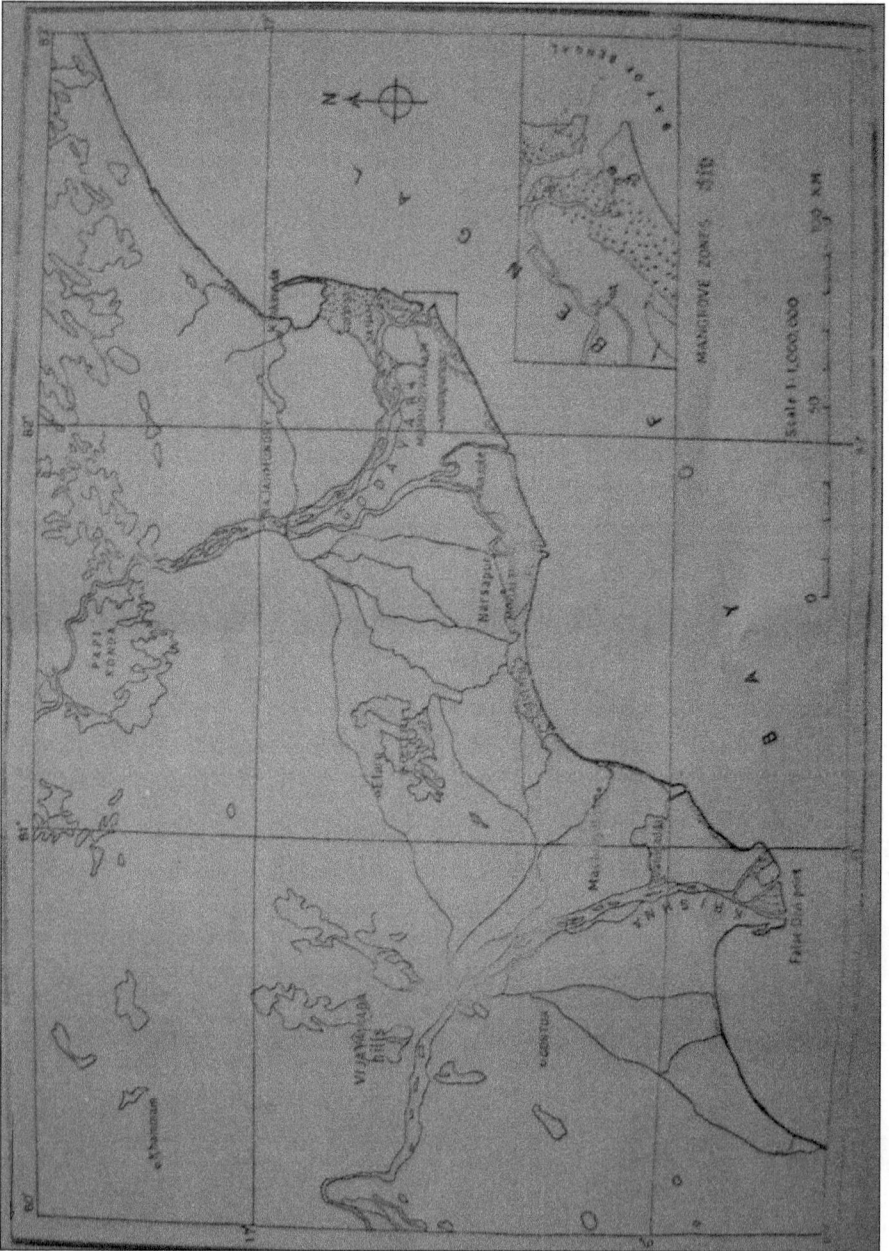

Figure 3.1

Mangroves of Vamsadhara estuary

Mangroves of Vishakhapatnam

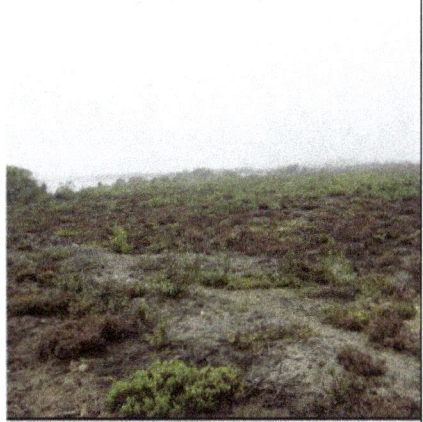

Mangroves of Sarada&Varaha estuarine complex

Plate 3.1

Mangroves of Godavari estuary

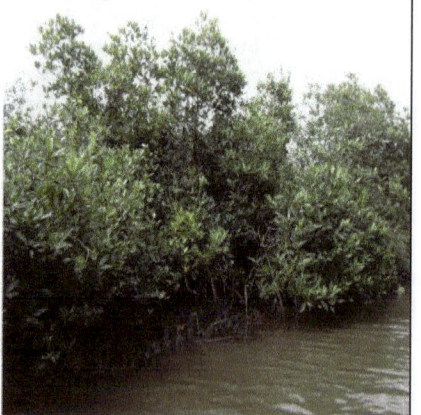

Mangroves of Godavari estuary

Mangroves of Krishna estuary

Plate 3.2

upward region Meghavaram is 10-12 kilometers, along this estuary mangrove forest occurs on either side of the river branch. From waterfront to barren zone mangrove forest spreads only 30 to 45 meters, transition zone comprises the halophytic plants only.

Materials and Methods

Five study sites were selected in Godavari estuary, three in Krishna estuary and in remaining regions single study site was selected. Hydrographical parameters such as air and water temperature, salinity and pH were collected from all study sites from February 2007 to January 2008. Surface water samples were taken form the center of the creek or channel. Temperature, pH and salinity were measured with a thermometer, portable pH meter, and salinometer respectively in all the sampling stations. Water transparency was determined by Secchi disc. Dissolved oxygen was estimated by the method given by Strickland and Parsons (1972), sediment analysis of the soil samples in mangroves habitats was carried out by using the pipette method (Carver,1971).

Data were collected by placing transect from the shore to the interior of the forest and extending transect up to the barren zone of the forest to obtain the real status of the forest cover. Quadrant (4 x 4m) was marked along the transect line at 5 m intervals from water front to inside the mangrove forest and up to end of the forest. A total of 84 transects (10 transects in Vamsadhara, 5 in each station of Visakhapatnam and Sarada and Varaha estuarine complexes, 24 in Krishna estuary and 40 in Godavari estuary) were used for collection of data in all five stations from February 2007 to January 2008. The plants present in each sample plot (16 m²) were counted. Quadrants samples collected from the different stations were analyzed, and densities of different species were estimated. GBH of all plants present in some quadrants was measured to estimate the relative abundance of different diameter classes of mangroves occurring in different stations of mangroves populations of Andhra Pradesh. Since larger tree species are less and many plants appear as small bushes, diameter of the plants was measured just above the ground level. During the course of investigation species of estuarine algae and sea grasses were collected from the pneumatophores and prop roots of the mangroves plants as well as from mud flats.

Results and Discussion

Hydrographical Studies of the Study Sites

Seasonal data collected on hydrographical features of different Mangrove regions/estuarine regions was presented in Tables 3.1A to 3.1E. Among these study sites, there is no much variation in all hydrographical parameters. Air temperatures in all three seasons ranged from 22.5°C to 32.5°C, minimum air temperature (22.5°C) was recorded in Meghadrigedda rain fed drain at Visakhapatnam in the North east monsoon season and higher temperatures reported in the Vamsadhara estuary during the pre-monsoon season (Table 3.1A to 3.1E). Similarly water temperature ranges from 21.5°C to 22.5°C during the period of study. Higher water temperature was recorded in pre monsoon season and lower temperature in the North east monsoon season. Secchi disc values show much variation in three seasons. Water turbidity was higher during the south west monsoon season from June to September months due to the high runoff of materials such as gravel and sand particles from near by areas, therefore low Secchi disc values were recorded. The highest Secchi disc values were recorded in pre monsoon season in all these study sites, when clear water was observed in the study zone (Tables 3.1A to 3.1E). Surface water salinity varies from 12.5‰ to 32.5‰ in the study sites during three seasons of the year. Lower salinities were recorded during the south west monsoons, whereas higher salinities were reported in the pre monsoon season period (Tables 3.1A to 3.1E). The pH of the surface waters varied seasonally and showed a positive relationship to the seasonal changes in the salinity. Minimum pH values were recorded in the Vamsadhara estuary and maximum values in the Krishna and Godavari estuary (Tables 3.1A to 3.1E). No seasonal trend was noticed in the content of the dissolved oxygen in all study sites of the present investigation. Seasonally dissolved oxygen content in the surface waters varied from 6.2 to 7.8 ml./L. In these study sites higher values of the hydrographical features were noticed during the pre monsoon period and lower values during the south west monsoon months.

Table 3.1A: Seasonal Changes in the Hydrographical Parameters of Vamsadhara estuary at Bhavanapadu

	Pre-monsoon (Feb-May)	SW Monsoon (June-Sep)	NE Monsoon (Oct-Jan)
Air Temperature ºC	32.5	29.5	24.0
Water Temperature ºC	25.5	27.0	24.5
Secchi disc (cm)	21	14	18
Salinity (per cent)	22.5	12.5	16.0
pH	7.1	6.9	7.1
D.O. (mg/l)	7.5	7.1	7.3

Table 3.1B: Seasonal Changes in the Hydrographical Parameters of Meghadrigedda Rainfed Drain at Mangrove Habitations, Vishakhapatnam

	Pre-monsoon (Feb-May)	SW Monsoon (June-Sep)	NE Monsoon (Oct-Jan)
Air TemperatureºC	30.5	27.5	22.5
Water Temperature ºC	24.0	23.5	22.0
Secchi disc (cm)	18	13	17
Salinity (per cent)	32.5	28.0	31.5
pH	7.2	7.0	7.2
D.O. (mg/l)	7.1	7.4	7.4

Table 3.1C: Seasonal Changes in the Hydrographical Parameters of Sarada and Varaha Estuarine Complex Near Mangrove Habitations

	Pre-monsoon (Feb-May)	SW Monsoon (June-Sep)	NE Monsoon (Oct-Jan)
Air Temperature°C	28.5	27.5	23.5
Water Temperature °C	24.5	23.0	21.5
Secchi disc (cm)	23	16	21
Salinity (per cent)	30.0	24.5	28.5
pH	7.0	7.2	7.1
D.O (mg/l)	7.5	7.1	7.3

Table 3.1D: Seasonal Changes in the Hydrographical Features of Different Stations of Gautami Godavari Estuary

	Pre-monsoon (Feb-May)	SW Monsoon (June-Sep)	NE Monsoon (Oct-Jan)
Air Temperature°C	29.5	26.5	24.5
Water Temperature °C	25.5	24.5	22.0
Secchi disc (cm)	25	12	22
Salinity (per cent)	28.0	22.5	26.5
pH	7.2	7.1	7.3
D.O (mg/l)	7.4	7.2	7.4

Table 3.1E: Seasonal Changes in the Hydrographical Features of Krishna Estuary

	Pre-monsoon (Feb-May)	SW Monsoon (June-Sep)	NE Monsoon (Oct-Jan)
Air Temperature°C	31.5	29.0	24.5
Water Temperature °C	25.5	24.5	23.5
Secchi disc (cm)	26	18	23
Salinity (per cent)	29.0	26.5	27.5
pH	7.2	7.3	7.3
D.O (mg/l)	7.7	7.8	7.5

Composition of Mangrove Species in Different Study Sites of Andhra Pradesh

Tables 3.2A to 3.2E shows the distribution of mangrove and associated plants species in different study sites of the present study. Of the 15 plant species reported in the quadrant samples of the mangroves populations of Vamsadhara estuary (Table 3.2A), only three plants are true mangrove species, 8 plant species are associated mangroves and remaining 4 plants are halophytes. Table 3.2B shows the plant species present in the Meghadrigedda rain fed drain of Visakhapatnam. Of the 11 plant species reported, two species are true mangroves, 5 species are associated mangroves and remaining four species are halophytes. In the Sarada and Varaha estuarine complex data collected from the quadrant samples reveals that presence of 2 mangrove species, six associated plants and four species belongs to the halophytes (Table 3.2C). Information collected from different stations of Godavari estuary aggregated together and presented in one table only (Table 3.2D). Of the 20 plant species recorded in the quadrants, 9 plants are true mangroves, 7 plants associated mangroves and remaining 4 species are halophytes. Table 3.2E shows the mangroves occur in three different stations of the Krishna estuary. Of the 19 plant species recorded, 9 plants are true mangroves, 5 are associated mangrove species and 5 species are the halophytes.

Table 3.2A: Density of Various Plant Populations Estimated from the Mangroves Forest of Vamsadhara Estuary

Sl.No.	Name of the Species	Family	Density/Individual/ha[1]
1.	*Acanthus ilicifolius* L.	Acanthaceae	473
2.	*Aegiceros corniculatus* (L.) Blanco	Myrsinaceae	24
3.	*Avicennia marina* (Forsk)Vierh	Verbenaceae	56
4.	*Avicennia officinalis* L	Verbenaceae	79
5.	Cynodon dactylon (L.) Pers	Poaceae	142
6.	*Dalbergia spinosa* (Dennst) Mabb.	Fabaceae	14
7.	*Derris trifoliata* Lour	Fabaceae	36
8.	*Excoecaria agallocha* L	Euphorbiaceae	224
9.	*Ipomoea tuba* (schlect.) G.Don	Convolvulaceae	45
10.	*Myriostachya wightiana* (Stend.) Hook.f.	Poaceae	12
11.	*Prosophis chilensis* (Molina) Stuntz.	Mimosaceae	362
12.	*Sesuvium porstulacastrum* (L) L.	Aizoaceae	471
13.	*Suaeda maritima* (L.) Dumm.	Chenopodiaceae	2175
14.	*Suaeda monoica* Forsk.ex. Gmel	Chenopodiaceae	1974
15.	*Suaeda nudiflora.* Moq	Chenopodiaceae	972

Density and Structure of Mangrove Populations

Density of the individual plant species was estimated based on the collected quadrant samples from the different study sites of the mangrove habitats. Station wise total number of quadrant samples in all transects were pooled and calculate the number of plant species per hectare. Density of mangroves varies from station to station (Tables 3.2A to 3.2E). Data on density of mangrove plant populations at Vamsadhara was presented in Table 3.2A. In this station, mangrove forest was dominated by halophytes and density of true mangroves was low (Plate 3.1). They distributed near to the water front only. Plant cover occurs on either side of the estuary in form of the thin strips only. Along the estuary most of the region converted into the aqua forms for commercial production of shrimp and fish. In Visakhapatnam

(Table 3.2B), mangrove forest was dominated by the *Suaeda* populations and maximum density of the forest biomass was covered with this vegetation. In this station most of the forest was converted into industrial and other purposes (Plate 3.1). Vegetation was reduced when comparing with earlier studies (Venkateswarlu *et al.*, 1972 and Venkanna *et al.*, 1989). In the Sarada and Varaha estuarine complex, (Table 3.2C), maximum density 3264 plants per hectare was reported for the *Suaeda monoica*, followed by *Suaeda maritima* with 2846 plants per hectare. Mangrove species density was very minimum (86 plants per hectare) for *Avicennia officinalis*. In the Godavari estuary (Table 3.2D), most of the forest cover was dominated by the plans *Excoecaria*

Table 3.2B: Density of Various Plant Populations from the Mangrove Regions of Visakhapatnam

Sl.No.	Name of the Species	Family	Density/Individual/ha[1]
1.	*Acanthus ilicifolius* L.	Acanthaceae	612
2.	*Aeluropus lagopoides*(L.)Trin. Ex Thw.	Poaceae	580
3.	*Avicennia marina* (Forsk)Vierh	Verbenaceae	78
4.	*Avicennia officinalis* L	Verbenaceae	66
5.	Arthocnemum indicum(Willd.)Moq.	Chenopodiaceae	924
6.	Cressa cretica L.	Convolvulaceae	421
7.	*Derris horrid (Dennst.)* Mabb.	Fabaceae	96
8.	*Excoecaria agallocha* L	Euphorbiaceae	328
9.	*Prosophis chilensis* (Molina) Stuntz.	Mimosaceae	872
10.	*Suaeda maritima* (L.) Dumm.	Chenopodiaceae	2853
11.	*Suaeda monoica* Forsk.ex. Gmel	Chenopodiaceae	3115

Table 3.2C:Density of Various Plant Populations Estimated at Sarada and Varaha Estuarine Complex

Sl.No.	Name of the Species	Family	Density/Individual/ha[1]
1.	*Acanthus ilicifolius* L.	Acanthaceae	964
2.	*Arthocnemum indicum*(Willd.)Moq.	Chenopodiaceae	1278
3.	*Avicennia marina* (Forsk)Vierh	Verbenaceae	98
4.	*Avicennia officinalis* L	Verbenaceae	86
5.	*Clerodendron inerme* (L.) Gaertn.	Verbenaceae	124
6.	*Cressa cretica* L.	Convolvulaceae	76
7.	*Derris horrid (Dennst.)* Mabb.	Fabaceae	84
8.	*Excoecaria agallocha* L	Euphorbiaceae	218
9.	*Prosophis chilensis* (Molina) Stuntz.	Mimosaceae	1452
10.	*Suaeda maritima* (L.) Dumm.	Chenopodiaceae	2846
11.	*Suaeda monoica* Forsk.ex. Gmel	Chenopodiaceae	3264
12.	*Sesuvium porstulacastrum* (L) L.	Aizoaceae	2388

agallocha, Suaeda maritima, Suaeda monoica and *Myrrostachya wightiana*. And these
species are found to be dominant forms and it was nearly 62 per cent of the total
population (Plate 3.2). Minimum density was estimated for the *Bruguiera gymnorrhiza*
with 6 plants per hectare. Similarly in Krishna estuary (Table 3.2E) plants such as
Excoecaria agallocha, Suaeda maritima, Suaeda monoica, Myrrostachya wightiana and
Acanthus ilicfolius are found to be dominant forms in the mangrove forest of Krishna
estuary. Maximum plant density was reported for the species *Suaeda monoica* (5122
plants per hectare) and minimum for the plant *Rhizophora mucronata* with density of
10 plants per hectare (Plate 3.2).

Table 3.2D: Density of Various Plant Populations Estimated from the Five Study Sites of the
Godavari Estuary

Sl.No.	Name of the Species	Family	Density/Individual/ha[1]
1.	*Acanthus ilicifolius* L.	Acanthaceae	2872
2.	*Aegiceros corniculatus* (L.) Blanco	Myrsinaceae	876
3.	*Avicennia marina* (Forsk)Vierh	Verbenaceae	784
4.	*Avicennia officinalis* L	Verbenaceae	2896
5.	*Bruguiera gymnorrhiza(*L.)Lamk.	Rhizophoraceae	06
6.	*Clerodendron inerme* (L.) Gaertn.	Verbenaceae	794
7.	*Ceriops decandra* (Griff.) ding Hou	Rhizophoraceae	374
8.	*Derris trifoliata* Lour	Fabaceae	832
9.	*Dalbergia spinosa* (Dennst) Mabb.	Fabaceae	36
10.	*Excoecaria agallocha* L	Euphorbiaceae	6680
11.	*Ipomoea tuba* (schlect.) G.Don	Convolvulaceae	264
12.	*Lumnitzera racemosa* Willd.	Combretaceae	492
13.	*Myriostachya wightiana* (Stend.) Hook.f.	Poaceae	1742
14.	*Rhizophora apiculata* Blume	Rhizophoraceae	18
15.	*Rhizophora mucronata* Lamk.	Rhizophoraceae	22
16.	*Sonneratia apetala*	Sonneratiaceae	1286
17.	*Sesuvium porstulacastrum* (L) L.	Aizoaceae	3864
18.	*Suaeda maritima* (L.) Dumm.	Chenopodiaceae	7458
19.	*Suaeda monoica* Forsk.ex. Gmel	Chenopodiaceae	4682
20.	*Suaeda nudiflora*. Moq	Chenopodiaceae	2438

GBH Classes of the Mangrove Species in Different Study Sites

The relative abundance of diameter classes of mangroves, associated flora and
halophytes of the different stations (Table 3.3) reveals that nearly 50 per cent plants
growing in the Andhra Pradesh are under 0-10 cm diameter class. Plants such as
*A.ilicifolius, D.trifoliate, M. wightiana, S. maritima, S.monoica, S.porstulacastrum, P.chilensis,
A. indicum, C. cretica, I. tuba, C. dactylon, A.lagopoides and D. horrid* are included in
minimum diameter class (0-10 cm) category. Only 7 species *A. officinalis, A. marina,*

B. gymnorrhiza, E. agallocha, S. apetala, R. apiculata and *R.mucronata* are larger tree species and their average diameter was more than 20 to 30 cm and few plants with more than 30 cm diameter class. Few species like *A. corniculatum, C. decandra, C. inerme, L. racemosa, D. spinosa* having the diameter classes of 0-10 and 10-20 cms. But one species *Clerodendoun inerme* shows the diameter classes of 0-10, 10-20 and 20-30 cms. These observations indicated that mangrove forest is mostly medium and bushy type. In general tree species of larger diameter classes were more abundant in the Godavari estuary and Krishna estuary; whereas in the remaining three stations plants with lesser diameter and bushy type vegetation is present.

Table 3.2E: Density of Various Plant Populations Estimated from the Three Study Sites of the Krishna Estuary

Sl.No.	Name of the Species	Family	Density/Individual/ha[1]
1.	*Acanthus ilicifolius* L.	Acanthaceae	2356
2.	*Aegiceros corniculatus* (L.) Blanco	Myrsinaceae	686
3.	*Avicennia marina* (Forsk)Vierh	Verbenaceae	654
4.	*Avicennia officinalis* L	Verbenaceae	1284
5.	*Bruguiera gymnorrhiza*(L.)Lamk.	Rhizophoraceae	14
6.	*Clerodendron inerme* (L.) Gaertn.	Verbenaceae	94
7.	*Ceriops decandra* (Griff.) ding Hou	Rhizophoraceae	424
8.	*Derris trifoliata* Lour	Fabaceae	32
9.	*Dalbergia spinosa* (Dennst) Mabb.	Fabaceae	218
10.	*Excoecaria agallocha* L	Euphorbiaceae	4676
11.	*Lumnitzera racemosa* Willd.	Combretaceae	328
12.	*Prosophis chilensis* (Molina) Stuntz.	Mimosaceae	856
13.	*Rhizophora apiculata* Blume	Rhizophoraceae	12
14.	*Rhizophora mucronata* Lamk.	Rhizophoraceae	10
15.	*Sonneratia apetala*	Sonneratiaceae	1016
16.	*Sesuvium porstulacastrum* (L) L.	Aizoaceae	254
17.	*Suaeda maritima* (L.) Dumm.	Chenopodiaceae	4421
18.	*Suaeda monoica* Forsk.ex. Gmel	Chenopodiaceae	5122
19.	*Suaeda nudiflora.* Moq	Chenopodiaceae	142

Estuarine Algae and Sea Grasses

During the course of study estuarine algae such as *Bostrychia tenella, Caloglossa leprieurii, Catenella impudica, Chara baltica, Enteromorpha was* collected from the different estuaries of Andhra Pradesh. In the Godavari and Krishna estuary good growth of the all estuarine algae listed in the Table 3.4 was observed. Their growth in terms of biomass was optimum in the months of the December and January. In Vamsadhara, Sarada and Varaha estuaries abundance of estuarine algae was poor and occurs here and there on mud flats and leaves of the mangrove plants. In the station 2 estuarine algae was not observed, mud flats covered by oil spills and pnuematophores also covered by the oil spills and contaminants. This may be due to the water pollution and other industrial activities in the area. In the present study for the first time *Chara baltica* was reported in the estuarine regions of Godavari and Krishna estuaries. Besides the estuarine algae two sea grasses such as *Halophila ovalis* and *Halophila baccarii* were also reported from the Godavari and Krishna estuaries.

Table 3.3: Percentage Frequency of Different Diameter Classes of Mangroves in the Study Sites of the Andhra Pradesh

Plant	Site 1, 2 and 3		Site 4				Site 5		
Diameter (cm)	0-10	10-20	0-10-	10-20	20-30	>30	0-10	10-20	20-30
A. officinalis	62	38	24	18	37	33	31	35	34
A. marina	58	42	30	28	23	19	34	32	34
B. gymnorrhiza	–	–	26	31	18	25	29	33	38
E. agallocha	47	53	32	38	25	5	36	38	26
S. apetala	–	–	26	22	23	24	34	28	38
R. apiculata	–	–	31	24	17	28	28	33	39
R. mucronata	–	–	28	27	21	24	39	33	28
A. corniculatum	68	32	47	43	10	–	61	39	–
C. decandra	–	–	48	31	21	–	57	29	14
C. inerme	58	42	62	38	–	–	67	33	–
L. racemosa	–	–	57	43	–	–	64	36	–
D. spinosa	76	24	72	28	–	–	65	35	–
A. ilicifolius	100	---	100	---	---	---	100	---	---
D. trifoliata	100	---	100	---	---	---	100	---	---
M. wightiana	100	---	100	---	---	---	–	---	---
S. maritima	100	---	100	---	---	---	100	---	---
S. nudiflora	100	---	100	---	---	---	100	---	---
S. monoica	100	---	100	---	---	---	100	---	---
S. porstulacastrum	100	---	100	---	---	---	100	---	---
P. chilensis	100	---	100	---	---	---	100	---	---
A. indicum	100	---	–	---	---	---	–	---	---
C. cretica	100	---	–	---	---	---	–	---	---
I. tuba	100	---	100	---	---	---	–	---	---
C. dactylon	100	–	–	–	–	–	–	---	---
A. lagopoides	100	–	–	–	–	–	–	---	---
D. horrid	100	–	–	–	–	–	–	–	---

Sediment Analysis in Different Stations of the Study

Mangroves live in a habitat where the soils are high concentration of salt and high water content. The soils have a low oxygen content and abundant hydrogen sulphide. Mangrove sediments are characterized by mud, sandy mud and clayey mud. Data on sediment analysis of soil samples in Vamsadhara, Visakhapatnam, Sarada and Varaha estuarine complex shows the similar results with high percentage of sand 35-45, low percentage of silt 10-29 and clay 20-31 per cent.In remaining stations such as Godavari and Krishna estuaries sand content varies from 5-18 per cent, silt content 11-61 per cent and clay content 17-87 per cent. Dense mangrove vegetation was observed in the regions where silt content is >50. Relatively high content of sand has been observed at the land ward region corresponding to their water front region. In general sand content in mangrove sediment is less than 10 per cent. Results of the present study agree with the earlier studies of Bhaskara Rao *et al.,*.

Table 3.4: Estuarine Algae and Sea Grasses Present in Different Estuaries of the Andhra Pradesh

Algae	Station 1	Station 2	Station 3	Station 4	Station 5
Bostrychia tenella	+	–	+	+	+
Caloglossa leprieurii	+	–	+	+	+
Catenella impudica	–	–	–	+	+
Chara baltica	–	–	–	+	+
Enteromorpha sps.	–	–	–	+	+
H. ovalis	–	–	–	+	+
H. baccarii	–	–	–	+	+

Zonation of Mangrove Plants in the Study Area

Zonation is a universal phenomenon where plants occupy different zones based on their adaptability to various environmental parameters. Blasco (1975) and Chapman (1976) observed the Zonation in mangrove ecosystems. Based on the pattern of Zonation real mangroves occupy the water front region and near by areas, followed by the mangroves and associated mangrove plants. Next to the associated mangroves halophytic populations were present. Species of the halophytic population covers the maximum region next to the mangroves and associated mangrove zone. After the halophytic vegetation, no plant population grows; only barren zone is present with out any flora. Twenty to thirty meters from the barren zone terrestrial plant population occurs. Due to the human settlements and urbanization, zonation within the mangrove ecosystem disturbs. In this context, other terrestrial plants also associated with mangrove vegetation. In Vamsadhara estuary, mangrove vegetation is in form of the strips on either side of the estuary and also on the channels of estuary, only three real mangrove species associated with other halophytes and associated mangrove plants. Most of the halophytic region and barren zone converted into the aqua farms for the production of shrimp and fish. Due to these human interference and activities halophytes also associated with true mangrove plants and occur along the waterfront zone (Plate 3.1). In Visakhapatnam, human activities such as construction of industries and other related units in the mangrove zone was completely alter the ecosystem and further spoil the forest (Plate 3.3). Oil spills, loading and unloading operations in the

Algal Blooms in drain

Pollution in mangroves

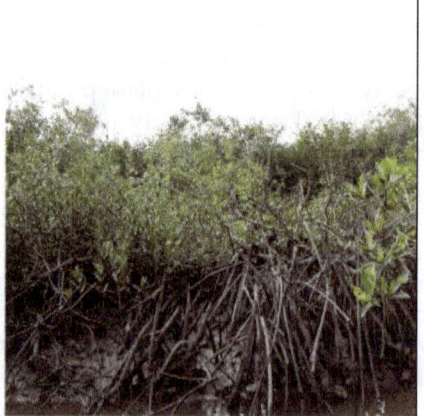

Illegal cuttings of mangroves

Plate 3.3

Aqua farms

Erosion

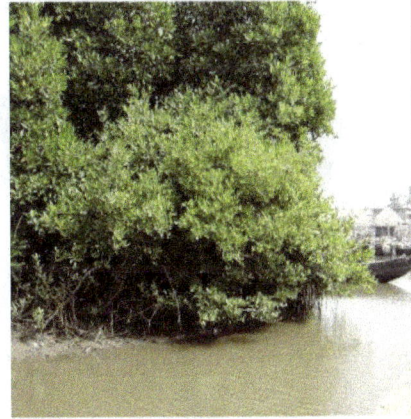

Human settlements in mangroves

Plate 3.4

Prayers by local people to plants

Plate 3.5

port may cause further degradation of the mangrove of this region. As whole, there is no zonation in the mangroves of the Visakhapatnam. Similarly in mangroves of Sarada and Varaha estuarine complex, most of the forest transform into the aqua farms for production of the shrimp and salt production with out any clear zonation as mentioned by Chapman (1976). Ultimately mangroves remained as patches of vegetation here and there. In the Godavari estuary most of the study sites with marked zoning of the plants in systematic manner. Maximum height and growth of the plants were reported in this estuary only. *Rhizophora, Bruguiera, Avicennia, Sonneratia* occupies the water front region. Plants like *Derris, Luminitzera, Clerodendron* and *Ceriops* occupy the next zone followed by the halophytes. In different stations of the Krishna estuary similar trend was observed. Activities such as aqua industry, manufacture of salt, erosion of the creeks (Plate 3.4) human settlements and illegal felling (Plate 3.3) of mangroves may be responsible for further destruction and deterioration of the mangroves in Godavari and Krishna estuaries. Prayers offered by the local inhabitants to the mangrove plants also seen in the Godavari estuary (Plate 3.5).

Mangroves are unique and special plants which grow in estuarine and regularly influenced by high tides and low tides of the sea. Transect studies in different stations of Andhra Pradesh reveals that in Godavari and Krishna estuaries, forest extends up to 60 to 70 meters and height of plants was more than 8 meters. In remaining stations forest extends 20-35 meters with average height of the plants. Intertidal region in the mangrove forest is briefly classified into three regions such as water front, land ward and barren regions. During the high tides of the sea water may touch the barren zone. Accumulation of salt is more in the barren zone, so these regions are converted into the manufacture of the salt. Two distinct zones are demarcated in mangrove forest, one is water front region with mangroves and associated mangroves and another one is land ward zone with halophytic populations. Formation of this halophytic population is due accumulation of more salt in the soil and poor inundation in this region.

Along the coast line of the Andhra Pradesh many rivers are merging into Bay Bengal, but all estuarine environments are not suitable for development and growth of the mangrove plants. That's why so many estuarine regions are remained without mangrove populations along Andhra coast. Salinity and percentage of the silt content in the soil favors the mangrove vegetation. Silt and sand composition in the sediments, soil salinity, sediment load from the rivers and organic matter in the waters may be responsible for development and good growth of mangroves

Acknowledgements

I express my sincere thanks to Head, Department of Botany for providing facilities and extend my grateful thanks to Dr. P.Prayaga Murthy of Botany Department for his timely help during this study. Financial assistance from the UGC-SAP is gratefully acknowledged.

References

1.   Banerjee, L.K. and Rao, T.A. (1990). *Mangroves of Orissa Coast and their Ecology.* Bishen Singh Mahendra Pal Singh, Dehradun, India, 1–118 pp.

2. Banerjee, L.K., Sastry, A.R.K and Nayar, M.P. (1989). *Mangroves in India: Identification Manual*. Botanical Survey of India, 1–113 pp.

3. Blasco, F. (1975). *The Mangroves of India*. French Institute, Fr. Sect. Sci. Tech., Pondicherry, 14: 1–175.

4. Bhaskara Rao, V., Narasimha Rao, G.M., Sarma, G.V. S. and Krishna Rao, B. (1992). Mangrove and its sediment characters in Godavari estuary, east coast of India. *Indian J. Mar. Sci.*, 21: 64–66.

5. Bunt, J.S. (1992). Introduction to mangrove ecosystem. In: *Tropical Mangrove Ecosystem*, (Eds.) A. I. Robertson and D.M. Alongi. American Geophysical Union, Washington, D.C., 1–6 pp.

6. Carver, R.E. (1971). *Sedimentary Petrology*. John Wiley and Sons Inc., London, 1–653 pp.

7. Chapman, V.J. (1976). *Mangroves Vegetation*. Cramer, Vaduz, 1–326 pp.

8. Duke, N.C. (1992). Mangrove floristics and biogeography. In: *Tropical Mangrove Ecosystem*, (Eds.) A.I. Robertson and D.M. Alongi. American Geophysical Union, Washington, D.C., 63–100 pp.

9. F.A.O. (2007). *The Worlds Mangroves 1980–2005*. FAO Forestry Paper 153, Rome, pp. 1–124.

10. International Union for Conservation of Nature and Natural Resources (2006). Mangroves for the future: Reducing vulnerability and sustaining livelihoods. *Coastal Ecosystems*, 1: 1–4.

11. Krishna Rao, B. and Narasimha Rao, G.M. (1992). Mangrove resources in Godavari and Krishna estuaries: Management and perspectives. *Indian J. Landscape Systems*, 15: 96–99.

12. Mathuda, G.S. (1959). Mangrove vegetation of Godavari delta. In: *Proceedings of the Mangrove Symposium*, Calcutta, p. 66.

13. Narasimha Rao, G.M. (1989). Ecological studies on some estuarine and marine algae. *Ph.D. Thesis*, Andhra University, Andhra Pradesh, p. 1–129.

14. Narasimha Rao, G.M. (1995). Seasonal growth, biomass, and reproductive behavior three species of red algae in Godavari estuary. *India J. Phycol.*, 31(2): 209–214.

15. Narasimha Rao, G.M. (2008). Mangrove populations of Visakhapatnam and Sarada and Varaha estuarine complex, India. *International Journal of Plant Sciences*, 3(2): 686–687.

16. Narasimha Rao, G.M. and Dora, S.V.V.S.N. (2009). An experimental approach to mangroves of Godavari estuary. *Indian Jour. For.*, 32(3): 263–265.

17. Narasimha Rao, G.M. and Vanilla Kumari (1997). Eco-physiological studies on *Bostrychia tenella* in Sarada and Varaha estuarine complex. *Phykos*, 36(1–2): 89–92.

18. Narasimha Rao, G.M. and Venkanna, P. (1996). Macro algae of the Sarada and Varaha estuarine complex. *Indian Jour. For.*, 19: 157–158.

19. Narasimha Rao, G.M. and Murthy, P.P. (2010). Mangrove populations of Vamsadhara estuary. *International Journal of Plant Sciences*, 5: (In Press).

20. Narasimha Rao, G.M., Rao, B.N., Reddy, K.S.N. and Varma, D. (2000). Ecology and physiology of *Halophila ovalis* (R.Br.) Hk.F. at Pandi back waters, Andhra Pradesh. *Ecol. Env. and Cons.*, 6(3): 297–300.

21. Raju, D.C.S. (1968). The vegetation on West Godavari: A study of tropical delta. In: *Proceedings of the Symposium on Recent Advances in Tropical Ecology, Part 1.* Banaras Hindu University, Varanasi, 348–358 pp.

22. Rao, R.S. (1959). Observations on the mangrove vegetation of the Godavari estuary. In: *Proceedings of the Mangrove Symposium*, Kolkata, 36–44 pp.

23. Saenger, P., Hegerl, E.J., and Davie, J.D.S. (1983). *Global Status of Mangrove Ecosystems.* Commission on Ecology Papers No. 3. Gland, Switzerland, World Conservation Union (IUCN).

24. Sidhu, S.A. (1963). Studies on the mangroves of India. I. East Godavari region. *Indian Forester*, 86: 337–351.

25. Spalding, M., Blasco, F. and Field, C. (1997). *World Mangrove Atlas.* International Society for Mangrove Ecology, World Conservation Monitoring Centre.

26. Strickland, J.D.H. and Parsons, T.R. (1972). A practical handbook of sea water analysis. *Fish. Res. Bd. Can. Bull.*, 167: 1–311.

27. Twilley, R.R. (1998). Mangrove wetlands. In: *Southern Forested Wetlands: Ecology and Management,* (Eds.) M.G. Messina and W.H.Conner. Lewis Publisher, Boca Raton, 445–473.

28. Umamaheswara Rao, M. and Narasimha Rao, G.M. (1988). Mangrove populations of the Godavari delta complex. *Indian J. Mar. Sci.*, p. 326–329.

29. Valiela, I., Bowen, J.L. and York, J.K. (2001). Mangrove forests: one of the world's threatened major tropical environments. *Bio. Sci.*, 51: 807–815.

30. Venkanna, P. (1991). Present status of the estuarine flora of the Godavari and the Krishna. *J. Bombay Nat. Hist. Soc.*, 88(1): 47–54.

31. Venkanna, P.P. and Narasimha Rao, G.M. (1993). Distribution pattern of the Mangroves in the Krishna estuary. *Indian Jour. For.*, 16(1): 48–53.

32. Venkanna, P., Buchi Raju, J. and Narasimha Rao, G.M., (1989). Mangrove and associated flora of Visakhapatnam. *Geobios New Reports*, 8: 157–158.

33. Venkateswara Rao, Y. and Narasimha Rao, G.M. (2008). *Mangrove Populations of Krishna Estuary.* National seminar on river Krishna through ages, Hrinkara Tirtha, Acharya Nagarjuna University, Nov.8th–9th, Guntur.

34. Venkateswarlu, J., Murthy, P.V.B. and Rao, P.N. (1972). *The Flora of Visakhapatnam.* A.P. Academy of Sciences, Hyderabad, 1–126 pp.

Biodiversity of Aquatic Reseources (2012)
Editors: Mamta Rawat & Sumit Dookia
Published by: DAYA PUBLISHING HOUSE, NEW DELHI

Pages 50-59

Chapter 4

Analysis of Growth Forms of the Macrophytes in Kharungpat Lake Manipur, North-East India

K. Khelchandra Singh*, B. Manihar Sharma
and Khuraijam Usha

Ecology Laboratory,
Department of Life Sciences, Manipur University,
Canchipur, Imphal – 795 003, Manipur

ABSTRACT

Kharungpat Lake is situated in the Thoubal District of Manipur at a distance of about 30 km from Imphal Town. During the study period 54 (fifty four) macrophytic species were recorded from Kharungpat Lake. The vegetation of the Lake is classified into (i) Submerged (ii) Rooted with floating leaves (iii) Free floating, and (iv) Emergent. Out of the 54 macrophytic species found in the lake, 7 species were recorded under submerged group, 6 species were recorded as rooted with floating leaves, 8 species in free floating and 33 species were recorded in the emergent group. These marcrophytes belong to 14 growth forms classes out of the 23 growth forms recognized by Hogeweg and Brenkert (1969). The 14 growth-forms are *Ceratophyllids, Charids, Eichhorniids, Helophytes, Lemnids, Magnolemnids, Marsileids, Nymphaeids, Parvopotamids, Pseudohydrophytes, Rhizopleustohelophytes, Trapids, Utriculariids* and *Vallisneriids*. Out of the 14 growth

* Corresponding Author E-mail: k_khel@yahoo.com

forms recorded Helophytes, Rhizopleustohelophytes, Nymphaeids and Magnolemnids has maximum number of species, *i.e.*, 20, 9, 5 and 4 respectively. Pseudohydrophytes recorded 3 species whereas Utriculariids, Parvopotamids, Eichhorniids and Lemnids reported 2 species each Ceratophyllids, Marsileids, Trapids, Charids and Vallisneriids recorded one species each.

Keywords: Growth forms, Macrophytes, Kharungpat Lake, Manipur.

Introduction

Freshwater macrophytes play a vital role in determining the productive nature of the water body and they are recognised as the most productive plant communities. These macrophytic species play a significant role in regulating the structure of a lake ecosystem. The macrophytes are also responsible for the regulation and stabilization of mineral cycling in the water bodies and hence they serve as indicators for the possible degree of damage in the ecosystem (Pieczynska and Ozimek, 1976). The macrophytes are differentiated into the various growth forms on the basis of their morphological differences on the atmospheric, aquatic or edaphic condition of their relative size and anatomical structure (stem and leaves). A particular growth form comprises plants of comparable structure and similar relations to their physical environment.

The structure of a stand of vegetation is defined as a complexity of the growth form spectrum in combination with the horizontal and vertical distribution of the growth forms represented in the stand (Segal, 1966). Misra (1968) recognised several serial communities which were termed as 'associes'. Hartog and Segal (1964) proposed a number of categories of growth forms for aquatic plants in the Netherlands. Hogeweg and Brenkert (1969) made efforts to identify the growth forms classification of Hartog and Segal (1964) with the Indian aquatic vegetations with slight alterations. The literature on the growth forms of the plants was critically reviewed in sufficient details by Hutchinson (1975) where he modified the system by including the emergent vegetations.

Materials and Methods

Kharungpat Lake is situated in Thoubal district of Manipur at a distance of about 30 km from Imphal town. At present, the lake is in enhanced eutrophic state due to heavy encroachment and piscicultural activities carried out by the people living in the vicinity of the lake. For detailed study, the lake is divided into four study sites, *viz.*, as Panchao (Site I), Pangalpat (Site II), Kambong Lairem (Site III) and Kharungpat Khong (Shamu Lanpham) (Site IV).

The macrophytic samples were collected at regular monthly intervals from different study sites (Site 1 to IV) in the lake for a period of two years from January 2008 to December 2009. The floristic composition along with seasonal distribution of the species in the different study sites were assessed by Quadrat method (Curtis 1959, Misra 1968). Growth form determination of the vegetation was preceded by detailed floristic studies. The growth form classification was adopted after the method given by Hogeweg and Brenkert (1969).

Results and Discussion

In the present investigation a total of 54 (fifty four) macrophytic species belonging to 28 (twenty eight) families were found distributed in the lake. Floristic survey of the lake has shown the presence of 32 (thirty two) species in Site-I, 27 (twenty seven) species in Site-II, 34 (thirty four) species in Site-III and 25 (twenty five) species in Site-IV. The floristic composition of the lake is furnished in Table 4.1. *Alternanthera philoxeroides, Azolla pinnata, Ceratophyllum demersum, Echinochloa stagnina, Eichhornia crassipes, Enhydra fluctuans, Ludwigia adscendens, Pistia stratiotes, Salvinia cucullata* and *Zizania latifolia* were found to occur in all the study sites. *Euryale ferox, Nelumbo nucifera, Nymphaea stellata, Oenanthe javanica, Potamogeton crispus* etc. were present only in site-III.

Table 4.1: Floristic Composition of Kharungpat Lake, Manipur

Sl.No.	Name of Species	Family	Site I	Site II	Site III	Site IV
1.	*Alisma plantago aquatica* Linn.	Alismataceae	+	−	−	−
2.	*Alternanthera philoxeroides* (Mart) Griseb.	Amaranthaceae	+	+	+	+
3.	*Alternanthera sessiles* (Linn.) R.Br.	Amaranthaceae	+	−	−	+
4.	*Azolla pinnata* R.Br.	Azollaceae	+	+	+	+
5.	*Brachiaria mutica* (Forsk). Stapf.	Poaceae	+	+	+	+
6.	*Ceratophyllum demersum* Linn.	Ceratophyllaceae	+	+	+	+
7.	*Ceratopteris thalictroides* (Linn.)	Ceratopteridaceae	−	+	+	−
8.	*Chara zeylanica* Willd.	Characeae	+	−	−	+
9.	*Commelina bengalensis* Linn.	Commelinaceae	+	−	−	−
10.	*Cymbopogon nardus* ((Linn.) Rendle.	Poaceae	+	−	+	−
11.	*Cyperus corymbosus* Rottb.	Cyperaceae	+	+	−	−
12.	*Cyperus distans* Linn.f.	Cyperaceae	+	−	+	+
13.	*Echinochloa stagnina* (Retz.) P. Beauv.	Poaceae	+	+	+	+
14.	*Eichhornia crassipes* (Mart.) Solms.	Pontederiaceae	+	+	+	+
15.	*Enhydra fluctuans* Lour.	Asteraceae	+	+	+	+
16.	*Euryale ferox* Salisb.	Nymphaeaceae	−	−	+	−
17.	*Hydrilla verticillata* (Linn. F.). Royle.	Hydrocharitaceae	−	+	+	+
18.	*Hygroryza aristata* (Retz.) Nees.	Poaceae	+	+	+	+
19.	*Imperata cyllindrica* (Linn.)	Poaceae	+	−	−	+
20.	*Ipomoea aquatica* Forsk.	Convolvulaceae	+	+	+	−
21.	*Ipomoea fistulosa* Mart.	Convolvulaceae	+	−	−	+
22.	*Kyllinga tenuifolia* Steud.	Cyperaceae	−	−	−	+
23.	*Lemna minor* Linn.	Lemnaceae	−	−	+	−
24.	*Leersia hexandra* Swartz.	Poaceae	+	+	−	−
25.	*Ludwigia adscendens* (Linn.) Hara	Onagraceae	+	+	+	+
26.	*Marsilea quadrifoliata* Linn.	Marsileaceae	+	−	+	+

Contd...

Table 4.1–Contd...

Sl.No.	Name of Species	Family	Site I	Site II	Site III	Site IV
27.	*Monochoria hastata* (Linn.) Solms.	Pontederiaceae	–	+	–	–
28.	*Nelumbo nucifera* Gaertn.	Nelumbonaceae	–	–	+	–
29.	*Neptunia prostrata* Bail.	Mimosaceae	–	–	+	–
30.	*Nymphoides cristatum* (Roxb.) O. Kuntze	Menyanthaceae	–	+	+	–
31.	*Nymphaea pubescens* Willd.	Nymphaeaceae	–	+	–	–
32.	*Nymphaea stellata* Willd.	Nymphaeaceae	–	–	+	–
33.	*Oenanthe javanica* (Bl) D.C.	Apiaceae	–	–	+	–
34.	*Oryza officinalis* Wall en Watt.	Poaceae	–	+	–	–
35.	*Oryza rufipogon* Griff.	Poaceae	+	+	–	–
36.	*Polygonum glabrum* Willd	Polygonaceae	+	+	–	–
37.	*Polygonum hydropiper* Linn.	Polygonaceae	–	–	+	+
38.	*Pistia stratiotes* Linn.	Araceae	+	+	+	+
39.	*Phragmites karka* (Retz.) Trin. Ex Stand.	Poaceae	+	–	+	+
40.	*Potamogeton crispus* Linn.	Potamogetonaceae	–	–	+	–
41.	*Pseudoraphis minuta* (Mez) Pilger	Poaceae	+	+	+	–
42.	*Pseudoraphis spinescens* (R.Br.) Vickery	Poaceae	–	–	–	+
43.	*Ranunculus scleratus* Linn.	Ranunculaceae	+	–	–	+
44.	*Rumex maritimus* Linn.	Polygonaceae	+	+	–	–
45.	*Sacciolepsis myosuroides* (R.Br.) A. Camus	Poaceae	–	–	+	–
46.	*Saccharum procerum* Roxb.	Poaceae	+	–	–	+
47.	*Sagittaria sinensis* Linn.	Alismataceae	+	–	–	–
48.	*Salvinia cucullata* Roxb.	Salvinaceae	+	+	+	+
49.	*Salvinia natans* Hoffm	Salvinaceae	–	–	+	–
50.	*Trapa bispinosa* Roxb.	Trapaceae	–	–	+	–
51.	*Utricularia flexuosa* Vahl.	Lentibulariaceae	–	+	+	–
52.	*Utricularia exoleta* R.Br.	Lentibulariaceae	–	+	–	–
53.	*Vallisnaria spiralis* Linn.	Hydrocharitaceae	–	–	+	–
54.	*Zizania latifolia* (Griseb.) Stapf.	Poaceae	+	+	+	+
	Total number of species		32	27	34	25

'+' and '–' sign indicate the presence and absence of species.

The group wise classification of macrophytes in the present lake is presented in Table 4.2. Generally the vegetation of Kharungpat Lake has been classified into four category *viz.* (i) Submerged (12.96 per cent) (ii) Rooted with floating leaves (11.11 per cent) (iii) Free floating (14.81 per cent) (iv) Emergent (61.11 per cent). These 54 macrophytes belonged to 14 Growth forms classes out of the 23 growth forms recorgnised by Hogeweg and Brenkert (1969). The 14 growth-forms are *Ceratophyllids,*

*Charids, Eichhorniids, Helophytes, Lemnids, Magnolemnids, Marsileids, Nymphaeids,
Parvopotamids, Pseudohydrophytes, Rhizopleustohelophytes, Trapids, Utriculariids* and
Vallisneriids. The growth form categories of the macrophytic plant species of
Kharungpat Lake are depicted in Table 4.3.

Table 4.2: Data on Groupwise Classification of the Macrophytes in Kharungpat Lake,
Manipur

Sl.No.	Group/Category	Name of Species	Number of Species	Percentage of Composition
1.	Submerged species	Ceratophyllum demersum Chara zeylanica Hydrilla verticillata Potamogeton crispus Utricularia exoleta Utricularia flexuosa Vallisnaria spiralis	7	12.96
2.	Rooted with floating leaves species	Euryale ferox Nelumbo nucifera Nymphoides cristatum Nymphaea pubescens Nymphaea stellata Trapa bispinosa	6	11.11
3.	Free-floating species	Azolla pinnata Ceratopteris thalictroides Eichhornia crassipes Lemna minor Neptunia prostrata Pistia stratiotes Salvinia cucullata Salvinia natans	8	14.81
4.	Emergent species	Alisma plantago aquatica Alternanthera philoxeroides Alternanthera sessiles Brachiaria mutica Commelina bengalensis Cymbopogon nardus Cyperus distans Eichinochloa stagnina Enhydra fluctuans Hygroryza aristata Imperata cyllindrica Ipomoea aquatica Ipomoea fistulosa Kyllinga tenuifolia Leersia hexandra Ludwigia adscendens Marsilea quadrifoliata Monochoria hastata Neptunia prostrata Oenanthe javanica Oryza officinalis	33	61.11

Contd...

Table 4.2–Contd...

Sl.No.	Group/Category	Name of Species	Number of Species	Percentage of Composition
		Oryza rufipogon		
		Polygonum glabrum		
		Polygonum hydropiper		
		Phragmites karka		
		Pseudoraphis minuta		
		Pseudoraphis spinescens		
		Ranunculus scleratus		
		Rumex maritimus		
		Sacciolepsis myosuroides		
		Saccharum procerum		
		Sagittaria sinensis		
		Zizania latifolia		

Out of the total fourteen (14) growth forms, Helophytes, Rhizopleustohelophytes, Nymphaeids and Magnolemnids represented the dominant growth forms over the others. The helophytes category are represented by twenty (20) emergent species *Viz. Brachiaria mutica, Echinochloa stagnina, Phragmites karka, Pseudorphis minuta, Zizania latifolia* etc. Rhizopleustohelophytes form a compact aerial floating mass on the water surface and this growth form included nine (9) species *viz. Alternanthera philoxeroides, Enhydra fluctuans, Hygroryza aristata, Ipomoea aquatica, Ludwigia adscendens* etc.

The Nymphaeids consisted of five (5) *species viz. Euryale ferox, Nelumbo nucifera, Nymphaea pubescens, Nymphaea* and *Nymphoides cristatum* which are rooted, submerged with floating-blades. Magnolemnids includes four (4) free-floating species *viz. Neptunia prostrata, Pistia stratiotes, Salivinia cucullata* and *Salivinia natans.* Utriculariids (*Utricularia flexuosa, Utricularia exoleta*), Parvopotamids (*Hydrilla verticillata, Potamogeton crispus*), Eicchorniids (*Eichhornia crassipes, Ceratopteris thalictroides*) and Lemnids (*Azolla pinnata, Lemna minor*) are represented by two (2) species each. The lowest number of species *i.e.* only one species each is recorded in various growth forms *viz.* Ceratophyllids (*Ceratophyllum demersum*), Marsileids (*Marsilia quadrifoliata*), Trapids (*Trapa bispinosa*), Charids (*Chara zeylanica*) and Vallisneriids (*Vallisneria spiralis*).

When compared with the findings of Hogeweg and Brenkert (1969) who reported 23 different growth-form classes of Macrophytes in the various regions of India, the present finding of 14 growth-forms classes in the Kharungpat Lake, Manipur is found to be comparatively low with the one recorded by Devi (2002) in Ikop Lake with sixteen (16) growth forms. Similar observation of 14 growth forms was recorded by Devi (2008) (14 growth forms classes) in Oksoipat Lake, Manipur. But it is found to be higher when compared with the findings of Devi (1993) and Devi and Sharma (2008) (8 growth form classes) in the Loktak Lake, Manipur; Devi (1993) (13 growth-form classes) in Waithou Lake, Manipur; Devi and Sharma (2008) in Laisoipat Lake, Manipur and Usha (2002) in Poiroupat Lake (11 growth forms).

The luxuriant growth of macrophytes in the present study indicated that the lake is eutrophic in nature. Therefore it is high time to take up conservative measures before further deterioration leading to extinction of the lake takes place.

Table 4.3: Growth Forms of Aquatic Macrophytes of Kharungpat Lake, Manipur

Sl.No.	Growth form Category	Relation to (Solid) Substratum	Relation to Aquatics Aerial Environment	Habit form	Name of Species	No. of Spcies (per cent)
1.	Cerato-phyllids	Free Floating attached	Completely submerged	Whorls of finely dissected foliage submerge flowering	*Ceratophyllum demersum*	1 (1.85 per cent)
2.	Charids	Attached	Completely submerged	Branched (verticillate) thallus	*Chara zeylanica*	1 (1.85 per cent)
3.	Eichhor-niids	Fee-floating or rooting soft partly organic sediment	Leaves emerged (at least the greater part of leaf and petiole)	Stoloniferous rosettes with floating to emerged 'petiolate' leaves	*Eichhornia crassipes,* *Ceratopteris thalictroides*	2 (3.70 per cent)
4.	Helophytes	Rooted	Emergent	When full-grown, vegetative parts and generative parts (almost) completely emerged as a rule.	*Brachiaria mutica,* *Cymbopogon nardus,* *Cyperus corymbosus,* *Cyperus distans,* *Echinochloa stagnina,* *Imperata cylindrica,* *Kyllinga tenuifolia,* *Leersia hexandra,* *Oryza officinalis,* *Oryza rufipogon,* *Polygonun glabrum,* *Polygonum hydropiper,* *Phragmites karka,* *Pseudoraphis minuta,* *Pseudoraphis spinescens,* *Ranunculus scleratus,* *Rumex maritinus,* *Sacciolepsis myosuroides,* *Saccharum procerum,* *Zizania latifolia*	20 (37.03 per cent)

Contd...

Table 4.3—Contd...

Sl.No.	Growth form Category	Relation to (Solid) Substratum	Relation to Aquatics Aerial Environment	Habit form	Name of Species	No. ol Spcies (per cent)
5.	Lemnids	Free-floating	Floating on the surface	Small to minute, sometimes not differentiated into leaved and stems ('corm' plantlets)	*Azolla pinnata, Lemna minor*	2 (3.70 per cent)
6.	Magnolem-nids	Free floating, sometimes secondary + rooting	Floating on the surface	Stoloniferous or branching rosette or stem like structure with floating or emergent sessile leaves	*Pistia stratiotes, Salvinia cucullata, Salvinia natans*	3 (5.55 per cent)
7.	Marsileids	Rooted	Floating leaf-blades or emerged	Rhizomatous, stemless; sometimes during life-cycle completely submerged with undivided leaves	*Marsilea quadrifoliata*	1 (1.85 per cent)
8.	Nymphaeids	Rooted	Submerged but the large leaf blades floating on the surface or emerging	Leaves in rosette or from short rhizome on long petioles	*Euryale ferox, Nelumbo nucifera, Nymphaea stellata, Nymphaea pubescens, Nymphoides cristatum*	5 (9.26 per cent)
9.	Parvo-potamids	Rooting during atleast a considerable part of the life cycle	Submerged	Usually developing long stems with undissected narrow leaves.	*Hydrilla verticillata, Potamogeton crispus*	2 (3.70 per cent)
10.	Pseudo-hydrophytes	Rooted	Emergent	When full-grown helophytic, but when younger with different in submerged foliage adopted to an aquatic habitat and emerged aerial organs	*Alisma plantago aquatica, Monochoria hastale, Sagittaria sinensis*	3 (5.55 per cent)

Contd...

Table 4.3–Contd...

Sl.No.	Growth form Category	Relation to (Solid) Substratum	Relation to Aquatics Aerial Environment	Habit form	Name of Species	No. of Spcies (per cent)
11.	Rhizopleusto-helophytes	Rooted in bank	Emergent	Stems floating on or just below the surface producing aerial leaves or leaf-bearing shoots	Alternanthera philoxeroides, Alternanthera sessiles, Commelina bengalensis, Enhydra fluctuans, Hygroryza aristata, Ipomoea aquatica, Ipomoea fistulosa, Ludwigia adscendens, Neptunia prostrata, Oenanthe javanica	10 (18.18 per cent)
12.	Trapids	Rooted or unattached	Floating leaves, sometimes foliage emergent	Strong, branched, long 'stems' producing tufts of leaves floating to emergent leaves; divided chlorenchymatic adventitious roots	Trapa bispinosa	1 (1.85 per cent)
13.	Utriculariids	Free floating never rooting	Submerged	Finely dissected leaves, emerged inflorescens	Utricularia flexuosa, Utricularia exoleta	2 (3.70 per cent)
14.	Vallisneriids	Rooted	Submerged	Broader flabby leaves from rosette	Vallisneria spiralis	1 (1.85 per cent)

References

1. Curtis, J.T. (1959). *The Vegetation of Wisconsion*, Univ. Wisconsin Press, Madison.

2. Devi, N. Beenakumari (1993). Phytosociology, primary production and nutrient status of the macrophytes of Loktak Lake, Manipur. *Ph.D. Thesis*, Manipur University, Manipur.

3. Devi, Ch. Nivanonee. (2002). Vegetational structure and primary production of the macrophytes of Ikop lake, Manipur. *Ph.D. Thesis*, Manipur University, Manipur.

4. Devi, S. Umeshori. (2008). Ecological analysis of the macrophytes in Oksoipat lake (Bishnupur), Manipur. *Ph.D. Thesis*, Manipur University, Manipur.

5. Devi N.B. and Sharma, B.M. (2008). Growth forms and seasonal variations of the macrophytes of the Loktak lake, Manipur, India. *Frontier Botanist (Special issue)*, 65–70 pp.

6. Hartog, C. den and S. Segal (1964). A new classification of water plant communities. *Acta. Bot. Neerl.*, 13: 367–393.

7. Hogeweg, P. and A.L. Brenkert (1969). Structure of aquatic vegetation; A comparison of aquatic vegetation in India. *Netherlands and Czechoslovakia. Trop. Ecol.*, 10(1): 139–162.

8. Hutchinson, G.E. (1975). *A Treatise on Limnology, Vol. 3: Limnological Botany*. John Wiley and Sons, New York.

9. Misra, R. (1968). *Ecology Workbook*. Oxford and IBH Publ. Co., New Delhi.

10. Segal, S. (1966). Oecology. Van hogere water Plankton. *Vakbal. Biol.*, 46: 138–149.

11. Usha, K. (2002). Macrophyte ecology of Poiroupat Lake, Manipur. *Ph.D. Thesis*, Manipur University, Manipur.

Biodiversity of Aquatic Reseources (2012)
Editors: *Mamta Rawat & Sumit Dookia*
Published by: DAYA PUBLISHING HOUSE, NEW DELHI

Pages 60–74

Chapter 5

Phytosociological Studies of the Aquatic Macrophytes in Poiroupat Lake, Manipur (North-East India)

Kh. Usha*, B. Manihar Sharma and K. Khelchandra Singh

Ecology Laboratory,
Department of Life Sciences, Manipur University,
Canchipur, Imphal – 795 003, Manipur

ABSTRACT

Poiroupat lake is situated in the Imphal East district of Manipur at a distance of 15 Km from the Imphal town. The lake is located about 881m above mean sea level with an area of 0.16 sq km. During the study period 30 macrophytic species were recorded from Poiroupat lake and they have been classified under submerged (3 species), rooted with floating leaf (5 species), free floating (6 species) and emergent (16 species) groups respectively. In all the study sites, maximum frequency was shown by *Alternanthera philoxeroides*, (30.0 per cent in winter to 90.0 per cent in summer) *Utricularia flexuosa* (50.0 per cent to 90.0 per cent), *Hydrilla verticillata* (20.0 per cent to 90.0 per cent) also exhibited maximum frequency only in sites IV and II. *Alternanthera philoxeroides, Hydrilla verticillata, Enhydra fluctuans, Nymphoides indicum, Ludwigia adscendens* were present in all the study sites. The maximum density was found in *Ceratophyllum demersum* with values ranging

* Corresponding Author E-mail: khuraijamusha36@gmail.com

from 19.2 to 72.0 plants m^{-2}. *Alternanthera philoxeroides* had density values ranging from 3.2 to 57.8 plants m^{-2}; *Eichhornia crassipes*, 3.2 to 40.0 plants m^{-2}; *Hydrilla verticillata*, 8.0 to 40.0 plants m^{-2}; *Ludwigia adscendens*, 11.2 to 38.4 plants m^{-2}. The abundance values of the different species showed wide variations, Maximum abundance was shown by *Ludwigia adscendens* (4.8–96.0 plants m^{-2}) in all the study sites among the regular occurring macrophytes. In all the study sites the highest value of IVI during the entire study period was recorded in *Ceratophyllum demersum* with a maximum value of 50.8 in site II. This was followed by *Ludwigia adscendens* with a value of 20.7 in site I. The maximum values of frequency, density, abundance and IVI of the different species were recorded during rainy season, which influence favourable climatic conditions for the successful and luxuriant growth of the macrophytes.

Keywords: *Phytosociology, Macrophyte, Poiroupat Lake, Manipur, Northeast India.*

Introduction

Phytosociology is the branch of ecology dealing with the origin, structure and composition as well as classification of the plants. It is the study of the vegetations occurring in a particular ecosystem (Ambasht, 1969). The qualitative and quantitative characteristics of the different communities are determined by a large number of environmental factors. Various inter-related parameters such as water depth, substrate texture, turbidity, nutrient concentrations and siltation rate etc. are the important factors for the growth and distribution of aquatic plants in a lake (Nichols, 1992). Investigation of structural aspects of the plant species in the aquatic ecosystem has become a pre-requisite for understanding the ecological dynamics of the ecosystem.

Vegetation represents the sum total growth of various forms of plant populations in the areas whereas communities constitute the ecosystem. One of the important aspects of the community dynamics is the periodical variation in the vegetation composition and structure mainly attributed to their interactions with the environmental variables (Tansley, 1935). Thus phytosociological studies are vital for the better understanding of the structure and dynamics of the vegetations.

Study Area

Poiroupat is situated in the Imphal East district of Manipur at a distance of 15km from the Imphal town. It lies between 24°40'6.24"N to 24°40'6.71"N latitude and 93°58'9.82" E to 93°58'10.25" longitude. The lake is about 881 m above mean sea level with an area of 0.16 sq. km. It is a much aged, and eutrophic and it is one of the endangered lakes of Manipur. Rain and groundwater are the only source of water to this lake. At present, the lake is threatened to extinction due to artificial eutrophication which is attributed to encroachment and pisciculture.

The depth of water has a minimum value of 0.79m and maximum of 1.92m, during the different seasons of the study period. The mean depth of the lake is 1.42m. The lake is more or less oval in shape. The basin is saucer shaped. The slope of the basin is found to be gentle. The total volume of the lake is about 0.00023 km^3. The shore line measures about 1.67km and the shoreline development is 1.17km. The development of volume is estimated as 2.22. The morphometry as well as bathymetric characters of Poiroupat lake is presented in Table 5.1.

Table 5.1: Morphometry of Poiroupat Lake

Altitude(m)	881
Latitude	24°40'6.24"N to 24°40'6.71"N
Longitude	93°58'9.82" E to 93°58'10.25"E
Maximum length (km)	0.53
Maximum breadth (km)	0.44
Surface area (km²)	0.16
Maximum depth (m)	1.92
Minimum depth (m)	0.79
Mean depth(m)	1.42
Mean depth/maximum depth ratio	0.73
Basin shape	Saucer
Basin slope	Gentle
Bottom texture	Silted
Shore-line (km)	1.67
Shore-line development	1.17
Volume of lake (km³)	0.00023
Surface area/volume ratio	695.65
Development of volume	2.22
Mean maximum temperature (°C)	22.5-39.5
Mean minimum temperature (°C)	2.0-22.0
Mean annual rainfall (mm)	254.7

For the present study, the lake is divided into four sites representing as site I, II, III and IV which are named as Sabam, Kabui Panung, Thambou Kom and Thaba Konjin respectively (Figure 5.1). Site I–Sabam: This site has an area of 0.02 sq km and water depth ranged from 82.4 to 186.0 cm during the whole study period. 18 macrophytes were recorded in this site in the different seasons. Site II–Kabui Panung: This site is just adjacent to site I with an area of 0.06 sq km. The depth of the water ranged from 85.4 to 190.0 cm. In this site 17 macrophytes were found during the study period. Site III–Thambou Kom: This is the largest site with an area of 0.07 sq. km and depth of water ranged from 82.5 to 192.0 cm during the whole study period. Maximum number of macrophytes (21 species) was recorded in this site in the different seasons. Site IV–Thabakonjin: This is the smallest site of the lake. It has an area of 0.01 sq. km. The depth of the water during the whole study period ranged from 79.0 to 172.3 cm. Here 17 macrophytes were recorded during the study period.

Materials and Methods

For detailed study and investigation, the entire lake was divided into four study sites *viz.* Site I (Sabam), Site II (Kabui Panung), Site III (Thambou Kom) and Site IV

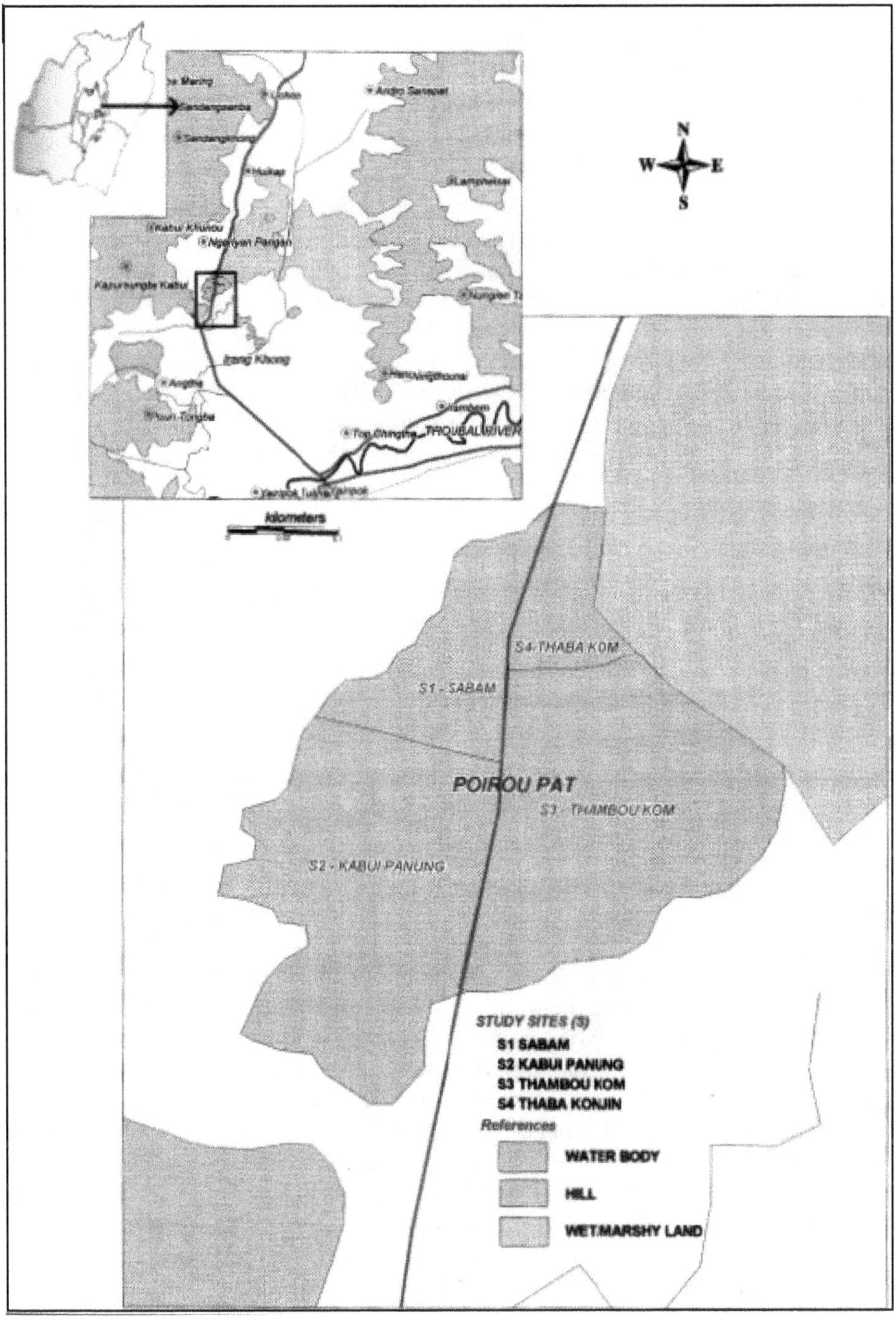

Figure 5.1: Map of Poiroupat Lake

(Thaba Konjin). The macrophytic plant samples were collected from each site on monthly basis during the entire study period. Using Quadrat method described by Curtis (1959) and Misra (1968), the macrophytes were collected on monthly basis for two years *i.e.* June 2000 to May 2002. For every month on each visit to the study sites, adequate numbers of quadrates were laid down randomly (Ambasht, 1970). Assessment of Frequency, Density, Abundance and Importance Value Index (IVI) comprised the quantitative analysis.

Frequency gives an account of the sampling units (per cent) in which a particular species occurs. Density represents the number of individuals per unit area while abundance measures the number of individuals per unit area of occurrence of the species. Density and abundance give an idea of the distribution pattern of the species. Even though the quantitative values of each of frequency, density, abundance and basal cover has its own significance, the composite picture of ecological importance cannot be obtained by anyone of the analysis. So, the complete picture of ecological importance of a species with respect to the community structure can be assessed by adding up the percentage values of relative frequency, relative density and relative abundance and the value out of 300 is termed as the Importance Value Index (IVI) of the species. The IVI reveals the overall picture of sociological structure of the species in a community.

Results and Discussion

The present study reveals the presence of 30 macrophytes. Generally the vegetation of Poiroupat lake has been classified into four classes *viz.* (i) Submerged (10 per cent) (ii) Rooted with floating leaves (16.7 per cent) (iii) Free floating (20 per cent) and (iv) Emergent (53.3 per cent).

Frequency

Variations in frequency range of the different *macrophytes* in all the study sites are shown in Table 5.2. In all the study sites, maximum frequency was shown by *Alternanthera philoxeroides,* (30.0 per cent in winter to 90.0 per cent in summer) *Utricularia flexuosa* (50.0 per cent to 90.0 per cent), *Hydrilla verticillata* (20.0 per cent to 90.0 per cent) also exhibited maximum frequency only in sites IV and II. *Alternanthera philoxeroides, Hydrilla verticillata, Enhydra fluctuans, Nymphoides indicum, Ludwigia adscendens* were present in all the study sites. For *Alternanthera philoxeroides* the value ranged from 30.0 per cent to 90.0 per cent. Highest value *i.e.* 90.0 per cent was recorded in June in site I and for *Ludwigia adscendens,* range of frequency values was recorded as 20.0 per cent to 80.0 per cent. This species was found in all the study sites. Lowest value of 50.0 per cent was 20.0 per cent to 80.0 per cent. This species was found in all the study sites. Lowest value of 50.0 per cent was observed in site I. *Hydrilla verticillata* was recorded to occur only in site II and III throughout the study period. *Nymphoides indicum* was present only in site I throughout the study period. *Imperata cylindrica* was found only in site IV during the study period. *Lemna perpusilla* and *Marsilea quadrifoliata* were present only in site III and I respectively. Trends of variations in the values of frequency were observed to be more or less identical in all study sites throughout the study period. The dominant species found in the Poiroupat lake,

Manipur were *Ceratophyllum demersum, Alternanthera philoxeroids, Eichhornia crassipes, Hydrilla verticillata, Ludwigia adscendens* etc. Highest frequency was recorded in *Alternanthera philoxeroides* (90.0 per cent) in site I. For *Ceratophyllum demersum*, the maximum value was found in site II and IV with a frequency percentage of 80 per cent. *Hydrilla verticillata* was also recorded in site III with its maximum frequency (90 per cent).

Table 5.2: Ranges of Frequency (per cent) of the Macrophytes in Different Study Sites

Sl.No.	Name of Species	Site I	Site II	Site III	Site IV
1.	Alternanthera philoxeroides	30.0–90.0	30.0–80.0	50.0–80.0	60.0–80.0
2.	Alternanthera sessiles	–	–	20.0–70.0	30.0–40.0
3.	Azolla pinnata	–	20.0–50.0	30.0–40.0	40.0–50.0
4.	Ceratophyllum demersum	30.0–60.0	30.0–80.0	50.0–70.0	60.0–80.0
5.	Commelina benghalensis	20.0–70.0	–	20.0–80.0	–
6.	Cyperus difformis	–	–	20.0–50.0	–
7.	Echinochloa stagnina	20.0–70.0	–	20.0–80.0	–
8.	Eclipta Prostrata	20.0–70.0	30.0–80.0	40.0–80.0	–
9.	Eichhornia crassipes	20.0–80.0	–	–	30.0–90.0
10.	Enhydra fluctuans	30.0–60.0	30.0–40.0	30.0–50.0	40.0–50.0
11.	Euryale ferox	20.0–50.0	30.0–40.0	–	–
12.	Hydrilla verticillata	20.0–50.0	30.0–50.0	20.0–90.0	50.0–60.0
13.	Hygroryza aristata	20.0–40.0	30.0–40.0	20.0–60.0	–
14.	Ipomoea aquatica	–	–	20.0–40.0	30.0–80.0
15.	Imperata cylindrica	–	–	–	20.0–70.0
16.	Lema perpusilla	–	–	20.0–50.0	–
17.	Ludwigia adscendens	20.0–50.0	30.0–80.0	30.0–60.0	40.0–70.0
18.	Marsilea quadrifoliata	20.0–30.0	–	–	–
19.	Nelumbo nucifera	20.0–80.0	20.0–40.0	20.0–80.0	30.0–80.0
20.	Nymphaea stellata	–	–	20.0–50.0	–
21.	Nymphoides indicum	20.0–50.0	20.0–70.0	40.0–80.0	20.0–70.0
22.	Oenanthe javanica	–	–	20.0–70.0	20.0–30.0
23.	Oryza officinalis	30.0–50.0	20.0–60.0	–	–
24.	Pistia stratiotes	20.0–50.0	30.0–40.0	20.0–30.0	–
25.	Polygonum glabrum	20.0–40.0	20.0–50.0	–	–
26.	Sacciolepsis myosuroides	–	–	–	20.0–60.0
27.	Sagittaria trifolia	–	–	–	20.0–70.0
28.	Salvinia cucullata	40.0–70.0	–	–	20.0–80.0
29.	Salvinia natans	–	30.0–80.0	40.0–60.0	–
30.	Utricularia flexuosa	20.0–80.0	40.0–60.0	50.0–80.0	50.0–90.0

From the present study, it was found that the dominant species of the lake showed its maximum frequency during rainy season which influence the growth of the macrophytes and favouring good climatic conditions. Rai *et al.* (1982) reported that the rainy season is the most favourable season for the germination of buried seeds of the perennial emergents (*Cyperus* species) and other mud growing species (*Eclipta, Enhydra, Ipomoea,* Caesula sp.) etc. Warm temperature during the rainy season in the tropics favours the rich growth of the aquatic macrophytes (Hogeweg and Brenkert, 1969). Maximum frequency for *Hydrilla verticillata* (80 per cent) during July, *Eichhornia crassipes* (80 per cent) during September and *Ceratophyllum demersum* (70 per cent) during March in the river Ganga were reported by Shah and Abbas (1979) which is in conformity with the present findings. Ambasht (1970) and Misra (1989) reported, high percentage frequency for various species *viz. Azolla* (100 per cent), *Trapa, Utricularia* and *Spirodella* (65 per cent to 80 per cent) and *Hydrilla* (55 per cent) in the freshwater ponds of Banaras Hindu University, Varanasi. Devi, S.U. (2008) reported high values of frequency for *Ceratophyllum demersum* (25 to 70 per cent) from Oksoipat Lake, Manipur. Similar findings were also reported by Devi, N.B, 1993 for *Salvinia cucullata* (33.2 per cent to 100 per cent), *Hydrilla verticillata* (100 per cent) *Trapa natans* (6.67 per cent to 73.33 per cent), *Ceratophyllum demersum* (6.67 per cent to 73.33 per cent) and *Vallisnaria spiralis* (6.67 per cent to 80.0 per cent) in the non-phumdi areas of Loktak lake, the percentage frequency values for *Alternanthera philoxeroides* were 20.0 per cent to 86.67 per cent. Babalonas and Papastergiadou (1989) also reported maximum frequency for Salvinia sp. (100 per cent) in lake Kerkini, Greece. Devi, Ch. U. (2000) also reported highest frequency percentage of 100 per cent for *Azolla pinnata, Salvinia cucullata* in the Freshwater Ecosystems of Canchipur, Manipur. In Sanapat lake, the reported values for *Alternanthera philoxeroides* was 53.33 per cent to 66.67 per cent, *Nelumbo nucifera,* 60.0 per cent to 100 per cent) *Eichhornia crassipes,* 40.0–93.3 per cent (Devi, Ch. B., 2001). Teresa *et al.* (1999), reported 306 species on mitigation wetlands, 274 species in naturally occuring wetlands and a total of 365 plant taxa identified acrossed all 96 study sites of which *Phalaris arundinaceae,* was found on 89 sites (93 per cent of the total).

The present frequency values are very high when compared with the reports by Rai *et al.* (1982) in the Chaurs of North Bihar for *Ceratophyllum demersum* (0.55 per cent to 2.5 per cent), *Hydrilla verticillata* (6.0 per cent to 30.5 per cent), *Potamogeton crispers* (4.6 to 21.2 per cent) and *Najas graminae* (1.10 per cent to 9.3 per cent). Lower frequency values were also reported by Devi, O.I. (1993) for *Ceratophyllum demersum* (0.95 to 6.67 per cent), *Hydrilla verticillata* (4.76 to 17.14 per cent), *Nymphoides indicum* (4.76 per cent to 14.29) and *Ludwigia adscendens* (2.86 to 24.76 per cent) in Waithou lake, Manipur. Seshavatharam *et al.* (1982) reported lower values of *Eichhornia crassipes* (5.98 per cent), *Alternanthera sessiles* (3.4 per cent), *Nymphaea nouchali* (2.5 per cent) in Kollelu lake, Andhra Pradesh.

Density

The density values have been expressed in plants m^{-2} for all the species which found in the study period. Variations of density values of the *macrophytes* in the study sites are shown in Table 5.3. Similar trend of variations in density with the change of seasons was observed in all the study sites. For *Alternanthera philoxeroids* maximum

Table 5.3: Ranges of Density (Plants m^{-2}) of the Macrophytes in Different Study Sites

Sl.No.	Name of Species	Site I	Site II	Site III	Site IV
1.	Alternanthera philoxeroides	8.0–67.2	3.2–38.4	20.0–57.8	20.8–51.2
2.	Alternanthera sessiles	–	–	11.2–41.6	12.8–40.0
3.	Azolla pinnata	–	1.6–32.0	6.4–36.8	3.2–28.8
4.	Ceratophyllum demersum	19.2–57.6	20.8–72.0	25.6–64.0	32.0–37.6
5.	Commelina benghalensis	3.2–38.4	–	6.4–41.6	–
6.	Cyperus difformis	–	–	8.0–19.2	–
7.	Echinochloa stagnina	3.2–27.2	–	6.4–24.0	–
8.	Eclipta Prostrata	1.6–41.6	3.2–36.8	11.2–38.4	–
9.	Eichhornia crassipes	3.2–33.6	–	–	4.8–40.0
10.	Enhydra fluctuans	4.8–28.8	6.4–22.4	17.6–22.4	19.2–22.4
11.	Euryale ferox	1.6–24.0	4.8–19.2	–	–
12.	Hydrilla verticillata	4.8–33.6	8.0–40.0	11.2–28.8	12.8–24.0
13.	Hygroryza aristata	4.8–40.0	4.8–24.0	12.8–40.0	–
14.	Ipomoea aquatica	–	–	1.6–27.2	12.8–24.0
15.	Imperata cylindrica	–	–	–	3.2–28.8
16.	Imperata cylindrica	–	–	3.2–30.4	–
17.	Ludwigia adscendens	12.2–38.4	12.8–35.2	22.4–28.8	28.8–33.6
18.	Marsilea quadrifoliata	1.6–4.8	–	–	–
19.	Nelumbo nucifera	4.8–35.2	8.0–40.0	8.0–33.6	25.6–28.8
20.	Nymphaea stellata	–	–	1.6–40.0	–
21.	Nymphoides indicum	1.6–44.8	3.2–38.4	22.4–27.2	27.2–28.8
22.	Oenanthe javanica	–	–	2.8–11.2	3.2–8.0
23.	Oryza officinalis	9.6–19.2	4.8–16.0	–	–
24.	Pistia stratiotes	11.2–24.0	12.8–32.0	17.6–22.4	–
25.	Polygonum glabrum	3.2–25.6	4.8–12.8	–	–
26.	Sacciolepsis myosuroides	–	–	–	1.6–16.0
27.	Sagittaria trifolia	–	–	–	1.6–19.2
28.	Salvinia cucullata	1.6–33.6	–	–	3.2–36.8
29.	Salvinia natans	–	3.2–40.0	6.4–20.8	–
30.	Utricularia flexuosa	4.8–57.6	8.0–56.0	20.5–44.8	33.6–38.4

density was observed in September and the minimum density during March. Maximum density for *Ceratophyllum* was recorded in the months of June and October and the minimum density during April. *Hydrilla verticillata* was present in all the study sites throughout the study period which showed its maximum density during February and March. *Ludwigia adscendens* was found in all the study sites throughout the study period. It showed its maximum and minimum density during winter and summer seasons respectively. *Nymphoides indicum* and *Utricularia flexuosa* were present throughout the year in site IV. Other species were found occasionally during a few months of the study period.

In the present study, maximum density was found in *Ceratophyllum demersum* with values ranging from 19.2 to 72.0 plants m^{-2}. *Alternanthera philoxeroides* had density values ranging from 3.2 to 57.8 plants m^{-2}; *Eichhornia crassipes*, 3.2 to 40.0 plants m^{-2} *Hydrilla verticillata*, 8.0 to 40.0 plants m^{-2}; *Ludwigia adscendens*, 11.2 to 38.4 plants m^{-2}. For some dominant species, maximum density was observed in rainy season which favours good climatic conditions for the growth of the macrophytes. In the study of freshwater ponds of Varanasi, Ambasht (1970) and Misra (1989) found out that *Hydrilla, Eleocharis, Chara* and *Utricularia* (56.0 to 210.0 plants m^{-2}) had high density values with exceptionally high density for *Azolla* (3445.0 plants m^{-2}). Gradual increase in the values of density by the onset of spring season *i.e.* maximum density of 48,137 and 116 shoots m^{-2} during June at Haokersar, Shalbogh and Kranchu respectively in Kashmir were also reported by Handoo and Kaul (1982). The present findings are also in agreement with the findings of Hogeweg and Brenkert (1969) and Rai and Munshi (1982). High values of density in dominant species in the Phumdi areas of Loktak lake, Manipur were also reported by Devi (1993), Devi (1998) also reported high density values for *Ceratophyllum demersum* (48.28 to 154.72 plants m^{-2}); *Hydrilla verticillata* (16.0 to 73.12 plants m^{-2}) *Ludwigia adscendens* (20.0 to 61.28 plants m^{-2}) in Utrapat lake, Manipur.

Billore *et al.* (1998) reported that there was a gradual increase in the density of *Eichhornia crassipes* during November to February in a tropical natural wetland at Ujjain. The present findings exhibit higher density values for some macrophytes when compared with the findings made by Devi (1993) in Waithou lake, Manipur (*Ceratophyllum demersum*, 0.04 to 0.65 plants m^{-2}; for *Hydrilla verticillata* in the mid rainy season; 8.5–10.8 plants m^{-2} *Hydrilla vertcillata*, 0.03 to 35.47 plants m^{-2}, *Alternanthera philoxeroides*, 4.46 to 17.03 plants m^{-2} etc.). Shah and Abbas (1979) also reported density of 10.9 to 22.10 plants m^{-2} for *Eichhornia crassipes* during January–February and 8.89 plants m^{-2} for *Ceratophyllum demersum* during summer. Higher density values were reported by Devi (2000) in the Freshwater Ecosystems of Canchipur, Manipur ranging from 3.2 to 155.68 for *Eichnochloa stagnina* plants m^{-2}, 190.40 plants m^{-2} for *Salvinia cucullata*. Devi (2001), reported density values for *Alternanthera philoxeroides* ranging from 4.16 to 62.88 plants m^{-2}; *Ludwigia adscendens*, 2.08 to 44.8 plants m^{-2}; *Salvinia cucullata*, 2.08 to 61.78 plants m^{-2} in Sanapat lake, Manipur. The above findings are very much comparable with the present findings.

Abundance

Ranges of abundance of the macrophytes in different study sites have been presented in Table 5.4. Owing to the change of seasons, the abundance values of the different species showed wide variations, Maximum abundance was shown by *Ludwigia adscendens* (4.8–96.0 plants m^{-2}) in all the study sites among the regular occurring macrophytes. For *Ceratophyllum demersum*, the abundance values ranged from 32.0–91.2 plants m^{-2}. Maximum value was observed in October and the minimum during April. *Hydrilla verticillata* had a range of 9.6–80.0 plants m^{-2} and it occurred in all the study sites, *Imperata cylindrica* showed its maximum abundance with values ranging from 4.48–144.0 plants m^{-2} among the species which occurred occasionally.

Table 5.4: Ranges of Abundance (Plants m⁻²) of the Macrophytes in Different Study Sites

Sl.No.	Name of Species	Site I	Site II	Site III	Site IV
1.	Alternanthera philoxeroides	6.4–96.0	25.6–68.8	25.6–84.8	48–60.3
2.	Alternanthera sessiles	–	–	16–132.8	27.2–88
3.	Azolla pinnata	–	3.2–80.0	20.8–184.0	32.0–72.0
4.	Ceratophyllum demersum	32.0–91.2	64.0–96.0	64.0–89.6	68.8–72.0
5.	Commelina benghalensis	4.8–96.0	–	16.0–80.0	–
6.	Cyperus difformis	–	–	12.0–72.0	–
7.	Echinochloa stagnina	28.8–48.0	–	6.4–89.6	–
8.	Eclipta Prostrata	1.9–104.0	16.0–59.4	36.8–48.0	–
9.	Eichhornia crassipes	3.5–48.0	–	–	4.0–28.8
10.	Enhydra fluctuans	9.6–64.0	20.8–64.0	35.2–48.0	17.6–44.8
11.	Euryale ferox	5.28–80.0	16.0–41.6	–	–
12.	Hydrilla verticillata	9.6–80.0	9.6–64.0	25.6–56.0	56.0–57.6
13.	Hygroryza aristata	24.0–99.2	24.0–64.0	41.6–48.0	–
14.	Ipomoea aquatica	–	–	8.0–89.6	9.6–80.0
15.	Imperata cylindrica	–	–	–	4.48–144.0
16.	Lema perpusilla	–	–	16.0–88.0	–
17.	Ludwigia adscendens	4.8–96.0	32.0–84.0	48.0–57.6	41.6–57.6
18.	Marsilea quadrifoliata	16.0–24.0	–	–	–
19.	Nelumbo nucifera	24.0–64.0	35.2–176.0	40.0–49.6	41.6–64.0
20.	Nymphaea stellata	–	–	3.2–104.0	–
21.	Nymphoides indicum	8.0–144.0	16.0–56.0	27.2–64.0	33.6–54.4
22.	Oenanthe javanica	–	–	4.4–48.0	16.0–41.6
23.	Oryza officinalis	16.0–48.0	25.6–38.4	–	–
24.	Pistia stratiotes	35.2–105.6	36.8–112.0	40.0–73.6	–
25.	Polygonum glabrum	16.0–48.0	32.0–51.2	–	–
26.	Sacciolepsis myosuroides	–	–	–	24.0–56.0
27.	Sagittaria trifolia	–	–	–	16.0–64.0
28.	Salvinia cucullata	8.0–60.8	–	–	24.0–48.0
29.	Salvinia natans	–	8.0–49.6	24.0–64.0	–
30.	Utricularia flexuosa	9.6–180.0	19.2–72.0	33.6–56.0	48.0–49.6

There were variations in the maximum values of abundance for the macrophytes found in the lake owing to the change of seasons. Maximum values of abundance for *Alternanthera philoxeroides* were recorded in rainy season (6.4 to 96.0 plants m⁻²). Variations were recorded in other species *viz. Azolla pinnata* (3.2-184.0 plants m⁻²), *Ceratophyllum demersum* (32.0–96.0 plants m⁻²), *Ludwigia adscendens* (4.8 to 96.0 plants m⁻²), *Nymphoides indicum* (8.4 to 144.0 plants m⁻²) *Utricularia flexuosa* (9.6 to 180.0 plants m⁻²) etc. The maximum abundance values recorded in the present study were found lower when compared with those reported for *Azolla* (215.30 plants m^{-2}) by Ambasht (1970) and Misra (1989) in the freshwater ponds of Varanasi.

High values of abundance were also reported for phumdi species of Loktak lake, Manipur *viz. Cryptococcum* (136.0 to 992.0 plants m^{-2}), *Echinochloa* (168.0–618.72 plants m^{-2}), *Alternanthera* (25.0–260.48 plants m^{-2}) and *Oryza* (40.0–408.0 plants m^{-2}) by Devi (1993). Lower values of abundance were recorded in the non-phumdi site of Loktak Lake. (*Hydrilla*, 19.0 to 98.08 plants m^{-2}); *Salvinia*, 16.0 to 149.28 plants m^{-2}).

Devi, O.I. (1993) in Waithou lake (Manipur), reported abundance values for *Eichhornia crassipes* as 41.87 to 68.53 plants m^{-2} in the second year and 68.87 to 100.85 plants m^{-2} for the first year. For *Hydrilla verticillata* the values ranged from 24.0 to 104.57 plants m^{-2} in the first year and 1.98 to 33.79 plants m^{-2} in the second year and *Alternanthera philoxeroides* had values ranging from 10.2 to 48.67 plants m^{-2} and 4.65 to 15.28 plants m^{-2} in the first and second year respectively. The above findings have been found in agreement with the findings of the present study. Devi, K.I. (1998) also reported maximum abundance values for *Ceratophyllum demersum* (48.28 to 154.72 plants m^{-2}), *Hydrilla verticillata* (16.0 to 73.12 plants m^{-2}), *Utricularia flexuosa* (20.0 to 97.60 plants m^{-2}) and *Ludwigia adscendens* (20.0 to 61.28 plants m^{-2}) in Utrapat lake, Manipur.

Higher values of abundance were reported by Devi. Ch. U (2000) in the Freshwater Ecosystems of Canchipur, Manipur. *Salvinia cucullata*, 220.32; *Hydrilla verticillata, Echinochloa stagnina, Schoenoplectus grossus*, 168.0 to 192.0 plants m^{-2}). Devi Ch. B. (2001) also reported comparable abundance values for *Alternanthera philoxeroides* (32.0 to 157.28 plants m^{-2}), *Ludwigia adscendens* (20.0 to 67.2 plants m^{-2}), *Azolla pinnata* (36.0 to 137.6 plants m^{-2}) in Sanapat lake, Manipur.

Importance Value Index (IVI)

Ranges of Importance Value Index (IVI) of the macrophytes in the different study sites have been furnished in Table 5.5. The values of IVI have been expressed out of 300. In all the study sites the highest value of IVI during the entire study period was recorded in *Ceratophyllum demersum* with a maximum value of 50.8 in site II. This was followed by *Ludwigia adscendens* with a value of 20.7 in site I. For *Alternanthera philoxeroides* both the highest value (17.5) and the lowest value (5.5) were noticed in site I itself. For the irregularly occurring species, the values of IVI were found lower in all the study sites throughout the study period as compared to the regular species. *Ceratophyllum demersum* and *Ludwigia adscendens* had the highest IVI values. The luxuriant growth and highest IVI values of these species may be due to the favourable climatic conditions prevailing in the lake. Gopal *et al.* (1978) reported that, the physico-chemical characteristics of water also influenced the growth of aquatic macrophytes, besides the climatic factors. Low pH is found to be dangerous to the successful growth of some macrophytes like *Pistia stratiotes* and *Nymphoides* as reported by Mitra (1966). However, Das (1968) reported that *Spirodela polyrhiza* is distributed in water having pH values between 6.2 and 8.6.

Table 5.5: Ranges of Importance Value Index (IVI) of the Macrophytes in Different Study Sites

Sl.No.	Name of Species	Site I	Site II	Site III	Site IV
1.	Alternanthera philoxeroides	5.5–17.6	13.1–17.0	12.1–14.3	9.8–13.2
2.	Alternanthera sessiles	–	–	6.9–14.2	9.1–12.2
3.	Azolla pinnata	–	5.0–13.2	5.1–10.1	8.3–9.7
4.	Ceratophyllum demersum	10.1–32.0	7.8–50.8	10.7–16.5	16.4–24.4
5.	Commelina benghalensis	4.8–12.5	–	5.5–31.4	–
6.	Cyperus difformis	–	–	4.6–8.6	–
7.	Echinochloa stagnina	6.2–11.3	–	7.9–10.8	–
8.	Eclipta Prostrata	5.6–8.5	–	–	11.1–13.3
9.	Eichhornia crassipes	0.4–8.8	5.7–8.8	6.7–12.2	–
10.	Enhydra fluctuans	5.8–10.8	8.5–9.8	6.9–10.0	7.6–9.5
11.	Euryale ferox	3.8–9.1	4.5–8.3	–	–
12.	Hydrilla verticillata	7.4–10.0	7.5–9.3	8.1–9.5	12.5–13.1
13.	Hygroryza aristata	6.7–14.5	8.0–12.0	9.3–11.5	–
14.	Ipomoea aquatica	–	–	3.4–16.6	5.9–10.3
15.	Imperata cylindrica	–	–	–	6.1–12.1
16.	Lema perpusilla	–	–	5.7–12.8	6.9–11.6
17.	Ludwigia adscendens	6.7–20.7	7.1–31.0	10.1–12.7	9.8–13.8
18.	Marsilea quadrifoliata	3.4–12.0	–	–	–
19.	Nelumbo nucifera	4.1–17.1	5.7–14.8	12.2–14.1	9.8–14.0
20.	Nymphaea stellata	–	–	5.0–12.4	–
21.	Nymphoides indicum	3.4–12.8	6.0–15.6	10.2–12.4	11.6–11.8
22.	Oenanthe javanica	–	–	4.2–7.4	4.8–8.7
23.	Oryza officinalis	5.0–11.2	6.2–7.7	–	–
24.	Pistia stratiotes	5.4–33.2	8.1–15.8	8.0–9.6	–
25.	Polygonum glabrum	3.4–9.0	5.5–6.7	–	–
26.	Sacciolepsis myosuroides	–	–	–	3.8–10.6
27.	Sagittaria trifolia	–	–	–	5.0–8.8
28.	Salvinia cucullata	6.2–14.6	–	–	9.3–13.4
29.	Salvinia natans	–	4.4–9.6	5.7–19.5	7.7–12.6
30.	Utricularia flexuosa	7.4–17.5	9.6–12.2	13.2–17.5	9.7–13.6

In the present study, maximum IVI was recorded in *Ceratophyllum demersum* with values ranging from 7.8 to 50.8. *Alternanthera philoxeroides* had values ranging from 5.5 to 17.6 and *Ludwigia adscendens*, 6.7 to 31.0, *Nymphoides indicum*, 3.4 to 15.6 and *Utricularia flexuosa*, 7.4 to 17.5. The observed results in the present study have been found to agree with the findings of Sankhla and Vyas (1982) in Bhagela Tank, Udaipur. They reported highly varying values of IVI for the different macrophytes.

Billore and Vyas (1982) reported IVI for *Eichhornia crassipes* ranging from 24.4 to 51.5 in Pichhola lake, Udaipur.

The dominant species had high IVI values which are in agreement with the findings of high values of IVI in *Hydrilla verticillata, Salvinia cucullata, Alternanthera* etc. in the Loktak lake, Manipur(Devi, N.B., 1993). Devi, O.I. (1993) also reported similar values of IVI in some species like *Hydrilla, Eichhornia, Nymphoides* etc. in the Waithou lake, Manipur. Devi, K.I., 1998 also reported high values of IVI for *Ceratophyllum demersum* (39.86 to 108.11), *Hydrilla verticillata* (6.28 to 36.78) and *Ludwigia adscendens* (8.86 to 35.06) in the study of Utrapat lake, Manipur. IVI values for species like *Salvinia cucullata* (8.5 to 168.98), *Salvinia natans* (27.04 to 79.68) were reported by Devi, Ch. U (2000) in the Freshwater Ecosystems of Canchipur, Manipur. The present findings when compared with the findings of Devi, Ch. B. (2001) in Sanapat lake, Manipur are found to be lower (2.02 to 225.0). Devi, L.G. (2007) reported IVI value of *Ceratophyllum demersum* (30.57 to 95.18) from Awangsoipat Lake, Manipur.

In the present study, the maximum values of frequency, density, abundance and IVI of the different species were assessed during rainy season, which influence favourable climatic conditions for the successful and luxuriant growth of the macrophytes. Similar observations were made by Hogeweg and Brenkert (1969) in the aquatic macrophytes of the tropics and Rai and Munshi (1982) in the emergent macrophytes of India.

References

1. Ambasht, R.S. (1969). *A Textbook of Plant Ecology.* Students Friends and Co., Lanka, Varanasi.

2. Ambasht, R.S. (1970). Freshwater ecosystems. In: *EcologyL Study of Ecosystems,* (Ed.) R. Misra. A.H. Wheeler and Co. (P) Limited, Allahabad, pp. 124–160.

3. Babalonas, D. and Papastergiadou, E. (1989). The water fern *Salvinia natans* (L) in the Kerkini Lake (north Greece). *Arch. Hydrobiol.,* 116(4): 478–498.

4. Billore, D.K. and Vyas, L.N. (1982). Distribution and production of macrophytes in Pichhola lake, Udaipur (India). In: *Wetlands: Ecology and Management,* (Eds.) B. Gopal, R.E. Turner, R.G. Wetzeland and D.F. Whigham. National Institute of Ecololgy and Internatiional Scientific Publications, India, pp. 45–54.

5. Billore, S.K., Bhardia, R. and Kumar, A. (1998). Potential removal of particulate matter and nitrogen roots of water hyacinth in a topical natural wetland. *Current Science,* 74(2): 152–156.

6. Curtis, J.T., 1959. *The Vegetation of Wisconsion.* Univ. Wisconsin Press, Madison.

7. Das, R.R (1968). Growth and distribution of *Eichhornia crassipes* and *Spirodela polyrhiza. Ph.D. Thesis,* Banaras Hindu University, Varanasi.

8. Devi, N. Beenakumari, 1993. Phytosociology, primary production and nutrient status of the macrophytes of Loktak Lake, Manipur. *Ph.D. Thesis,* Manipur University, Manipur.

9. Devi, Ch. Bebika (2001). Variations in species distribution and primary productionof the macrophytes in Sanapat lake, Manipur. *Ph.D. Thesis*, Manipur University, Manipur.

10. Devi Ch. Umabati (2000). Phytosociology and primary production of the macrophytes in the freshwater ecosystem of Canchipur, Manipur. *Ph.D. Thesis*, Manipur University, Manipur.

11. Devi, O. Ibeton (1993). Distribution, primary production and nutrient status of the macrophytic communities in Waithou Lake, Manipur. *Ph.D. Thesis*, Manipur Univeristy, Manipur.

12. Devi, K.I. (1998). Ecological studies of freshwater macrophytes in Utrapat lake, Manipur. *Ph.D. Thesis*, Manipur University, Manipur, India.

13. Devi, Ch. Nivanonee (2002). Vegetational structure and Primary production of the macrophytes of Ikop lake, Manipur. *Ph.D. Thesis*, Manipur University, Manipur.

14. Devi, L. Geetabli (2007). Studies on the vegatational dynamics and primary productivity of Awangsoiat Lake, Bishnupur (Manipur). *Ph.D. Thesis*, Manipur University, Manipur.

15. Devi, S. Umeshori (2008). Ecological Analysis of the macrophytes in Oksoipat lake (Bishnupur), Manipur. *Ph.D. Thesis*, Manipur University, Manipur.

16. Gopal, B., Sharma, K.P. and Trivedy, R.K. (1978). Studies on ecology and production in Indian freshwater ecosystems at primary producer level with emphasis on macrophytes. In: *Glimpses of Ecology*, (Eds.) J.S. Singh and B. Gopal. Int. Sci. Publs., Jaipur, pp. 349–376.

17. Handoo, J.K. and Kaul, V. (1982). Phytosociological and standing crop studies in wetlands of Kashmir. In: *Wetlands: Ecology and Management*, (Eds) B. Gopal, R.E. Turaer, R.G. Vetzel and D.F. Whigham. National Institute of Ecology and International Scientific Publications.

18. Hogeweg, P. and A.L. Brenkert (1969). Structure of aquatic vegetation: A comparison of aquatic vegetation in India, Netherlands and Czechoslovakia. *Trop. Ecol.*, 10(1): 139–162.

19. Misra, R. (1968). *Ecology Workbook*. Oxford and IBH Publ. Co., New Delhi.

20. Misra, K.C. (1989). *Manual of Plant Ecology*, 3rd Edition. Oxford and IBH, New Delhi.

21. Mitra, E. (1966). Contribution to our knowledge of India freshwater plants. V. on the morphology, reproduction and autecology of *Pistia stratiotes* Linn. *J. Asiatic Soc.*, 8: 115–135.

22. Nichols, S.A. (1992). Depth-substrate and turbidity relationships of some wiscons in lake plants. *Transactions, Wisconsin Academy of Science, Arts and Letters*, 80: 91–119.

23. Rai, D.N. and Munshi, J.S. Datta (1982). Ecological characteristics of Chaurs of North Bihar. In : *Wetlands: Ecology and Management*, (Eds.) B. Gopal, R.E. Turner, R.G. Wetzel and D.F. Whigham. International Scientific Publications and National Institute of Ecology, pp. 89–95.

24. Seshavatharam, V., Dutt, B.S.M. and Venu, P. (1982). An ecological study of the vegetation of the Kolleru lake. *Bull. Bot Surv., India*, 24(1–4): 70–75.

25. Shah, J.D. and Abbas, S.G. (1973). Seasonal variation in frequency, density, biomass and rate of production of some aquatic macrophytes of the river Ganges at Bhagalpur (Bihar). *Trop. Ecol.*, 20(2): 127–134.

26. Sankhla, S.K. and Vyas, L.N. (1982). Observations on the moist bank community of Baghela Tank, Udaipur (India). In: *Wetlands: Ecology and Management*, (Eds.) B. Gopal, R.E. Turner, R.G. Wetzel and D.F. whigham. National Institute of Ecology and International Scientific Publications, India, pp. 197–206.

27. Tansley, A.G. (1935). The use and abuse of vegetational concepts and terms. *Ecology*, 16: 284–307.

28. Teresa, K. Magee, Ted L. Ernst, Mary E. Kentula and Kathleen A. Dwire (1999). Floristic comparison of freshwater wetlands in an urbanizing enviroment. *Wetlands*, 19(3): 517–534.

Biodiversity of Aquatic Reseources (2012)
Editors: **Mamta Rawat & Sumit Dookia**
Published by: **DAYA PUBLISHING HOUSE, NEW DELHI**

Pages **75–89**

Chapter 6

Assessment of Environmental Flow Requirements of Tungabhadra River Basin

S. Srikantaswamy* and B.K. Harish Kumara

Department of Studies in Environmental Science,
University of Mysore, Manasagangotri
Mysore – 570 006, Karnataka

ABSTRACT

Environmental flow determination requires decisions on total volume of flow to be allocated for the environment is depending on spatial and temporal of the flow. Generally it is beneficial to maintain the natural variation and timing in flows, as many ecological processes are dependent on flow variations and also on implicit links to other triggers such as temperature, rainfall etc. These linkages are not always understood and are best preserved by maintaining a natural flow pattern. In this research work an attempt has been made to understand the existing river flow in the Tungabhadra (TB) River and significance of the ecosystem services for the local livelihood. Increased water drawn from the TB Dam for the agricultural, domestic and industrial uses have gradually altered the natural flow. This has an impact on the socio-economic condition of the downstream dependent population of the river. The downstream of the river bed for about a 100 km has shrunken in its size. The study also identify that, the livelihood support has

* Corresponding Author E-mail: srikantas@hotmail.com

gradually declined during last decade leading in shifting of the occupation and migration of the community. The main aim of this research work is identifying the degradation of the natural ecosystem and its effects on the fishermen communities.

Keywords: *Environmental flows, Ecosystem services, Downstream ecosystem, Tungabhadra river, Water for environment.*

Introduction

In India, as elsewhere in the world, freshwater and freshwater-dependent ecosystems provide a range of services for humans, including fish, flood protection, wildlife, etc. (Revenga *et al.*, 1998). Water needs to be allocated to ecosystems to maintain these services, as it is allocated to other users like agriculture, power generation, domestic use and industry. Balancing the requirements of the aquatic environment and other uses is becoming critical in many of the world's river basins as population and associated water demands increase. India is no exception, on the other hand, the assessment of water requirements of freshwater-dependent ecosystems represents a major challenge due to the complexity of physical processes and interactions between the components of the ecosystems. Environmental requirements are often defined as a suite of flow discharges of certain magnitude, timing, frequency and duration for day-to-day management of particular river. These flows ensure a flow regime capable of sustaining a complex set of aquatic habitats and ecosystem processes and are referred to as *"environmental flows", "environmental water requirements", "environmental flow requirements", "environmental water demand"*, etc. (Dyson *et al.*, 2003). In recent years many methods for determining these requirements have been emerged. They are known as *environmental flow assessments* (EFA). The mean annual sum of estimated environmental flows represents a total annual water volume, which could be allocated for environmental purposes.

Hydraulic civilizations have highly valued importance of water in human life and in the entire ecosystem. Degradation of ecosystem services has resulted in both social and economic costs. Affected people are largely from poorer sections. Recognizing the full value of the ecosystem services and investing in them accordingly can safeguard livelihoods in the future, which can help to achieve sustainable development goals. In this regard, if failing to do so may cause serious damages to the ecosystems (Dyson *et al.*, 2003, Millennium Ecosystem Assessment 2005, Pearce 2007). It is estimated that more than 60 per cent of the world's rivers are fragmented by hydrological alterations. This has led to widespread degradation of aquatic ecosystems (Millennium Ecosystem Assessment, 2005; Dyson *et al.*, 2003; Postel and Richter, 2003; Revenga *et al.*, 2005). The Natural Flow Paradigm where the natural flow regime of a river is recognized as vital to sustaining ecosystems has now been widely accepted. This recognition of flow as a key driver of aquatic ecosystems has led to the development of the environmental flows concept (Dyson *et al.*, 2003).

"Environmental flows (EF)" refers to the water considered sufficient for protecting the structure and function of an ecosystem and its dependent species. These flow requirements are defined both by, the long-term availability of water and its variability and are established through environmental, social, and economic assessment (King *et al.*, 2003; Dyson *et al.*, 2008, Christer *et al.*, 2005). Environmental Flows describes the quantity, quality and timing of water flows required to sustain freshwater and estuarine ecosystems, human livelihoods and well-being that depend on these ecosystems (Brisbane declaration, 2007).

According to the World Bank's recently approved Water Resources Sector Strategy, "the environment is a special 'Water-using sector' in that most environmental concerns are a central part of overall water resources management, and not just a part of a distinct water using sector" (Richard and Hirji, 2003). Until recently, in Asian countries, the concept of environmental flows have related primarily to the flows required to flush the river systems and restore water quality. In these regions, the western concept of 'Environmental Flows' is nowadays represented by the terms 'ecological and environmental water requirements' (EEWR), 'ecological water demand' and 'eco-environmental water consumption' (Song and Yang, 2003).

Materials and Methods

The main aim of this research work is to carryout an evaluation of the present environmental condition such as flow rates and its impact on aquatic organism like fishes in Tungabhadra River and assess the potential benefits/consequences and examine whether environmental flows have been maintained or not.

Many early applications of environmental flow assessments (EFA) were focused on single species. As a result, environmental flows were set to maintain critical levels of habitat for these species. However, managing flows without consideration for other ecosystem components may fail to capture the system processes and biological community interactions that are essential for creating and sustaining the habitat and well-being of that target species. Recent advances in EFAs reflect this knowledge and EFA methodologies increasingly take a holistic approach (King and Brown 2006). Table 6.1 shows the overview of EF assessment methods. In the present study EFA has been carried out by using two modules for the assessment of the EF in study area, as follows:

1. Biophysical assessment.
2. Socio-economic assessment.

Biophysical Assessment

Flow rates were analyzed for the thirty year data of inflow and outflow from the TB dam and compared the flow status with the benchmarking of available methods. Flow status was compared with the existing flow status with French fisheries law method and Montanna method of environmental flow assessment. This analysis indicates the physical status of the river flow over a period of time.

Socio-economic Assessment

Socio-economic studies have been carried out on the river course by analyzing common-property users for subsistence, and the river-related health profiles of these people and their livestock. Questionnaires, both closed and open-ended and focus group discussions were used to collect household-wise and community level information on ecosystem services, water utilization and its conservation. A survey was conducted in the downstream of TB dam, which covered 10 villages along the bank of the river. All these studies are linked to the flow, with the objective of predicting how the people have been affected by specified river changes.

Table 6.1: Overview of Environmental Flow Assessment Methods

Organization	Categorization of Methods	Sub-category	Example
IUCN (Dyson et al., 2003)	Methods	Look-up tables	Hydrological (*e.g.* Q95 index) ecological (*e.g.* tennant method)
		Desk-top analyses	Hydrological (*e.g.* Richter method) hydraulic (*e.g.* wetted perimeter method)
		Functional analyses	BBM, expert panel assessment method, benchmarking methodology
		Habitat modeling	PHABSIM
	Approaches		Expert team approach, stakeholder approach (expert and non-expert)
	Frameworks		IFIM, DRIFT
World Bank (King and Brown, 2003)	Prescriptive approaches	Hydrological index methods	Tennant method
		Hydraulic rating methods	Wetted perimeter method
		Expert panels	
		Holistic approaches	BBM
	Interactive approaches		IFIM DRIFT
IWMI (Tharme, 2003)	Hydrological index methods		Tennant method
	Hydraulic rating methods		Wetted perimeter method
	Habitat simulation methodologies		IFIM
	Holistic methodologies		

Study Area

River Tungabhadra is the largest tributary of the river Krishna, contributing an annual discharge of 14,700 million m³ at its confluence point to the main river. The river is transboundary and flows about 531km from its origin in Karnataka state, before it confluence with river Krishna at Sanghameshwaram, near Kurnool in the neighboring state of Andhra Pradesh (Figure 6.1). The TB Sub-basin (TBSB) stretches over an area of 48,827km² in both the riparian states of Karnataka (38,790 km²) and Andhra Pradesh (9037km²) and finally joins Krishna that flows into Bay of Bengal.

TB covers seven districts[1] and twenty-seven talukas[2] in Karnataka and four districts in Andhra Pradesh[3], the sub-basin is mostly rainfed, dominated by red soils

---

1   Shimoga, Chikamaglur, Davanagere, Haveri, Bellary, Koppal and Raichur.

2   Chikamagalore, Tarikere, Koppa, Sringeri, N R Pura, Shimoga, Bhadravathi, Chennagiri, Harihar, Ranebennur, Hospet, Siruguppa, Koppal, Raichur, Bellary, Gangavathi, Manvi, Sindhanur, Harappanahalli, HB Halli, Honnali, Haveri, Davanagere, Theerthahalli, Mundaragi.

3   Karnool, Cudappa, Ananthpur and Mahaboob Nagar.

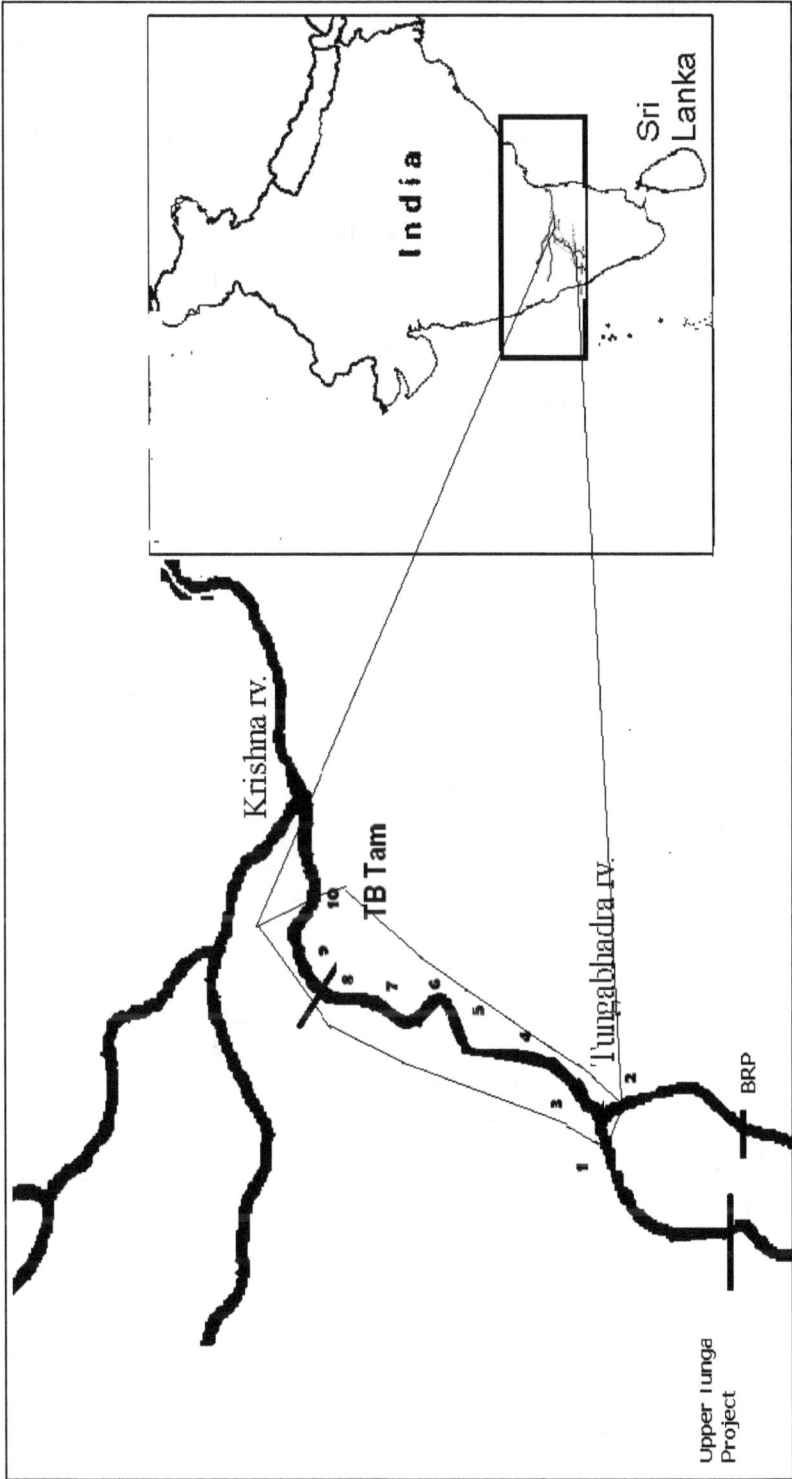

Figure 6.1: Map of Study Area, TB River Basin

with an average annual rainfall of 1200 mm. The upper basin of TB is characterized by undulating terrain with high rainfall while the lower portion of the basin is characterized by much lower rainfall, drought conditions and mainly plain terrain. Agriculture is the major occupation across the basin. The major crops grown are paddy, jowar, sugarcane, cotton and ragi (finger millet). The river catchment includes a number of large and small-scale units supporting industrial activities. Fishing is the next major activity that supports more than 10,000 families. Brick making, pottery, etc are other livelihood options practiced along the basin.

In the present study EFA has been carried out of Tungabhadra river basin at TB dam. TB dam was constructed across the river TB River at Munirabad, Hospet to irrigate the Bellary and Raichur districts in the year 1953. Reservoir water storage capacity of the TB reservoir is reducing due to siltation from mining activities, dust, soil erosion etc. Conflicts within and across sectors are common apart from interstate disputes. Population pressure and increased urbanization have added on. With issues being complex at various levels, the impacts have been serious resulting in land use changes and pollution affecting human health, agriculture and livelihoods of marginal fishing communities.

Results and Discussion

In the study area, population growth detail has shown that urban population is increasing, adding more pressures on the infrastructure. Particularly between 1991 and 2001, there is 36 per cent increase while the rural population shows a decreasing trend (Table 6.2). There are 12 anicuts on left and right banks of the river to facilitate the water extraction for irrigation at downstream of TB Dam. It is observed that the water flow regime in downstream has undergone extreme changes and even in some years it has insignificant flow.

Table 6.2: Population Growth of the River Basin

Year	Population	
	Urban	Rural
1961	3,184,585	2,444,369
1971	3,851,098	3,005,054
1981	5,456,334	3,905,472
1991	6,162,941	4,727,507
2001	9,669,701	5,402,166

Source: Census of India, 1961–2001.

Agriculture is the main occupation in the TB command area. The water consumption for agriculture has drastically increased in the study area and also with Groundwater exploitation. Recent trends show that surface water usage is on decline. Cultivation of water demanding crops like sugarcane and paddy and spread of bore wells are causative factors. Both surface and surface water meet drinking water requirements for various human settlements across the basin. The requirement of water across the various towns are ranged from 70 to 135 litre per capita per day (LPCD) (Table 6.3). The sewage from these ULBs is directly enters into the river system or agricultural fields for villages located in the river basin; the main source of water being the groundwater, which is supplied through mini water supply schemes.

Table 6.3: Water Supply in Urban Local Bodies

Name of the ULBs	Surface Water		Groundwater		LPCD
	Source	Volume Supplied (in TMC)	No. of Wells	Volume Supplied (in TMC)	
Tarikere	Bhadra canal	0.057	10	0.02	100
Sringeri	Tunga river	0.005	–	–	70
Koppal	Tunga river	0.006	14	0.10	70
Gadag Batageri	TB river	0.035	27	0.15	135
Mundaragi	TB river	0.034	10	0.50	100
Byadagi	TB river	0.043	12	0.10	100
Haveri	TB river	0.093	10	0.50	100
Ranebennur	TB river	0.150	14	0.50	100
Shimoga	TB river	0.619	32	2.00	135
Thirthahally	TB river	0.017	10	0.10	70
Honnali	TB river	0.018	12	0.10	70
Davanagere	TB river	0.820	22	1.00	135
Harihara	TB river	0.125	10	0.20	100
Bellary	TB river	0.710	35	2.00	135
Hosapet	TB canal	0.369	15	0.20	135
Kamalapura	TB power canal	0.036	15	0.10	100
Kampli	TB river	0.059	15	0.15	100
Siruguppa	TB river	0.071	10	0.10	100
Tekalkopta	TB canal	0.039	10	0.10	100
Huvinahadagali	TB river	0.039	15	0.12	100
Sindhanur	TB canal	0.102	10	0.10	100
Gangavathi	TB river	0.210	15	0.20	135
Koppa	TB river	0.094	15	0.20	100
Manvi	TB river	0.064	15	2.00	100
Lakeshmeshwar	TB river	0.050	15	2.00	100
Harapanahalli	TB river	0.068	20	0.20	100
Bhadravathi	Bhadra river	0.362	20	0.25	135
Chennagiri	Bhadra river	0.030	15	0.10	100
Total		4.325	423	13.09	

Source: KWSDB, 2005.

The average flow status from the TB reservoir of 30 years from 1975-2005 is shown in Table 6.4 and Figure 6.2, compared with the standards with the different methods. It is clearly indicated that the flow varies over the years. During the monsoon season it flows in the moderate level. The utilization from the reservoir is high. It is interested to note that during non-monsoon periods, while inflows are nil, outflows from the reservoir is high (due to the commitments with the downstream users). Natural Flow Regime in TB River was completely changed as early as 1960s. The earlier agricultural pattern of cultivating semi arid crops has now been shifted to water intense crops where ever is possible, in the entire basin. Bore wells are another important feature.

Ecosystem Services in the River Basin

Aquatic ecosystems, such as rivers, provide a great variety of benefits to people. These include 'goods' such as clean drinking water, fish, fibre, and 'services' such as water purification, flood mitigation and recreational opportunities. Healthy rivers and associated ecosystems also have an intrinsic value to people that may be expressed, in terms of cultural significance, particularly for indigenous cultures. One of the important indicators of water quality is the impact on fish and other aquatic life (Manikya and Venkateshwaralu, 1986). Fish kills is a serious issue resulting from pollution, specifically industrial pollution, which occurred more during summer season due to minimum flow and minimum dilution (Reddy and Venkateswarlu, 1987). In TBSB, fish kills are observed at an average of 3 to 4 times in a year. In 1984, a large-scale fish kill was observed in the TB River downstream because of the effluent discharge from Birla industries. The Karnataka State Pollution Control Board (KSPCB) pointed out that, uncontrolled discharge of industrial effluents was the primary cause and issued a notice, to stop releasing effluents to the concerned industries. Sediment load also increased from mining activities and less flow in the river (Krishnaswamy *et al.*, 2006).

Another occurrence was in March 1994 in the downstream of TB to a stretch of 25-30 km affecting many villages. This was one of the massive fish kill ever observed where an estimated 2.5 tons of fish were dead. This case was narrated in the study by the National Environmental Engineering Research Institute (NEERI), Nagpur, in response to the 1994 fish kill, the Tungabhadra Parisara Samiti, held regular protests and processions. The study showed that the BOD levels in the raw effluents released into the river were 1,000 mg/l. The effects of pollution were felt 40 kms downstream during the summers.

During the field study many fishermen expressed concern over the increasing use of chemical fertilizers in agriculture. Several stretches of the river are polluted affecting around 75 villages, 47 per cent of the people were of the opinion that the pollution of water affected their income and resulted in fish kills, while only 26 per cent people reported health related problems. A small group of 9 per cent of the people opined that water pollution had resulted in loss of income, fish kills and had negative health implications too. 44 per cent of the fishermen felt that it was the responsibility of the government to clean the river while 20 per cent felt it should be the responsibility of the polluters.

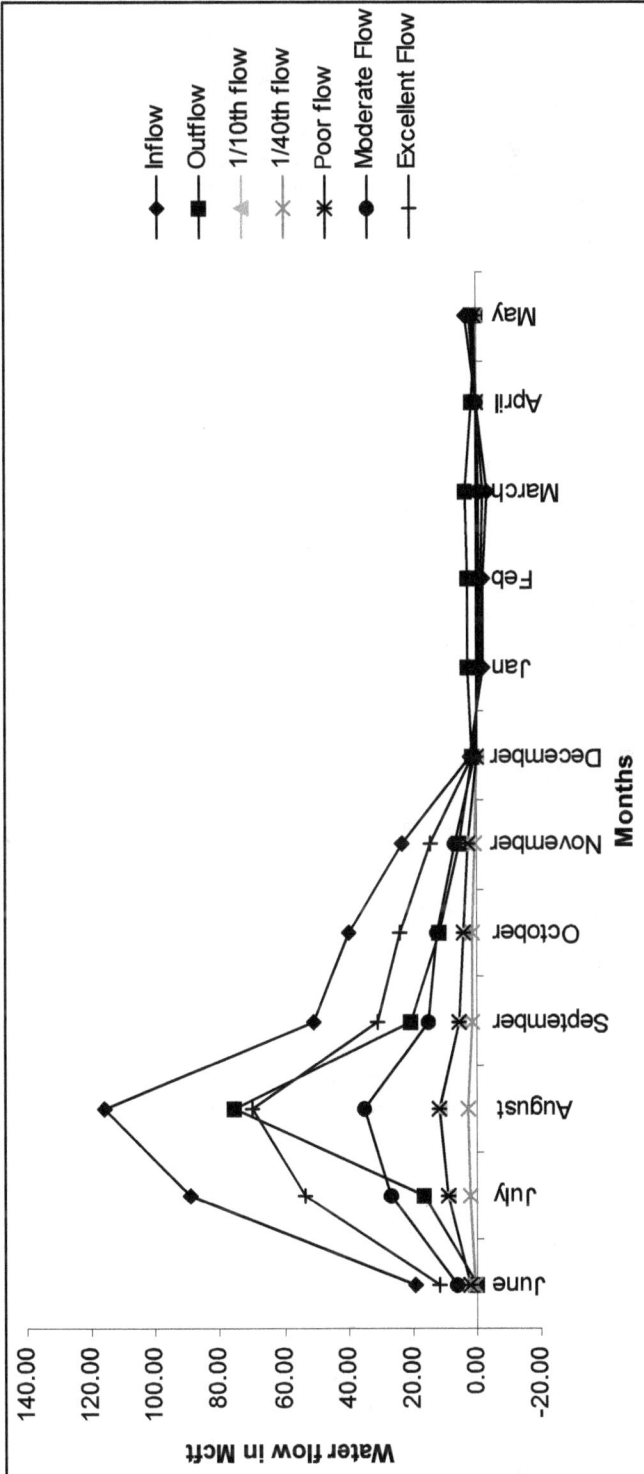

Figure 6.2: Average Flow from the TB Dam

Table 6.4: Quantum of EF Required from TB Dam

Month	Inflow	Outflow	French Fisheries Law		Montana Method			Remarks
			1/10th Flow	1/40th Flow	Poor Flow	Moderate Flow	Excellent Flow	
June	19.39	0.22	1.938853	0.484713	1.938853	5.81656	11.63312	Less flow
July	89.25	16.46	8.924952	2.231238	8.924952	26.77485	53.54971	Less than moderate flow
August	116.05	75.32	11.60506	2.901265	11.60506	34.81518	69.63037	Excellent flow
September	50.98	20.29	5.097742	1.274435	5.097742	15.29323	30.58645	Moderate flow
October	39.90	11.73	3.989503	0.997376	3.989503	11.96851	23.93702	Moderate flow
November	23.32	5.16	2.33241	0.583102	2.33241	6.997229	13.99446	Less than moderate flow
December	2.04	1.20	0.204158	0.05104	0.204158	0.612474	1.224948	Moderate flow
Jan	-2.16	2.46	-0.2156	-0.0539	-0.2156	-0.64681	-1.29363	Excellent flow
Feb	-2.03	2.53	-0.20299	-0.05075	-0.20299	-0.60898	-1.21796	Excellent flow
March	-3.51	3.07	-0.35141	-0.08785	-0.35141	-1.05423	-2.10846	Excellent flow
April	-0.10	1.17	-0.00973	-0.00243	-0.00973	-0.0292	-0.05839	Excellent flow
May	3.31	0.26	0.33059	0.082648	0.33059	0.99177	1.98354	Less flow

Table 6.5: Ecosystem Services from the TB River

Service Category	Service Provided	Key Flow Related Function	Key Environmental Flow Component or Indicator
Production Services	Water for people–subsistence/rural and piped/urban	Water supply	Floodplain inundation
	Fish/shrimp/crabs (non-recreational)	Habitat availability and connectivity, food supply	Instream flow regime, floodplain inundation
	Soil fertility	Supply of nutrients and soil moisture	Floodplain inundation
	Wildlife	Habitat availability and food web	Floodplain inundation, flows sustaining riparian productivity
	Fibre/organic raw material for building/firewood/handicraft	Seasonality of moisture conditions in soils	Floodplain inundation, flows sustaining riparian vegetation
	Inorganic raw material for construction and industry (gravel, sand, clay)	Sediment supply, transportation and deposition (fluvial geomorphology)	Instream flow magnitude and variability
Regulation Services	Chemical water quality control (purification capacity)	Denitrification, immobilization, dilution	Instream flow regime,
	Physical water quality control	Flushing of solid waste	Instream flow regime
	Groundwater replenishment (low flow maintenance)	Groundwater (aquifer) replenishment	Floodplain inundation
	Health control	Flushing of disease vectors	Instream flow regime, water quality
	Pest control	Habitat diversity	Instream flow regime
	Erosion control (riverbank/bed and delta dynamics)	Healthy riparian vegetation	Flows sustaining riparian vegetation
	Prevention of saltwater intrusion (salinity control)	Freshwater flow, groundwater replenishment	Instream flow regime
	Microclimate stabilization	Healthy ecosystems	Floodplain inundation, flows sustaining riparian vegetation
Information Services	Recreation and tourism (incl. fishing and hunting)	Presence of wildlife, aesthetic significance, good water quality	Site specific
	Biodiversity conservation	Sustaining ecosystem integrity (habitat diversity and connectivity)	Natural flow regime
	Cultural/religious/historical/symbolic activities	Site specific	Site specific
Life support	The prior existence of healthy ecosystems	All	Natural flow regime

This intrinsic value is often overlooked as it is difficult to identify and quantify. Until now in Karnataka, 201 freshwater fish species belonging to 9 Orders, 27 Families and 84 genera have been recorded, among them 40 fish species are under 'threat' and urgent conservation measures need to implement to ensure their survival. The systematics of freshwater fish species recorded from the inland waters of the State is based on the work carried out by Jayaram (1999). There are 81 fish species from 8 orders with 14 families that are endemic to the TB River. Annual fish production in the state from the freshwater sector is about 1.2 lakh metric tonnes as against an estimated potential of 2.6 lakh metric tones. In order to boost the inland fish production and in turn meet the ever growing demand, the State Department of Fisheries has laid more emphasis on the culture of fast-growing Indian major carps like *Catla catla*, *Labeo rohita* and *Cirrhinus mrigala*, as also exotic fish–*Cyprinus carpio*, *Hypophthalmichthys molitrix* and *Ctenopharyngodon idella*. These introduced fish species have adopted well in the various types of freshwater bodies. Introduction of exotic fish species, in a way, has resulted in the decline in native fish population comprising of *Labeo*, *Cirrhinus*, *Puntius*, Catfish, Murrels, etc. The culture of African catfish–*Clarias gariepinus* undertaken by private entrepreneurs is going to damage the entire picture of indigenous fish fauna in particular if corrective measures are not implemented at the earliest. Fish farming offers good scope for rural development, apart from providing employment, income generation, and protection to livelihoods and also provides inexpensive protein rich food.

Ecosystems provide a wide range of services to people (Costanza, 2003, Emerton and Bos, 2005, Millennium Ecosystem Assessment, 2005, Pearce *et al.*, 2006). The services provided by environmental flows, may either be provided directly by flow (*e.g.* flushing of sediments, salinity control) or indirectly via ecosystem functions. The alteration of natural water flow regimes has had the most pervasive and damaging effects on river ecosystems and species, ecological impacts associated with flow alteration (Poff *et al.*, 1997, Postel and Richter 2003). Many species are highly dependent upon lateral and longitudinal hydraulic connectivity, which can be broken through flow alteration; and the invasion of exotic and introduced species in river systems can be facilitated by flow alteration (Bunn and Arthington, 2002). The extent to which ecosystem functions create ecosystem services depends on the cultural, socio-economic and technical setting. Thus, the list of services given in Table 6.5 is not entirely determined by the suite of ecosystem functions, but also by human ingenuity in deriving benefits.

Conclusion

The immediate impact of flow regime change was observed in the vicinity of the river in the study area, where farmers observed that their consumption of inorganic fertilizers has increased and their fields were getting enriched by the silt. The groundwater table has declined along the river course and quality of river water has decreased which indicate reduction in the diversity factor. Agricultural and industrial water demand has increased the stress on the water resources of the river basin. Several incidents had occurred in the basin due to less flow. Local community had failed to link the effects due to the less flow of runoff. In many places on the river bank, forest cover has clearly vanished. Agricultural activities on the river bed during lean months are commonly seen in the study area. It clearly indicates that the flow varied over the years. During the monsoon season it flows at a moderate level. The utilization from the dam was high. It is interesting to note that during non-monsoon periods, while inflows were nil, outflows from the reservoir were high.

Local communities have observed negative impacts from the TB dam, when large amounts of water were released from the reservoir, causing massive downstream flooding at Hampi, Kampli. Rainy season flooding, believed to have been caused by water releases from the dam, has damaged agricultural crops and flooded villages along the TB River every year, although floods in the upper part of the basin have been much less severe than those in lower parts.

Communities have been impacted by irregular dry season water level fluctuations, which are characterized by extreme highs and lows, and rapid changes in water levels. Massive surges of water over 2 metres high have caused serious damage downstream, including large amounts of riverbank erosion. Dry seasons gardens have been flooded, and a number of other dry season activities were fishing etc., have been severely disrupted. The water quality in the TB River has seriously deteriorated. However, local communities opined that the water quality problems originated with the TB dam, which may be contaminated with toxic elements. The river has become turbid and smells bad. Irregular fluctuations in the TB River have seriously affected riverine vegetation, birds, reptiles and various aquatic life forms whose lifecycles are dependent on the natural rhythm of the TB River.

Native fish, fish habitat and riverine fisheries have been severely impacted by changes in the hydrological regime and water quality. Villagers, who are highly dependent on fishing, for food and income, are badly affected, owing to the drastic decline in fish catches. All fish species have been impacted; apparently, large fish have been affected more. Fish diseases have also increased. The rapidly rising waters, which occur without warning, have washed away large numbers of fishing boats. It is impossible to imagine a river in the virgin condition, but it is necessary to maintain river in pristine condition by allocating minimum water release regularly to the downstream of the river.

Aquatic species have evolved life history strategies, such as their timing of reproduction, in direct response to natural flow regimes, which can be desynchronized through flow alteration. Endemism is very high in the river basin, both in the terrestrial ecosystem and aquatic ecosystem. Agricultural water demand and industrial water demand has increased the stress on the water resources of the river basin. Several incidents have occurred in the basin due to less flow. It is impossible to imagine a river in to the virgin condition, but it is necessary to maintain a river in pristine condition, by allocating minimum water release regularly to the downstream of the river.

References

1. Brisbane Declaration (2007). The Brisbane declaration: Environmental flows are essential for freshwater ecosystem health and human well-being. In: *Declaration of the 10th International River Symposium and International Environmental Flows Conference*, Brisbane, Australia, 3–6 September.

2. Bunn, S.E. and Arthington, A.H. (2002). Basic principles and ecological consequences of altered flow regimens for aquatic biodiversity. *Environmental Management*, 30: 492–507.

3. Christer Nilsson, Catherine, A.R., Dynesius, M. and Revenga, C. (2005). Fragmentation and flow regulation of the world's large river systems. *Science*, 308(5720): 405–408.

4. Costanza, R. (2003). Social goals and the valuation of natural capital. *Environmental Monitoring and Assessment*, 86: 19–28.

5. Dyson, M., Bergkamp, G. and Scanlon, J. (2008). *Flow: The essentials of Environmental Flows*, 2nd Edition. Gland, Switzerland: IUCN. Reprint, Gland, Switzerland: IUCN.

6. Dyson, M., Bergkamp, M. and Scanlon, J. (2003). *Flow: The Essentials of Environmental Flows*. IUCN, Gland, Switzerland and Cambridge, U.K.

7. Emerton, L. and Bos, E. (2005). *Value. Counting Ecosystems as an Economic Part of Water Infrastructure*. IUCN, Gland, Switzerland and Cambridge, U.K.

8. Jayaram, K.C. (1999). *The Freshwater Fishes of the Indian Region*. Narendra Publishing House, New Delhi, p. 551.

9. Krishnaswamy Jagdish, Milind Bunyan, Vishal K. Mehta, Niren Jain, K. Ullas Karanth (2006). 'Impact of iron ore mining on suspended sediment response in a tropical catchment in Kudremukh, Western Ghats, India'. *Forest Ecology and Management*. 224: 187–198

10. Karnataka Urban Water and Sewage Development Board (KUWSDB) Report 2005, Government of Karnataka, Bangalore.

11. King, J., Brown, C. and Sabet, H. (2003). A scenario-based holistic approach to environmental flow assessments for rivers. *River Research and Applications*, 19: 619–639.

12. King, J. and Brown, C. (2006). Environmental flows: Striking the balance between development and resource protection. *Ecology and Society*, 11(2): 26.

13. Manikya, R.P. and Venkateswarlu, V. (1986). Ecology of algae in the paper mill effluents and their impact on the river TB. *Journal of Environmental Biology*, 7: 215–224.

14. Millennium Ecosystem Assessment (2005). *Millennium Ecosystem Assessment Synthesis Report*. Island Press, Washington DC.

15. Pearce, D., Atkinson, G. and Mourato, S. (2006). *Cost–Benefit Analysis and the Environment: Recent Developments*. OECD Publishing.

16. Pearce, F. (2007). *When the Rivers Run Dry: What Happens When our Water Runs Out?* Transworld Publishers, London.

17. Poff, N.L., Allan, J.D., Bain, M.B., Karr, J.R., Prestegaard, K.L., Richter, B.D., Sparks, R.E. and Stromberg, J.E. (1997). The natural flow regime: A paradigm for river conservation and restoration. *Bio Science*, 47: 769–784

18. Postel, S. and Richter, B. (2003). *Rivers for Life: Managing Water for People and Nature.* Island Press, Washington, DC.

19. Reddy, P.M. and Venkateswarlu (1987). Assessment of water quality and pollution in the river Tungabhadra near Kurnool, Andhra Pradesh. *Journal of Environmental Biology*, 8: 109–119.

20. Revenga, C., Murray, S., Abramovitz, J. and Hammond, A. (1998). *Watersheds of the World.* World Resources Institute, Washington, D.C., USA.

21. Revenga, C., Campbell, I., Abell, R., De Villiers, P. and Bryer, M. (2005). Prospects for monitoring freshwater ecosystems towards the 2010 targets. Philosophical transactions of the Royal Society of London, Series B, *Biological Sciences*, 360: 397–413.

22. Richard, D. and Hirji, R. (2003). Water resources and environment technical note C.2: Environmental flows: Case studies. The World Bank, Washington, D.C.

23. Song, B.Y. and Yang, J. (2003). Discussion on ecological use of water research. *Journal of Natural Resources*, 18: 617–625.

24. Tharme, R.E. (2003). A global perspective on environmental flow assessment: Emerging trends in the development and application of environmental flow methodologies for rivers. *River Research and Applications*, 19: 397–441.

Biodiversity of Aquatic Reseources (2012)
Editors: **Mamta Rawat & Sumit Dookia**
Published by: **DAYA PUBLISHING HOUSE, NEW DELHI**

Pages **90~104**

Chapter 7

Aquatic Biodiversity of Rana Pratap Sagar Lake, Rajasthan

P.C. Verma[1]*, L.L. Sharma[2] and A.G. Hegde[1]

[1]*Environmental Studies Section,*
Health Physics Division, BARC, Mumbai, Mh.
[2]*College of Fisheries, MPUAT, Udaipur, Rajasthan*

ABSTRACT

Aquatic biodiversity has enormous economic and aesthetic value and is largely responsible for maintaining and supporting overall aquatic environmental health. Humans have long depended on aquatic resources for food, medicines and materials as well as for recreational and commercial purposes such as fishing and tourism. Aquatic organisms also rely upon the great diversity of aquatic habitats and resources for food materials and breeding ground. The aquatic biodiversity of a water body gets significantly affected by several factors such as contamination due to anthropogenic wastes from industry or other human activities including sewage disposal into the reservoirs. Aquatic biodiversity of Rana Pratap Sagar lake is presented in this chapter.

Keywords: Aquatic biodiversity, Thermal profile, Ponderal index, Thermal tolerance.

---

* Corresponding Author E-mail: pcverma@barc.gov.in

Introduction

Aquatic biodiversity may be defined as the variety of species and their abundance in the aquatic ecosystem. The aquatic ecosystem provides home to several aquatic life species including phytoplanktons, zooplanktons, fishes and other abiotic entities like sediment, weeds etc. The interaction among all these biotic species and abiotic matter forms an aquatic ecosystem. Besides several other factors, the temperature has many fold effects on biodiversity as it can adversely impact no growth of an organism, distribution of animals, physiological processes like oxygen consumption, reproduction, rate of embryonic development and behavioral changes in organisms. It also affects the biochemical, metabolic and physiological activities of fishes. The growth rate, reproduction and enzyme activity increases with increase in temperature up to a certain limit, beyond that it creates adverse effect on the organisms. The critical temperature range is that over which a significant disturbance in the normal behaviour of the fishes may occur *i.e.* sign of stress.

Rana Pratap Sagar is one of the largest reservoirs of Rajasthan. On its eastern bank there exists Rawatbhata Site, comprising of multi–nuclear facilities. There are four PHWR units of RAPS which are in operations, two are under commissioning stages and another two are under advanced stage of planning. In addition to nuclear power plants the site also houses a Heavy Water Plant and other allied facilities such as cobalt facility and waste management facilities.

RAPS draws water from RPS lake through a 300 m long conduit pipe located at lake bottom about 20 m below the surface. Duly treated low level radioactive liquid effluents from RAPS facilities are injected to the warmed condenser outlet and then allowed to discharge to RPS in a controlled manner. The warm water is likely to remain at the surface and get mixed with lake water and cooled due to dilutions, evaporation from the lake surface and wind currents.

The heat release to the RPS lake through condenser outlet may affect the microbiological and water quality parameters, planktonic biodiversity, fish productivity etc. and thus it is imperative to conduct the thermal ecological studies pertaining to this reservoir.

Thermal ecological studies were carried out at Rana Pratap Sagar lake under DAE-BRNS project sanctioned to College of Fisheries of Maharana Pratap University of Agriculture and Technology, Udaipur in collaboration with Environmental Survey Laboratory of Bhabha Atomic Research Centre, Mumbai.

This chapter presents the results of thermal ecological studies conducted at Rana Pratap Sagar (RPS) lake. This study provides the details on collected data over three years on planktonic biodiversity, productivity and fisheries aspects of RPS reservoir. It includes identification of aquatic species, evaluation of Ponderal index and thermal tolerance study in respect of RPS fishes. Attempts have also made to estimate the harvesting age of commercial fishes of RPS lake.

Methodology

Description of the Study Area

RPS lake is the largest man made reservoir of Rajasthan. The reservoir forms part of the great Ganga basin and constructed on the river Chambal. The reservoir has water spread area of 19600 ha and the average depth is about 25 meters. The salient features of the lake are given in Table 7.1.

Table 7.1: Salient Features of Rana Pratap Sagar Lake

Maximum level	352.8 m MSL (1976)
Minimum level	344.4 m MSL (1985)
Gross Storage	2.9×10^9 m³
Submergence area	198.5 km²
Minimun draw level	342 m MSL
Annual fish catch	~ 550 MT

RPS has rocky bed and practically no silt load.

Sampling Locations

The outline map of RPS lake is shown in Figure 7.1. In order to complete the study five off-shore locations as shown in Figure 7.1, were chosen at RPS lake for sampling namely one is *Control* which is located at about 5km upstream, near *Intake* jetty of RAPS, *Discharge* area of RAPS, *Dam* side which is located about 6 km down stream and the last one is *Sentab* which is about 9 km down stream away from the discharge area of RAPS. In addition to off-shore sampling locations, samples were also collected from another five locations on the downstream and three locations on the upstream side of RAPS discharge point along the RPS shore.

Samples of water, sediment, lake biota etc. were periodically collected from the identified locations. Equipments such as Kemmerer sampler for collection of depth wise water samples, Ekman dredge for collection of bottom sediments, hand held WTW make pH meter with temperature probe for depth wise measurements and GPS for locating the precise position of the sampling were used in a motorized boat for carrying out sampling work.

Besides thermal profile, measurements on selected water quality parameters were monitored at all the sampling stations. Studies were also conducted depth wise at all the five off-shore sampling locations. The water quality parameters were analysed as per the standard methods (APHA, 1985). Collection, identification and density estimations for planktonic, benthic and aquatic macrophyte communities were also made using standard keys (Cole, 1979). In view to assess the seasonal effects the sampling were carried out regularly on alternate months. In view to quantify the abundance of biota, various diversity indices were obtained.

The RPS fishes were collected from landing centers and identified. The length and weight data in respect of fishes were used to calculate the Pondral index, which indicate general well being of the fishes. Thermal tolerance and oxygen consumption

N

Scale

0   1   2   3 Cm.

0   1.27   2.54   3.8 Km.

Year 2003-06

J.S. Dam

Kota Barrage

Borabas

Baroli

Bhagatpura

Sub Bridge

Township Colony

Padajar

Saddle Dam

Phase II Colony

R.P.S. Colony

Dippura

Amarpura

5

4

Phase I Colony

Chainpura   Hempura

2

1

R.A.P.P.

3

Kalgarh

Raheliya

Lake Rana Pratap Sagar

River Chambal

LEGEND

Sampling Station

Village

Submerge area

Dam Site

Township

Atomic Power Plant

River course

Gandhi Sagar Dam

1. Plant site (Intake canal)
2. Plant site (Discharge canal)
3. Plant sites (Control)
4. Dam site
5. Downstream site

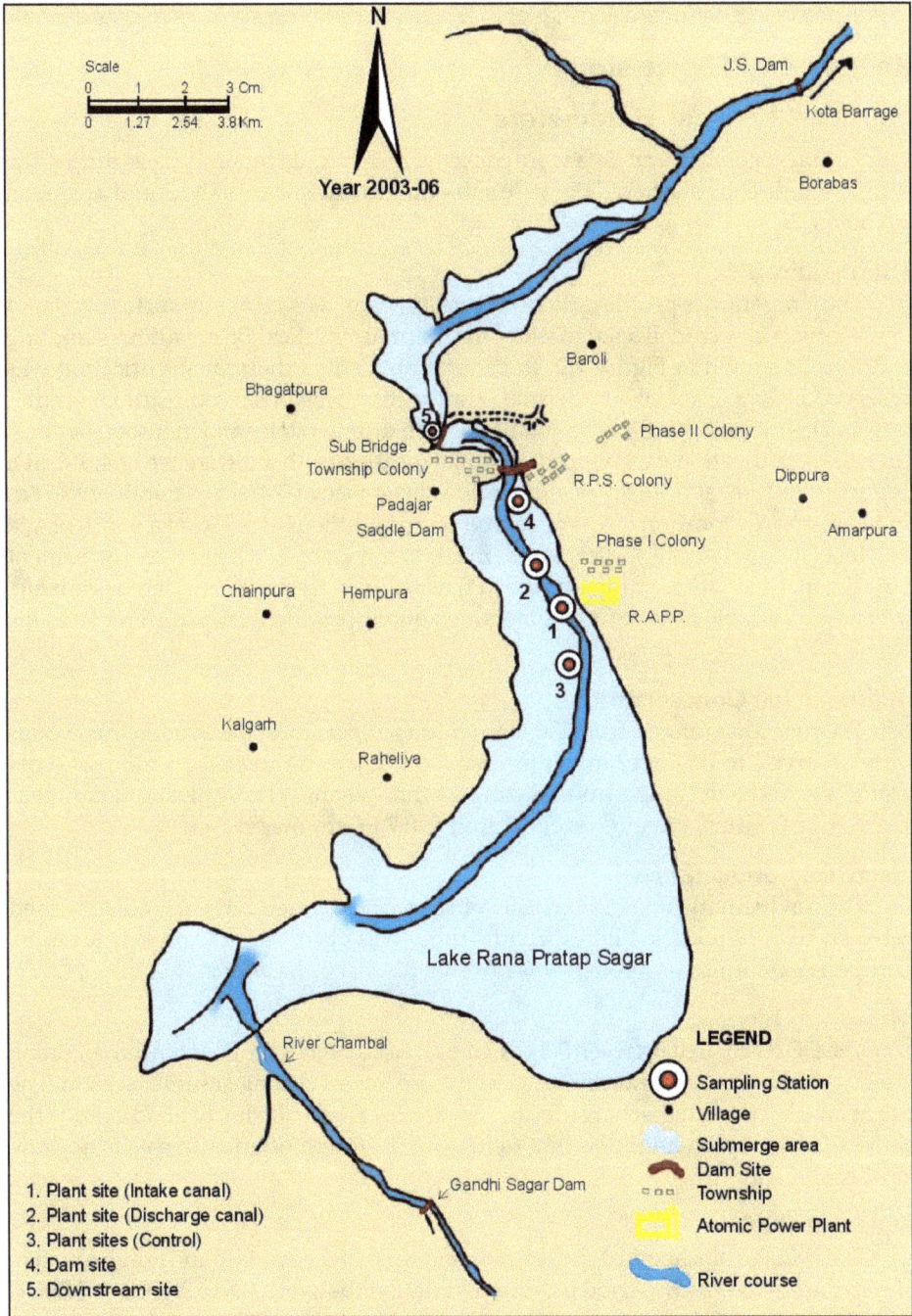

Figure 7.1: Outline Map of Rana Pratap Sagar Lake

of common fishes were also studied. Attempts have also made to estimate the harvesting age of commercial fishes of RPS lake using their scale rings measurements.

Results and Discussions

Physico-Chemical Parameters

The physico-chemical data were collected on alternate monthly basis during the study period. The collected data were subjected to statistical analysis and are given in Table 7.2.

Thermal Profile

The temperature profile data from surface water as well as sub surface water at 2 meter interval were collected to study the thermal stratification at various sampling locations as shown in Figure 7.2. In general no distinct thermal stratification was observed at RPS lake. Weak thermal stratification was observed during summer season. Thermal stratification was totally lacking during winter and monsoon periods. Thermal stratification was noticed at Intake location with a difference of 4.5°C at a depth of 8 to 10m, at Control location with a difference of 3.4°C at a depth of 6 to 8m and at discharge location it was observed with a gradient of 4.5° C at a depth of 4 to 6 meters. Thermal profile indicates polymictic behavior of RPS lake water. As evident from the thermal profile study that the RPS water mass frequently undergoes mixing, which is a desirable feature for augmenting aquatic productivity at different trophic levels.

Hydrogen Ion Concentration

The pH value ranged from 7.6 to 9.4. Like temperature, the average pH values were observed to be lower in winter than summer and monsoon period. The pH profile showed a declining trend during all the seasons. The vertical gradient was found to be 0.2 during winter as compared to 0.7 in summer season.

Electrical Conductance

The conductivity ranged from 260 to 462 µS. The average vertical gradient varied between 10 to 14 µS. The highest conductivity was observed at Dam side location. Compared to control location, it was lower at discharge locations.

Dissolved Oxygen

The DO fluctuated between 0.2 to 12 mg/l. As expected the DO showed declining trend with the depth of lake and observed to be lowest during summer season. The DO at intake jetty and discharge location was observed to be fairly good even at the bottom layers which might be due to luxuriant growth of aquatic weeds at these locations.

BOD and COD

The average biochemical oxygen demand for surface, middle and bottom waters of RPS during the study period indicated variations between 0.6 to 3.0, 0.6 to 2.8 and 1.4 to 2.9 mg/l in winter, summer and monsoon respectively. Moreover, higher average BOD value was seen for winter as compared to summer and monsoon. The BOD also showed a decreasing trend as in case of Dissolved Oxygen.

Table 7.2: Seasonal Average of Physico-Chemical Parameters in Rana Pratap Sagar Lake

Sample Collected from	Temperature (°C)			pH			Dissolved Oxygen (mg/l)			BOD (mg/l)			COD (mg/l)			EC (μS)			Total Dissolved Solids (mg/l)		
	M	W	S	M	W	S	M	W	S	M	W	S	M	W	S	M	W	S	M	W	S
Surface	30.2	21.0	30.8	8.5	8.1	8.5	9.6	10.6	8.4	2.9	3.0	2.8	29	48	64	299	337	325	133.0	118.0	137.7
Middle	30.0	20.9	25.7	8.3	8.0	8.0	7.5	8.0	4.1	2.3	1.1	0.8	35	29	50	284	321	332	123.2	92.5	126.6
Bottom	29.0	20.6	22.1	7.9	7.9	7.8	4.2	4.2	1.1	1.4	0.6	0.6	69	39	50	298	323	335	121.0	94.0	125.6

Sample Collected from	Bicar. Alkalinity (mg/l)			Total Alkalinity (mg/l)			NO_2 Nitrogen (μg/l)			NO_3 Nitrogen (μg/l)			NH_3 Nitrogen (μg/l)			Orthophosphate (mg/l)			Total Phosphorus (mg/l)		
	M	W	S	M	W	S	M	W	S	M	W	S	M	W	S	M	W	S	M	W	S
Surface	99	99	104	101	99	104	6.0	9.0	15.0	2.0	4.0	2.0	2.0	6.0	4.0	0.16	0.11	0.07	0.69	0.39	0.36
Middle	100	104	106	100	104	106	3.0	8.0	27.0	2.0	7.0	5.0	0.0	7.0	0.0	0.20	0.07	0.09	0.55	0.78	0.50
Bottom	104	106	112	104	106	112	3.0	13.0	7.0	2.0	9.0	15.0	27.0	107	3.0	0.22	0.11	0.08	0.69	0.67	0.44

Sample Collected from	Dis. Silicate (mg/l)			Sulphate (mg/l)			Ca Hardness (mg/l)			Mg Hardness (mg/l)			Total Hardness (mg/l)			Water Color (Hazen units)			Free CO_2 (mg/l)		
	M	W	S	M	W	S	M	W	S	M	W	S	M	W	S	M	W	S	M	W	S
Surface	5.3	8.1	7.1	8.0	8.3	9.2	50.2	44.9	55.3	31.7	37.7	12.7	81.9	82.6	68.0	14.2	10	22.5	8.50	11.14	13.17
Middle	5.2	9.5	9.0	8.0	8.2	9.3	49.0	37.6	62.6	30.3	46.8	15.1	79.3	84.4	77.7	14.4	11	15.0	7.30	12.60	13.20
Bottom	5.0	9.5	9.7	8.4	9.4	10.2	50.0	52.4	63.4	35.0	32.8	6.3	85.0	85.2	79.7	16.9	11	15.0	10.8	14.40	15.46

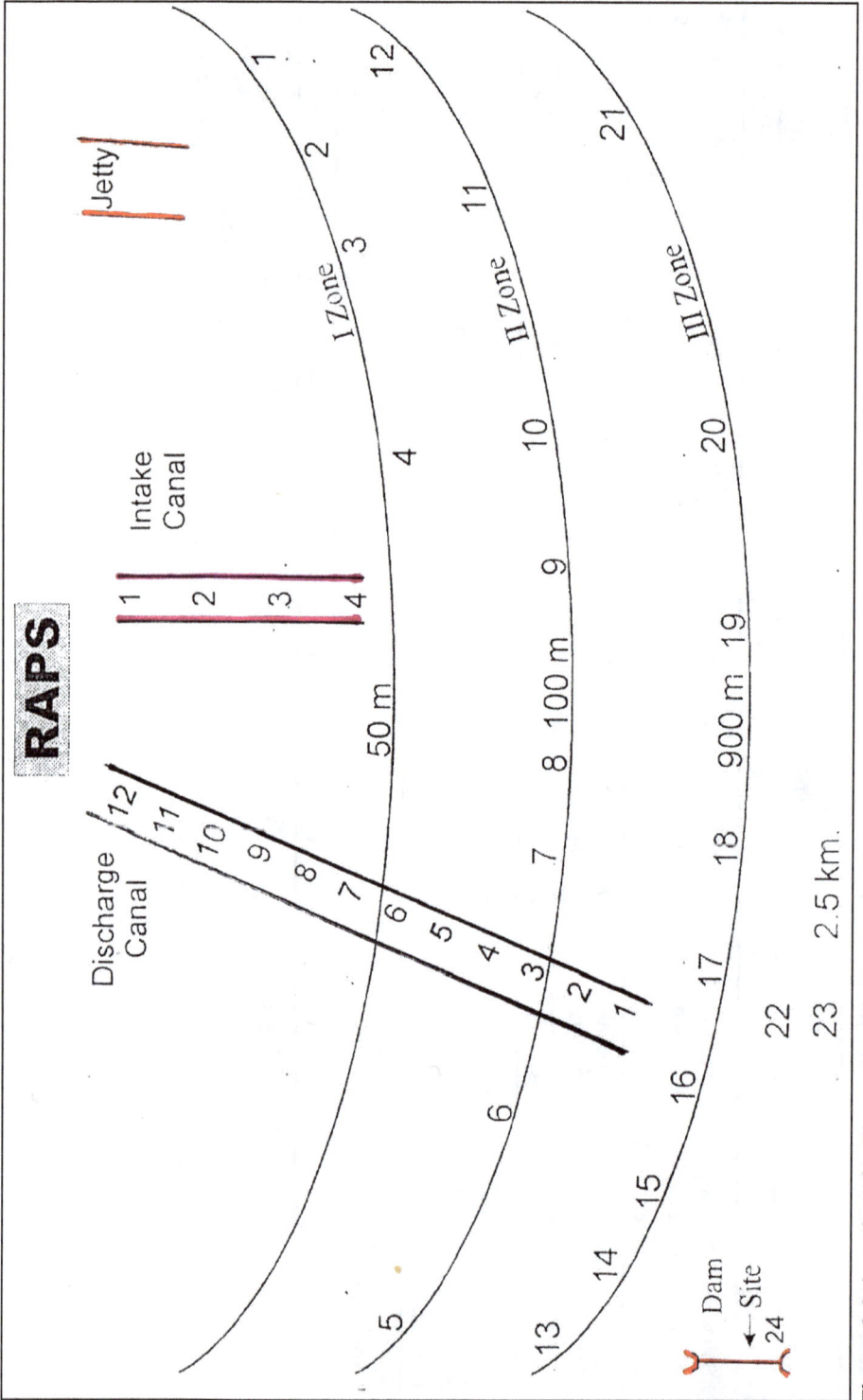

Figure 8.2: Locations for Thermal Profile Studies in RPS

The average COD values are found to be higher in summer than monsoon and summer periods which may be due to excessive evaporation in summer season and resulting higher dissolved contents.

Water Colour and Transparency

The water colour ranged between 14.2 to 22.5 Hazen units. The over all oscillations in average transparency were 20 to 287 cm during the study period. The average transparency (excluding monsoon) was observed to be 221 cm.

Alkalinity

The alkalinity was mainly due to bicarbonates. The carbonate alkalinity was almost negligible and was observed to be present during June to September months whenever carbon dioxide was absent. The seasonal average alkalinity showed ascending trend with the depth of lake. The total alkalinity was observed to be lowest during monsoon and was comparatively higher during summer season. The vertical gradient was 6 to 8 mg/l during the study period.

Hardness

It ranged from 52 to 98, 40 to 100 and 74 to 104 mg/l in surface, middle and bottom layers of RPS waters respectively. The average value was found to be higher during winter and lower during summer season.

Total Dissolved Solids

The seasonal average TDS varied from 92 to 137.7 mg/l. The average values were observed to be lower during winter as compared to monsoon and summer. The vertical gradient ranged from 12 to 25 mg/l.

Sulphate

It ranged from 5.9 to 13.1 mg/l. The seasonal average value of sulphate showed ascending trend with the depth of RPS lake. It was observed to be lower in monsson and higher during summer. The vertical gradient varied from 0.4 mg/l in monsoon and 1.0 mg/l in summer.

Dissolved Silicates

The seasonal average value varied from 5.0 mg/l to 9.7 mg/l. Like hardness, the silicates also showed higher values during winter season. The vertical gradient varied from 0.3 mg/l to 2.6 mg/l.

Ortho-Phosphate and Phosphorus

The seasonal average total phosphorus values varied from 0.36 to 0.69 mg/l. No distinct trend in phosphorus contents could be observed. However, in general the ortho-phosphate was observed to be higher in bottom layers.

Nitrite and Nitrate Nitrogen

Like phosphorus no distinct trend in nitrogen parameters was observed. However, the average nitrogen values were higher in summer than winter and monsoon seasons. In general, the bottom layer maintained higher nitrate as compared to middle and surface water layers.

Ammonia Nitrogen

The seasonal average value of ammonia nitrogen was observed to be higher during winter as compared to summer and monsoon periods.

The nitrogen and phosphorus did not show any significant variation and remained well with in the limits prescribed for drinking water standard. The low concentrations of these nutrients may be due to rapid utilization by phytoplankton and microphyte communities, as reported elsewhere especially in case of eutrophic lake (Sreenivasan, 1968).

The monitored data on water quality were subjected to statistical analysis to evaluate the ecological risk (Verma, *et al.*, 2007). Using Fuzzy synthetic evaluation system on the basis of water quality guidance from WHO, BIS and ICMR with combination of expert perception the water quality belongs to desirable category during all the seasons of the year. The study reveals that there is no adverse effect on RPS water quality owing to receipt of the warmed effluents from RAPS (Verma *et al.*, 2007). Furthermore, it shows that RPS water is nearly homogeneous and shows weak thermocline and chemocline patterns. Based on monitoring data, the reservoir could be assigned as mild eutrophic status. The RPS water appears to be homogeneous as revealed from the parallel studies carried out on depth wise tritium measurements (Verma *et al.*, 2005).

Macrophytes

A total of 19 species of macrophytes have been recorded. The prominent macrophytes are Potamogeton *spp, Hydrilla, Vallisneria* and *Najans minor.* Macrophytes were noticed to grow luxuriously even at the depth of 6 to 8 meters. A typical specimen sheet is shown in Figure 7.3. The average biomass ranged between 0.86 to 4.58 kg/m^2 during the study period. The Odum's diversity index for macrophytes was found to be 594.6.

Macro Invertebrates

A total of 38 species of macro invertebrates have been recorded from RPS during the course of study. The *Mollusca, Diptera, Hemiptera, Coleptera, Odonata, Isopod, Zygoptera, Crutacea,and Oligochaeta* constitutes the main groups of the macro invertebrates. A typical specimen sheet is shown in Figure 7.4. The average density of benthos varied between 462 (summer) to 4288 N/m^2 (monsoon).The frequency and concentration of macro invertebrates varied from sampling location to location. Using average density of benthos, the Odum's diversity index works out to be 7.88.

Plankton and Biodiversity

The biodiversity of planktonic organisms in RPS has been observed fairly well. About 82 phytoplanktons and 63 zooplanktons have been observed and identified. During the study period the dominant phytoplankton identified in RPS are *Peridinium palatinum, Pediastrum simplex, Microcystis aeruginosa, Tabellaria, Fragilaria capucina, Alpanizomenom* and dominant zooplankton species are *Nauplius larvae, Keratella tropica, Mesocyclops, Keratella coochlearis,*

Hemiptera

Diplonychus annuletus

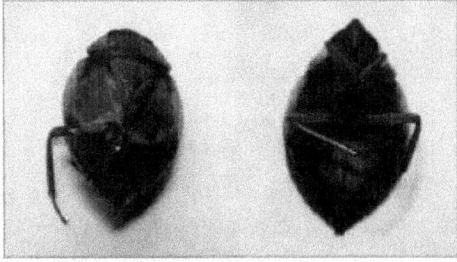

Diplonychus rusticus

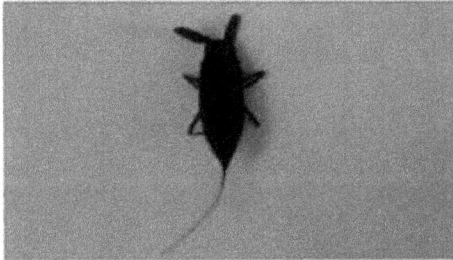

Laccotrephes griseus

Coleoptera

Cybister tocipunetatus asiatieus

Stemolophus rcufipes

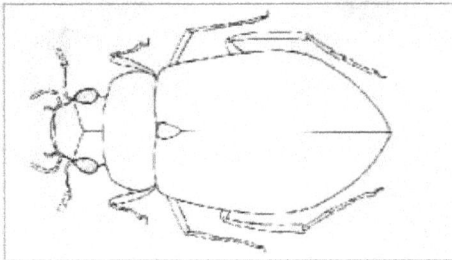

Berosus dolerosus

Figure 7.3: Macrophytes Specimens

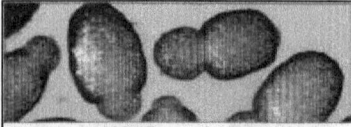

Wolfia arrhiza

Marsilea quadrifolia

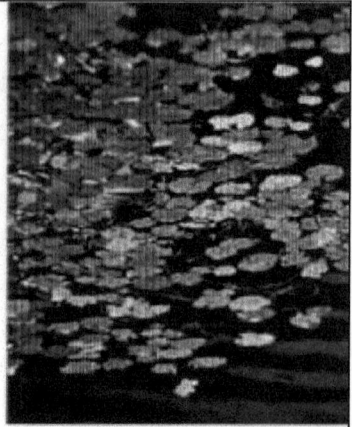

Nymphoides platatum

Najas graminea

Najas minor

Echinochloa

Polygonum hydropiper

Figure 7.4: Benthos Specimens

Table 7.3: Seasonal Average of Gross Primary Productivity, Net Primary Productivity and Community Respiration (gc/m³) in RPS

Location	Winter			Summer			Monsoon		
	GPP	*NPP*	*CR*	*GPP*	*NPP*	*CR*	*GPP*	*NPP*	*CR*
Control	0.312	0.062	0.187	0.325	0.162	0.325	0.375	0.062	0.312
Discharge Point	0.437	0.062	0.437	0.575	0.000	0.575	0.313	0.185	0.125
Intake Jetty	0.362	0.063	0.062	0.125	0.375	0.750	0.313	0.063	0.250
Dam Side	0.193	0.081	0.112	0.218	0.139	0.062	0.187	0.112	0.175
Down Stream	0.306	0.131	0.275	0.255	0.161	0.093	0.237	0.149	0.092

Brachionus caudatus, Collothecia cornata, Alleonella, Rotifera. The diversity indices for plankton were estimated in terms of Odum's index and Menhinick's index. They are found to be 165.26 and 3.68 for phytoplankton whereas 276.4 and 4.17 for zooplankton respectively.

Comparing biodiversity and density of plankton of Control location with other sampling locations especially with Discharge area and Dam side, it was observed that phytoplankton density was higher at Dam side with low species diversity. Discharge area had higher density of phytoplankton than Control with almost the same diversity at both the locations. In case of blue green algae *Microcystis aeruginosa* and *Phormidium* showed high density at Discharge area.

Among green algae *Cosmarium* showed higher density at Discharge area and was absent at Control and Dam side locations. Similar trend have been shown by *Fragilaria capucina* among diatoms. This could be due to prevailing relatively higher temperature at this site which may be favorable for proliferation of these biota.

Zooplankton also showed similar trends with higher density at Dam side in comparison to Discharge area and Control locations. The species diversity was almost same at Discharge area, Control and Dam side during winter while during summer Dam side showed an increase in the diversity than Discharge area and Control locations. From the observed number of species it would be seen that the biodiversity of plankton appears to be fairly higher than many water bodies of Rajasthan (Sharma and Durve, 1985).

Sawyer (1966) considered *Microcystics* as an indicator of eutrophy. The presence of phytoplankton such as *Fragilaria, Cymbella, Microcystics, Aphanizomenon, Volvox, Phormidium, Spirogyra, Ulothrix, Peridinium, Ceratium etc.* and zooplanktons such as *Daphnia, Mesocyclops, Brachionus, Rotaria, Filinia, Arcella, Vorticella, Difflugia etc.*, exhibits the eutrophic status of RPS reservoir (Kaushik, 2003 and Wetzel, 1975).

Primary Productivity

The seasonal average data related to Gross Primary Productivity (GPP), Net Productivity (NPP) and Community Respiration (CR) are given in Table 7.3. The GPP manifested by surface water of RPS during the study period indicated seasonal average value 0.362, 0.499 and 0.285 g C/m³/hr for winter, summer and monsoon respectively.

The higher GPP average was seen for the intake jetty location as compared to other locations whereas the minimum GPP values were seen at dam side location.

The chlorophyll 'a' showed an overall average value of 0.031 and 0.042 mg/m³ in monsoon and winter respectively. Location wise the highest chlorophyll 'a' value was exhibited at Dam side. The values of chlorophyll when compared to GPP and phytoplankton density, it may be seen that the dam side did not show appreciably high GPP. This may be due to the fact that the planktonic counts at dam side also comprises of the algae drifted from lentic zone and their skeletal material devoid of chlorophyll thus exhibiting moderate GPP. From the productivity data, it is evident that RPS passes the mild eutrophic status (Madhusudan, *et al.*, 1984).

Fish and Fisheries

A total of 40 fish species have been identified from RPS. Out of these, 18 species are of commercial values and regularly harvested from RPS and collected at landing stations. The length weight data in respect of 16 common fishes were used to calculate the Ponderal Index, the indicator for general well being of these fishes. The Poderal Index varied from 0.93 to 7.5 which indicate the satisfactory growth status of this water body. The principal major carp species have shown appreciably high values of Ponderal index especially for *Catla catla, Labeo rohita, Tor khudree* and *Labeo calbasu* fishes.

The harvestable sizes of five commercially important fishes of RPS were estimated from their scale rings measurements. A typical imagery of *Catla catla* scale is shown in Figure 7.5. It is estimated to harvest *Cirrhinus mrigla* and *L.Calbasu, Catla catla* and *L.rohita, and Tor khudree* respectively at the age of 3, 4 and 5 years respectively. Moreover, it is interesting to note that during the last 30 years the fish catch has shown consistent increase which can be co-related to heat inputs (Chaudhary and Juyal 2003).

Thermal tolerance and oxygen consumption studies on *Rasbora sp.* acclimated to temperatures *viz.* 20°C, 28°C and 36°C respectively were conducted using critical thermal methodology. The temperature tolerance polygon of *Rasbora sp.* is calculated as 449.12°C. It is observed that the temperature tolerance of the fishes changed with the change in acclimation temperature. The oxygen consumption rates increased in fishes acclimated to 28°C. However, the rates decreased at 36°C. The decrease in oxygen consumption rates may be due to the damage caused to the gills structures at high temperature.

The aquatic biodiversity based on of this study has been documented in the form of a compendium on flora and fauna of RPS lake. This document provides the details of identification and colored photographs of phytoplantons, zooplanktons, benthos, macrophytes and fishes of RPS Lake.

Conclusions

The aquatic biodiversity of RPS lake has been fairly good, as revealed by the study showing presence of 38 species of micro invertebrates, 19 species of macrophytes, About 82 phytoplanktons and 63 zooplanktons in the reservoir. It habitats about 40 species of the fishes. The fishes enjoy a healthy environment which

Scale of *Catla catla* (Age 4+ year class)

Figure 7.5: Imagery of *Catla catla* Scales

is evident from the Pondral index values as well as general morphology. The consistent increase in fish catch during last thirty years can also be closely related with beneficial impact of warmed waters released from RAPS into the reservoir without any adverse impact on fisheries. This indicates the balanced environmental status of RPS reservoir.

Acknowledgement

The authors are thankful to Shri H.S.Kushwaha, Direcror, Health, Safety and Environment Group, BARC, Mumbai, Shri M.L.Joshi, Ex Head, Health Physics

Division, BARC for their interest and encouragement for the study. We are also thankful to Dr. Pratap Singh, Director, Research, MPUAT, Udaipur and RAPS authorities for providing facilities to conduct this study. Our thanks are also due to BRNS for providing financial support. We are also thankful to ZSI, Kolkotta authorities for collaborating the proper identification of RPS fishes. The support and cooperation provided by Shri Anil Kumar and Shri Niranjan Sarang, Project JRF for carrying out field work is gratefully acknowledged

References

1.   APHA (1985). *Standard Methods for the Examination of Water and Wastewater*, 16th Edition, Washington, D.C.

2.   Chaudhary, S. and Juyal, C.P. (2003). Fisheries kingdom of RPS. *Fishing Chimes*, pp. 12–18.

3.   Cole, A.G. (1979). *Textbook of Limnology*, 2nd Edition. The C.V. Mosby Company, St. Louis, Toranto, pp. 426.

4.   Kaushik, B.D. (2003). *Manual for Short Course on Ecology of Freshwater* at MPUAT, Udaipur.

5.   Madhusudan, Sharma L.L. and Durve, V.S. (1984). Eutrophication of lake Pichola in Udaipur. *Raj. Poll. Res.* 3(920): 426.

6.   Sawyer, D.N. (1966). Basic concepts of eutrophication. *J. Wat. Poll. Cont. Fed.*, 38: 737–744.

7.   Sharma, M.S. and Durve, V.S. (1985). Trophic status and fishery potential of Rajasthan waters. In: *Proc. Nat. Sympos. Evalu. Environ. (Special Volume Geobios)*, (Eds.) S.D. Mishra, D.N. Sen and I. Ahmad, p. 180–186.

8.   Sreenivasan, A. (1968). The limnology and fish production in two ponds in Chinglepat (Madras). *Hydrobiologia*, 32: 131–144.

9.   Verma, P.C., Alpana Roy and Hegde, A.G. (2005). Shore and offshore monitoring of Rana Pratap Sagar Lake. *Environmental Geochemistry*, 8: 225.

10. Verma, P.C., Sharma, L.L., Venkataramani, B.and Hegde, A.G. (2007). Ecological risk assessment of RAPS wastewater disposal in RPS lake. In: *National Symposium on Limnology*, NSL–2007 at MPUAT, Udaipur during February.

11. Wetzel, R.G. (1975). *Limnology*. Saunders College Publishing, Philadelphia, pp. 526.

Biodiversity of Aquatic Reseources (2012)
Editors: **Mamta Rawat & Sumit Dookia**
Published by: **DAYA PUBLISHING HOUSE, NEW DELHI**

Pages **105–123**

Chapter 8

Rotifers of Hyderabad, Andhra Pradesh

S.V.A. Chandrasekhar

Freshwater Biology Regional Centre, Zoological Survey of India
Plot 366/1, Attapur (V), Hyderguda (P.O.),
Hyderabad – 500 048, A.P.

ABSTRACT

The rotifer fauna of Hyderabad and its neighborhood was studied from the plankton collections of seven water bodies, KBR National Park and in the context of published information on the ponds of Osmania University campus. The study has revealed the presence of 48 species of Rotifera belonging to 20 genera spread over 15 families and three orders. Details of material examined, description, status, distribution and source have been presented.

Keywords: Rotifers, Lake, Hyderabad.

Introduction

Hyderabad, the fifth largest city in India and also the capital Andhra Pradesh has a number of major and minor water bodies (approximately 170 in number) within its metropolitan limits. The Musi river passes through a distance of 15 kms through the heart of the city and is one of the major tributaries of the river Krishna. Presently, river Musi is heavily contaminated with domestic sewage and industrial effluents loaded with toxic chemicals and metals. The river is acting as a drain for sewage from the city and there is no regular flow of water in the river due to the construction of two reservoirs. Manjira lake, Osmansagar and Himayatsagar, are the main sources of drinking water for the city.

Studies on the rotifer fauna in Andhra Pradesh had earlier been carried out by Dhanapathi (1974, 1975, 1976a, 1978 and 2000); Dhanapathi and Rama Sarma, (2000); Chandra Mohan and Rao, 1976; Siddiqi and Chandrasekhar, 1995; Chandrasekhar and Siddiqi, 2005), Ahson Ahsan Md. (1980), Jaya Devi (1985), Chandrasekhar (1997) and Malathi *et al.* (2003). These are some of the major contributions on the ecological studies of the lakes in Hyderabad and its neighbourhood inventorying the rotifer fauna was presented. Some major contributions on the rotatorian fauna of some water bodies of Hyderabad in particular, have come from Md. Arshaduddin and Khan (1991), Chandrasekhar and Kodarkar (1994, 1995), Chandrasekhar (1996 a, 1996 b). Dhanapathi (1974) had partly discussed the rotatorian fauna of Hussainsagar lake in his studies on Rotifer fauna of Andhra Pradesh. Out of the 325 species available in India (Sharma, 1997), 91 have been reported from Andhra Pradesh (Dhanapathi, 2000).

Since there is no systematic documentation on the total Rotifer composition of the water bodies in and around Hyderabad, the author intends to take up the work and reported here 48 species belonging to 20 genera of 15 families in three orders. The study had been carried out from the plankton collections of the water bodies of Hyderabad and its neighborhood and also literature available in Freshwater Biological Station (Freshwater Biological Station, Zoological Survey of India.), Zoological Survey of India (ZSI), Hyderabad. These water bodies include Hussainsagar, Himayatsagar, Osmansagar, Mir alam tank, Manjira lake, Saroornagar lake, Medchal tank, water bodies in the Kasu Brahmanda Reddy National Park, and the ponds in the Osmania University campus of Hyderabad. Presently the ponds in the Osmania University campus do not exist (dried).

Hussainsagar

This lake situated between the twin cities of Hyderabad and Secunderabad, was excavated in 1562 AD mainly to store drinking water brought from Musi river by Balakpur canal. However, with the passage of time the lake lost its importance as a source of potable water, nevertheless, it was extensively used for washing, bathing and recreation. The lake gets its pollution load from surrounding industrial areas and settlements. Its water-spread area is 27.5 sq km., length 3.2 km; width 2.8 km and maximum depth is 12.5m

Himayatsagar

This water body was constructed in 1926 on the tributary of Musi river and its catchment (1,307 sq km) is made of rocky undulating ground. This water body is 19 km from the city on the southwestern side. This is also one of the major drinking water sources to the twin cities at present. Its surface area is 21 km, maximum depth 23.9 km., length 7.8 km and width is 4.2 km.

Osmansagar

This reservoir built in 1920 on Musi river system was mainly to control floods and provide drinking water to the twin cities. This ecosystem is about 18 km away on the western side of the city. It is one of the drinking water sources to the twin cities even today. The catchment of lake has scrub jungles and the forest of Anantagiri hills at its origin and aerable agricultural land along the course of the water body. Its catchment area is 740 sq km., surface area 22 sq km., maximum depth 31.7m, length 8.8 km and width 5.1 km.

Mir Alam Tank
Constructed in 1806, it was one of the oldest tanks created on Musi river system and the only multi-arch (21 arches) dam of its kind in the world. It is about 7 km southwest of the city. The water body was used as a source of drinking water up to 1960 by the Rajendranagar municipality. Presently it serves as the main source of water to the Nehru Zoological Park and its surroundings. Its surface area is 1.7 sq km., maximum depth 13.41m, and catchment area of 16.5 sq km. The catchment is made of rocky undulating terrain.

Saroornagar Lake
Impounded in 1626 AD Saroornagar lake was meant for agricultural and drinking purposes and today is one of the major aquatic ecosystems about 8 km on the eastern side of Hyderabad city. Its water spread area is 35 hectare, maximum depth 6.1 m. In the last two decades, due to growing urbanization the catchment has undergone drastic modification with the consequent effects on the morphometry and limnology of the lake ecosystem. Scientifically speaking, the lake has got aquaculture, ecological and recreational potential.

Manjira Lake
This water body is situated near the village, Kalabgur near Sangareddy town in Medak District of Andhra Pradesh and in about 50 km North-West side of Hyderabad city. After the construction of a barrage on Manjira Lake near Sangareddy town in 1965, this has become a reservoir and one of the main sources of drinking water to the twin cities. Its catchment area is about 1680 sq km., maximum depth is 4.0 m approximately and the catchment area is about 1896 hectares.

Medchal Tank
This is an irrigation tank situated in Medchal Mandal of Ranga Reddy District, at about 30 km. from Hyderabad city on its North-West direction on Nagpur High way.

Ponds of Osmania University
There are four small seasonal ponds in the Osmania University campus each of about half acre in area. Most of these ponds have water for around 4–6 months while few of them dry up at the on set of summer and again filled with water during monsoon. A couple of them also receive organic pollutants in the form of sewage.

Ponds of Kasu Brahmananda Reddy (KBR) National Park
With an area of 142.5 hectares, this park is situated in the hill rocks of prestigious Jubilee hills, in the heart of the city. It is named after the former Chief Minister of Andhra Pradesh, Shri Kasu Brahmananda Reddy and this park had been declared

as National Park in the year 1994. It houses 3 small ponds with an area of 0.5 to 1 hectare, one of which is comparatively big (one hectare) and perennial one.

Material and Methods

During the course of limnological investigations of water bodies of Hyderabad and its neighborhood, the scientists of Freshwater Biological Station, Zoological Survey of India (ZSI), Hyderabad, have collected plankton samples since 1979. The samples are deposited in the National Zoological Collections of the station.

These plankton samples were collected by towing the plankton net (No. 25) on the littoral zone of these water bodies and collected organisms were preserved in 5 per cent formaldehyde solution. The present study reveals the rotifer fauna of all the water bodies mentioned here except the ponds of Osmania University campus (Md. Arshaduddin and Khan, 1991). Species identification was carried out with the help of Sharma (1978a, 1978b, 1979,1980,1987,1997,1998) and Dhanapathi,1975, 1976, 1978, 2000). The status of each species is given with reference to the lakes of Hyderabad and its environs and the species occurring in different water bodies presented in this paper are given in a Table 8.1.

SYSTEMATIC ACCOUNT

Class : ROTIFERA

Subclass : MONOGONONTA

Order : PLOIMIDA

Family : BRACHIONIDAE

Anuraeopsis fissa (Gosse, 1851)

Material examined: Not recorded in the present study, but Md. Arshaduddin and Khan (1991) have reported from the ponds of Osmania University campus.

Description : Lorica with two thin plates, and ovate in outline; without caudal spine and foot.

Status : Rare

Distribution : Andhra Pradesh, Assam, Gujarat, Haryana, Kerala, Meghalaya, Orissa, Punjab, Rajasthan, and West Bengal.

Source : Sharma and Michael (1980).

Brachionus angularis Gosse, 1851

Material examined : SN, MT, HMS, (OU); Lorica length 0.09–0.20; maximum width 0.08–0.15; anterior spine 0.01–0.06 mm.

Status : Uncommon

Distribution : Andhra Pradesh, Assam, Delhi, Haryana, Kashmir, Madhya Pradesh, Maharashtra, Orissa, Punjab, West Bengal,

Source : Sharma (1980)

Table 8.1: Showing the Rotifer Fauna of Hyderabad and its Neighbourhood

Sl.No.	Species	Hussain-sagar	Saroor Nagar Lake	Manjira Lake	Osman-sagar	Medchal Tank	Himayat-sagar	Ponds of KBR Natl. Park	Mir Alam Tank	Ponds of OU
(I)	(II)	(III)	(IV)	(V)	(VI)	(VII)	(VIII)	(IX)	(X)	(XI)
1.	Anuraeopsis lissa (Gosse)	–	–	–	–	–	+	–	–	+
2.	Brachionus angularis Gosse	–	+	–	–	+	+	–	–	+
3.	B. bidentata Anderson	+	–	–	–	–	–	–	–	+
4.	B. calyciflorus Pallas	+	+	+	–	+	–	+	+	+
5.	B. caudatus Bar. and Daday	+	+	–	+	+	–	–	+	+
6.	B. diversicornis Daday	–	–	–	–	–	–	+	–	+
7.	B. durgae Dhanapathi	+	+	+	–	–	–	+	+	+
8.	B. falcatus Zacharias	+	–	+	+	+	–	+	+	+
9.	B. forticula Wierzejshi	–	+	–	+	–	–	–	–	+
10.	B. patulus Muller	–	–	–	–	–	–	–	–	+
11.	B. plicatilis Mueller	+	–	–	–	–	–	–	–	–
12.	B. quadridentatus Hermann	–	–	–	–	–	+	–	–	+
13.	B. rubens Ehrn.	+	+	–	–	–	–	–	–	–
14.	Keratella cochlearis Gosse	+	–	–	–	–	+	–	–	–
15.	K. procurva (Thorpe)	–	–	–	+	–	–	–	–	–
16.	K. tropica (Apstein)	+	+	+	–	+	+	+	–	+
17.	K. volga (Ehrn.)	+	–	–	–	–	–	–	+	–
18.	Euchlanis dilatata Ehren.	–	–	–	–	–	–	–	–	+
19.	Tripleuchlanis plicata (Lav.)	–	–	–	–	–	–	–	–	+

Contd...

Table 8.1–Contd...

Sl.No.	Species	Hussain-sagar	Saroor Nagar Lake	Manjira Lake	Osman-sagar	Medchal Tank	Himayat-sagar	Ponds of KBR Natl. Park	Mir Alam Tank	Ponds of OU
(I)	(II)	(III)	(IV)	(V)	(VI)	(VII)	(VIII)	(IX)	(X)	(XI)
20.	Trichotria tetractis (Ehren.)	–	–	–	–	–	–	–	–	+
21.	Mytilina ventralis (Ehrn.)	+	–	–	–	–	–	+	–	+
22.	Colurella obtuse (Gosse)	–	–	–	–	–	–	–	–	+
23.	Lepadella ehrenbergi (Perty)	–	–	–	–	–	–	–	–	+
24.	L. ovalis (Muller)	–	–	–	–	–	–	–	–	+
25.	L. rhomboides (Gosse)	–	–	–	–	–	–	–	–	+
26.	Lecane (L) curvicornis Mur.	–	–	–	–	–	–	–	–	+
27.	L. (L) hastate (Mur.)	–	–	–	–	–	–	–	–	+
28.	L. (L) liudgwigi (Eck.)	–	–	–	–	–	–	–	–	+
29.	L. (L) luna (Mul.)	–	–	–	–	–	–	–	–	+
30.	L. (L) papuana (Murray)	–	–	–	–	–	+	–	+	+
31.	L. (L) ploenensis (Voigt)	–	–	–	–	–	–	–	–	+
32.	Lecane (M) bulla (Gosse)	+	–	–	–	–	+	+	+	+
33.	L. (M) closterocerca (Sch.)	–	–	–	–	–	–	–	–	+
34.	L. (M) lunaris (Ehren.)	–	–	–	–	–	–	–	–	+
35.	L. (M) obtuse Murray	–	–	–	–	–	–	–	–	+
36.	Cephalodella gibba (Ehrn.)	–	–	–	–	–	+	–	–	+
37.	Scaridium longicaudum (Muller)	–	–	–	+	–	–	–	–	+
38.	Trochocerca rattus (Muller)	+	–	–	–	–	–	–	–	–
39.	Asplanchna intermedia (Hud.)	+	+	–	–	–	–	–	+	+

Contd...

Table 8.1–Contd...

Sl.No.	Species	Hussain-sagar	Saroor Nagar Lake	Manjira Lake	Osman-sagar	Medchal Tank	Himayat-sagar	Ponds of KBR Natl. Park	Mir Alam Tank	Ponds of OU
(I)	(II)	(III)	(IV)	(V)	(VI)	(VII)	(VIII)	(IX)	(X)	(XI)
40.	A. sieboldi (Sudzuku)	–	–	–	–	–	–	–	–	+
41.	Polyarthra vulgaris Carliu	–	–	–	+	–	–	–	–	–
42.	Limnias meliceria Weissa	–	–	–	–	–	–	–	–	+
43.	Hexarthra intermedia Wizneiski	+	–	–	–	–	–	–	–	–
44.	Filinia longiseta (Ehrn.)	+	+	–	+	–	+	–	+	+
45.	F. opoliensis Zacharias	+	–	–	–	–	–	–	–	+
46.	F. terminalis (Plate)	+	+	–	–	–	–	–	+	–
47.	Testudinella palina (Hermann)	–	–	–	–	–	–	+	–	+
48.	Rotaria neptunea (Ehren.)	–	–	–	–	–	–	–	–	+

Brachionus bidentata Anderson, 1889

Material examined : HS, (OU); Total length 0.16, Maximum width 0.12 mm.

Status : Rare

Distribution : Andhra Pradesh, Haryana, Punjab, Orissa and West Bengal.

Source : Sharma (1997)

Brachionus calyciflorus Pallas, 1776

Material examined: HS, SN, ML, MT, KBR, MAT and (OU); Lorica length 0.15–0.2; maximum width 0.14–0.2.

Status : Common

Distribution : Assam, Bihar, Kashmir, Maharashtra, Manipur, Meghalaya, Orissa, Punjab and Rajasthan,

Source : Sharma (1997)

Brachionus caudatus Barrois, 1894

Material examined: HS, SN, MT, MAT and (OU);

Measurements : Lorica length 0.08–0.1; maximum width 0.075–0.9; antero-lateral spine 0.015–0.03; antero-medium spine 0.075–0.09; posterior spine 0.04–0.06 mm.

Status : Common

Distribution : Andhra Pradesh, Assam, Bihar, Delhi, Maharashtra, Meghalaya, Orissa, Rajasthan and Tamil Nadu.

Source : Patil (2001)

Brachionus diversicornis (Daday, 1883)

Material examined : OS, KBR and (OU); Lorica Total length 0.24–0.35, Maximum width 0.13–0.24 mm.

Status : Uncommon

Distribution : Andhra Pradesh, Assam, Orissa, Punjab and West Bengal.

Source : Sharma (1980)

Brachionus durgae Dhanapathi, 1974

Material examined : HS, SN, Ml, KBR, MAT and (OU); Lorica length 0.18–0.2; maximum width 0.15–0.18; anterior spines 0.015–0.01mm.

Status : Common

Distribution : Andhra Pradesh.

Source : Dhanapathi (1974)

Brachionus falcatus Zacharias, 1898

Material examined : ML, OS,MT, KBR, MAT and (OU); Lorica length 0.09–0.12, maximum width 0.1–0.18; length of antero-lateral spin e 0.015–0.02; length of intermediate spine 0.12–0.15; length of median spine 0.007–0.02; posterior spine 0.1–0.4 mm.

Status : Common

Distribution : Andhra Pradesh, Assam, Bihar, Gujarat, Haryana, Kerala, Madhya Pradesh, Orissa, Punjab, Rajasthan and West Bengal.

Source : Dhanapathi (1974)

Brachionus forficula Wierzeski, 1891
Material examined : SN, OS, and (OU); Lorica length 0.1–0.15, maximum width 0.09–0.12, antero-lateral spine 0.015–0.02; antero-median spine 0.01–0.015; posterior spine 0.05–0.07 mm.

Status : Uncommon

Distribution : Andhra Pradesh, Gujarat, Orissa, Punjab and West Bengal.

Source : Dhanapathi (1974)

Brachionus patulus Muller, 1786
Material examined : Not recorded in the present study, but Md. Arshaduddin and Khan (1991) have reported from the ponds of Osmania University campus.

Status : Rare

Distribution : Andhra Pradesh, Gujarat, Kashmir, Maharashtra and West Bengal

Source : Edmondson and Hutchinson (1934)

Brachionus plicatilis Muller, 1786
Material examined : HS; Total length 0.18, maximum width 0.135 mm

Description : Lorica oval, anterior margin with six angular saw toothed spines; foot opening posterior.

Status : Rare

Distribution : Andhra Pradesh, Assam, Rajasthan, Punjab and West Bengal.

Source : Sharma (1980)

Brachionus quadridentatus Hermann, 1783
Material examined : HMS and (OU); Lorica length 0.08; maximum width 0.12; antero-lateral spines 0.02; antero-median spines 0.03; antero-intermediate spines 0.015; posterior spines 0.07 mm.

Status : Rare

Distribution : Andhra Pradesh, Bihar, Maharashtra, Meghalaya, Kerala, Orissa and West Bengal.

Source : Dhanapathi (1974)

Brachionus rubens Ehrenberg, 1838
Material examined : HS, SN ; Total length 0.19–0.2, maximum width 0.14–0.16 mm.

Status : Rare

Distribution : Andhra Pradesh, Assam, Haryana, Orissa, Punjab, Rajasthan and West Bengal.

Source : Sharma (1980)

Keratella cochlearis Gosse, 1851

Material examined : HS and HMS; Total length 0.18–0.2, maximum width 0.065–078 mm

Status : Rare

Distribution : Andhra Pradesh, Assam, Kashmir, Kerala, Ladak, Punjab, Rajasthan and West Bengal

Source : Dhanapathi (1974)

Keratella procurva (Thorpe, 1891)

Material examined : OS; Total length 0.15, maximum width 0.065 mm.

Status : Rare

Distribution : Andhra Pradesh, Kashmir, Kerala, Orissa and West Bengal.

Source : Sharma (1980)

Keratella tropica Apstein, 1907

Material examined : HS, SN, ML, MT, HMS, KBR, MAT and (OU); Total Length 0.19–0.25, maximum width 0.068–0.09; antero-lateral spine 0.015–0.018; antero-median spine 0.045–0.05; Right posterior spine 0.12–0.15; left posterior spine 0.04–0.05 mm.

Status : Common

Distribution : Andhra Pradesh, Gujarat, Kerala, Maharashtra and West Bengal.

Source : Dhanapathi (1974)

Keratella volga (Ehrenberg, 1834)

Material examined : HS; Lorica length 0.24, maximum width 0.72; Right posterior spine 0.15 mm, left posterior spine 0.07 mm.

Status : Rare

Distribution : Andhra Pradesh, Gujarat, Kerala, Maharashtra.

Source : Sharma (1998)

Family : EUCHLANIDAE

Euchlanis dilatata Ehrenberg, 1832

Material examined: This species was not observed by the author but Arshaduddin and Khan, 1991 found from the ponds of Osmania University campus.

Status : Rare

Distribution : Andhra Pradesh, Assam, Gujarat, Kashmir, Meghalaya, Orissa, Punjab and West Bengal.

Source : Sharma (1998)

Tripleuchlanis plicata (Lavender, 1894)

Material examined : The author did not find this species from his collections but Md. Arshaduddin and Khan (1991) found from the ponds of Osmania University campus.

Status : Rare

Distribution : Andhra Pradesh and West Bengal

Source : Sharma (1998)

Family : TRICHOTRIDAE

Trichotria tetractis (Ehrenberg, 1832)

Material examined : The author did not find this species from his collections but Md. Arshaduddin and Khan (1991) found from the ponds of Osmania University campus.

Status : Rare

Distribution : Andhra Pradesh, Assam, Gujarat, Kerala, Kashmir, Madhya Pradesh, Punjab, Tamil Nadu and West Bengal.

Source : Dhanapathi (1974)

Family : MYTILINIDAE

Mytilina ventralis Ehrenberg, 1832

Material examined : HS, KBR and (OU); Lorica length 0.17, maximum width 0.09; toes 0.06 mm.

Status : Uncommon

Distribution : Andhra Pradesh, Assam, Gujarat, Kerala, Kashmir, Madhya Pradesh, Punjab, Tamil Nadu and West Bengal.

Source : Dhanapathi (1974)

Family : COLURELLIDAE

Colurella obtusa (Gosse, 1886)

Material examined : The author did not find this species from his collections but Md. Arshaduddin and Khan (1991) found from the ponds of Osmania University campus.

Status : Rare

Distribution : Andhra Pradesh, Punjab and West Bengal

Source : Sharma (1998)

Lepadella ehrenbergi (Perty, 1850)

Material examined : The author did not find this species from his collections but Arshaduddin and Khan (1991) found from the ponds of Osmania University campus.

Status : Rare

Distribution : Andhra Pradesh, Assam, Meghalaya, Nagaland, Orissa,Punjab and West Bengal.

Source : Sharma (1978b)

Lepadella ovalis (Muller, 1786)

Material examined : The author did not find this species from his collections but Md. Arshaduddin and Khan (1991) found from the ponds of Osmania University campus.

Status : Rare

Distribution : Andhra Pradesh, Kashmir, Orissa, Punjab and West Bengal.

Source : Sharma (1978b)

Lepadella rhomboides (Gosse, 1886)

Material examined : The author did not find this species from his collections but Md. Arshaduddin and Khan (1991) found from the ponds of Osmania University campus.

Status : Rare

Distribution : Andhra Pradesh, N.E. India, Tamil Nadu, Gujarat and West Bengal.

Source : Sharma (1978b)

Family : LECANIDAE

Lecane (Lecane) curvicornis Murray, 1913

Material examined : The author did not find this species from his collections but Md. Arshaduddin and Khan (1991) found from the ponds of Osmania University campus.

Status : Rare

Distribution : Andhra Pradesh, Madhya Pradesh and West Bengal.

Source : Dhanapathi (1976)

Lecane (Lecane) hastata (Murray, 1913)

Material examined : The author did not find this species from his collections but Md. Arshaduddin and Khan (1991) found from the ponds of Osmania University campus.

Status : Rare

Distribution : Andhra Pradesh and West Bengal

Source : Sharma (1998)

Lecane (Lecane) ludgwigi (Eckstein, 1883)

Material examined : The author did not find this species from his collections but Md. Arshaduddin and Khan (1991) found from the ponds of Osmania University campus.

Status : Rare

Distribution : Andhra Pradesh, Orissa, Punjab and West Bengal.

Source : Dhanapathi (1976)

Lecane (Lecane) luna (Muller, 1776)
Material examined: The author did not find this species from his collections but Md. Arshaduddin and Khan (1991) found from the ponds of Osmania University campus.

Status : Rare

Distribution : Andhra Pradesh, Kashmir, Gujarat, N.E. India, Orissa, Punjab and West Bengal.

Source : Dhanapathi (1976)

Lecane (Lecane) papuana (Murray, 1913)
Material examined : HMS, MAT and (OU); Lorica length 0.14–0.16; maximum width 0.08–0.1; length of toes 0.03–0.04; length of claw 0.008–0.01 mm.

Status : Uncommon

Distribution : Andhra Pradesh, Mizoram, Tamil Nadu, Kashmir and West Bengal.

Source : Dhanapathi,1975

Lecane (Lecane) ploenensis (Voigt, 1902)
Material examined: The author did not find this species from his collections but Md. Arshaduddin and Khan (1991) found from the ponds of Osmania University campus.

Status : Rare

Distribution : Andhra Pradesh, Gujarat, Mizoram, Meghalaya, Punjab and West Bengal.

Source : Sharma (1998)

Lecane (Monostyla) bulla (Gosse, 1851)
Material examined : HS, HMS, KBR and (OU); Dorsum 0.1–0.14, ventral 0.12–0.14; dorsal width 0.079–0.09, ventral 0.76–0.08, toe 0.46–0.5.

Status : Common

Distribution : Andhra Pradesh, N.E. Region, Gujarat, Rajasthan, Punjab, Tamil Nadu, Kashmir, Orissa and West Bengal.

Source : Dhanapathi (1976)

Lecane (Monostyla) closterocerca (Schmarda, 1898)
Material examined : The author did not find this species from his collections but Md. Arshaduddin and Khan 1991 found from the ponds of Osmania University campus.

Status : Rare

Distribution : Andhra Pradesh, Gujarat, Kashmir, Orissa, Punjab, Rajasthan, Rajasthan Tamilnadu and West Bengal.

Source : Sharma (1978a,b)

Lecane (Monostyla) lunaris Ehrenberg, 1832

Material examined : The author did not find this species from his collections but Md. Arshaduddin and Khan (1991) found from the ponds of Osmania University campus.

Status : Rare

Distribution : Andhra Pradesh, Gujarat, Kashmir, N.E. Region and West Bengal.

Source : Sharma (1978a)

Lecane (Monostyla) obtusa Murray, 1913

Material examined : The author did not find this species from his collections but Md. Arshaduddin and Khan (1991) found from the ponds of Osmania University campus.

Status : Rare

Distribution : Andhra Pradesh and West Bengal.

Source : Sharma (1978a)

Family : NOTAMMATIDAE

Cephalodella gibba (Ehrenberg, 1832)

Material examined : HMS and (OU); Lorica length 0.25–0.27; toes 0.08–0.88 mm.

Status : Uncommon

Distribution : Andhra Pradesh, Gujarat, Kashmir and West Bengal.

Source : Sharma (1979)

Scaridium longicaudum (Muller, 1786)

Material examined : OS and (OU); Total length 0.04, foot 0.012, toe 0.013 mm.

Status : Uncommon

Distribution : Andhra Pradesh, Gujarat, Punjab and West Bengal

Source : Dhanapathi (1978)

Family: TRICHOCERCIDAE

Trichocerca rattus (Muller, 1776)

Material examined: HS; Lorica 0.17, left toe 0.013 mm

Status : Rare

Distribution : Andhra Pradesh, Kashmir, Gujarat, Punjab and West Bengal.

Source : Dhanapathi and Rama Sarma (2000)

Family : ASPLANCHNIDAE

Asplanchna intermedia Hudson, 1886

Material examined : HS, SN, MAT and (OU); Body length 0.36–0.4; maximum width 0.18–0.2 mm.

Status : Common

Distribution : Andhra Pradesh, Maharashtra and Tamil Nadu.

Source : Dhanapathi (1975)

Asplanchna sieboldi (Sudzuku, 1956)

Material examined : The author did not find this species from his collections but Md. Arshaduddin and Khan (1991) found from the ponds of Osmania University campus.

Status : Rare

Distribution : Andhra Pradesh, Tamil Nadu.

Source : Md. Arshaduddin and Khan (1991)

Family : SYNCHAETIDAE

Polyarthra vulgaris Carliu, 1943

Material examined: The author did not find this species from his collections but Jaya Devi, 1985 found from Osmansagar.

Status: Rare

Distribution: Andhra Pradesh, Assam, Orissa, Punjab and West Bengal.

Source: Sharma (1998).

Order : GNESIOTROCHA

Family : FLOSCULARIDAE

Limnias melicerta Weissa, 1848

Material examined : The author did not find this species from his collections but Md. Arshaduddin and Khan (1991) found from the ponds of Osmania University campus.

Status : Rare

Distribution : Andhra Pradesh and Tamilnadu.

Source : Md. Arshaduddin and Khan (1991)

Family : HEXARTHRIDAE

Hexarthra intermedia (Wizneiski, 1929)

Material examined : HS; Body length 0.3 mm.

Status : Rare

Distribution : Andhra Pradesh, Kashmir and Punjab.

Source : Dhanapathi and Rama Sarma (2000)

Family : FILINIDAE

Filinia longiseta Ehrenberg, 1834

Material examined : HS,SN, OS, MT, HMS, MAT and (OU); lorica length 0.18–0.2, lateral spine 0.1–0.25, posterior setae 0.1–0.17 mm.

Status : Common

Distribution : Andhra Pradesh, Assam, Gujarat, Haryana, Madhya Pradesh, Orissa, Punjab, Rajasthan and West Bengal

Source : Sharma (1998b)

Filinia opoliensis (Zacharias, 1898)

Material examined : HS and (OU); Lorica length 0.15–0.19, anterior setae 0.25–0.3,Posterior long setae 0.23–0.25, posterior short setae 0.05–0.06 mm.

Status : Rare

Distribution : Andhra Pradesh, Assam, Gujarat, Haryana, Madhya Pradesh, Orissa, Punjab, Rajasthan and West Bengal.

Source : Sharma (1998)

Filinia terminalis (Plate, 1886)

Material examined: HS, SN, OS and MAT; Lorica length 0.1–0.18, anterior spine 0.32–0.4, posterior spine 0.26–0.38 mm.

Status : Uncommon

Distribution : Andhra Pradesh, Gujarat, Tamil Nadu and West Bengal.

Source : Sharma (1998)

Family : TESTUDINELLIDAE

Testudinella patina (Hermann, 1783)

Material examined: KBR and (OU); lorica length 0.18–0.2, maximum width 0.16–0.18 mm.

Status : Uncommon

Distribution : Andhra Pradesh, Assam, Gujarat, Kashmir, Orissa, Punjab and West Bengal.

Source : Sharma (1998)

Order : BDELLOIDA

Family : PHILODINIDAE

Rotaria neptunea (Ehrenberg, 1832)

Material examined : The author did not find this species from his collections but Md. Arshaduddin and Khan (1991) found from the ponds of Osmania University campus.

Status : Rare

Distribution : Andhra Pradesh, Assam and West Bengal.

Source : Sharma (1998)

The varieties like *Brachionus calyciflorus* var. *hymani* Dhanapathi (1974), had been reported as new variety by Dhanapathi (1974) from a tank at Vizianagaram of Andhra Pradesh and *Brachionus calyciflorus* var. *dorcas* Gosse, 1851 had been reported by Dhanapathi (1974) from Hussainsagar and the author had also observed these two varieties from the plankton samples of the lakes in and around Hyderabad. Among the species mentioned here *Brachionus durgae* have been described as new species by Dhanapathi (1974) from Hussainsagar and also is endemic to Andhra Pradesh. Among the species reported in this paper, *Anuraeopsis fissa, Brachionus angularis, B. caudatus, B. rubens, B. calyciflorus, B. falcatus, Keratella tropica, Polyarthra vulgaris, Filinia longiseta, F. opoliensis* and *Rotaria neptunea* are quite common in alkaline waters and the species like *Euchlanis dilata* is of acidophilic in nature (Sharma, 1997). According to Khan and Seshagiri Rao (1981), *Brachionus falcatus, B. forficula* and *B. quadridentatus* are the indicator species for clean waters and *B. angularis, B. calyciflorus, Keratella tropica, Filinia longiseta,* and *F. terminalis* are indicator species for heavily polluted waters. It had been reported earlier that *Brachionus angularis, Filinia longiseta, F. opoliensis* and *Rotaria neptunea* occur under eutrophic to hypereutrophic conditions and *Brachionus caudatus, B. calyciflorus, Anuraeopsis fissa, Keratella tropica* and *Polyarthra vulgaris* commonly noticed in alkaline eutrophic waters. Presence of *Brachionus forficula, B. diversicornis* and *Keratella cochlearis* indicates the eutrophic soft and acidic waters. So, it is quite clear that most of the water bodies studied are eutrophic to hypereutrophic.

Acknowledgements

The author is thankful to Dr J.R.B. Alfred, Former Director of Zoological Survey of India (ZSI), Kolkata and Dr. C.A. Nageswara Rao, Officer-in-Charge, ZSI, Hyderabad for extending facilities in writing this paper.

References

1.   Ahsan Mohd. (1980). Ecology of freshwater zooplankton (Hussainsagar and Saroornagar lakes). *Ph.D. Thesis,* Osmania University, Hyderabad, 175 pp.

2.   Chandra Mohan, P. and Rao, R.K. (1976). Epizoic rotifers observed in odonata nymphs from Visakhapatnam. *Sci. Cult.,* 42: 527–528.

3.   Chandrasekhar, S.V.A. (1997). Ecological studies on Saroornagar lake, Hyderabad. *Ph.D. Thesis,* Osmania University, Hyderabad, 150 pp.

4.   Chandrasekhar, S.V.A. (1996a). An account of rotatorian and cladoceran fauna of Manjira lake, Andhra Pradesh with a note on their abundance and indicator value. *Proc. Academy of Envl. Biology,* 7(1): 27–30.

5.   Chandrasekhar, S.V.A. (1996b). Zooplankton diversity of Medchal irrigation tank (R.R. District, Andhra Pradesh) with a special reference to rotifera and cladocera. *J. Freshwater Biol.,* 4: 197–200.

6.   Chandrasekhar, S.V.A. and Kodarkar, M.S. (1994). Biodiversity of zooplankton from Saroornagar lake, Hyderabad, A.P. *J. Aqua. Biol.,* 9(1 and 2): 30–33.

7.   Chandrasekhar, S.V.A. and Kodarkar, M.S. (1995a). Studies on *Brachionus* from Saroornagar lake, Hyderabad. *J. Aqua. Biol.,* 10(1 and 2): 48–52.

8.   Chandrasekhar, S.V.A. and Kodarkar, M.S. (1995b). *Conservation of Lakes with Special Reference to Water Bodies in and Around Hyderabad*. Indian Association of Aquatic Biologists, 3: 82.

9.   Chandrasekhar, S.V.A. and Siddiqi, S.Z. (2005). Kondakarla lake, Andhra Pradesh: A Taxo-ecological Profile. *Rec. Zool. Surv. India*, 104 (3–4): 63–76.

10. Dhanapathi, M.V.S.S.S. (1975). Rotifers from Andhra Pradesh, India. *Zool. J. Linn. Soc. (London)*, 57: 85–94.

11. Dhanapathi, M.V.S.S.S. (1976). Rotifers from Andhra Pradesh, India–III, family Lecanidae including two new species. *Hydrobiologia*, 48: 9–16.

12. Dhanapathi, M.V.S.S.S. (1978). New species of rotifer from India belonging to the family Brachionidae. *Zool. Linn. Soc. (London)*, 62: 226–229.

13. Dhanapathi, M.V.S.S.S. (2000). *Taxonomic Notes on the Rotifers from India (from 1889–2000)*. Indian Association of Aquatic Biologists, Publ. 10.

14. Dhanapathi, M.V.S.S.S. and Sarma, Rama (2000). Futher studies on the rotifers from Andhra Pradesh, India, including a new species. *J. Aqua. Biol.*, 15(1 and 2): 1–12.

15. Edmondson, W.T. and Hutchinson, G.E., (1934). Report on rotatoria, Article IX. Yale North India Expedition. *Mem. Cann. Acad. Arts Sci.*, 10: 153–186.

16. Jaya Devi, M. (1985). Ecological studies on limnoplankton of three freshwater bodies of Hyderabad. *Ph.D. Thesis*, Osmania University, Hyderabad, 126 pp.

17. Khan, M.A. and Seshagiri Rao, I. (1981). Zooplankton in the evaluation of pollution. WHO workshop on biological indicators and indices of environmental pollution. *Cent. Bd. Prev. Cont. Poll.*, p. 35–148.

18. Malathi, D., Chandrasekhar, S.V.A. and Kodarkar, M.S. (2003). Ecological son Hussainsagar lake, Hyderabad with special reference to zooplankton communities. *Rec. Zool. Surv., India* (submitted for publication), 175 pp.

19. Md. Arshaduddin and Khan, M.A. (1991). Rotifer fauna of some seasonal ponds of Osmania University campus,Hyderabad (A.P.) India. *Indian J. Microbiol. Ecol.*, 2: 29–40.

20. Patil, S.G. (2001). Rotifera: Fauna of Nilgiri biosphere reserve. In: *Fauna of Conservation Area Series* No. 11: 25–28, Zoological Survey of India Publ.

21. Sharma, B.K. (1978a). Contribution to the rotifer fauna of West Bengal. Part I. Family Lecanidae. *Hydrobiologia*, 57: 143–153.

22. Sharma, B.K. (1978 b). Contribution to the rotifer fauna of West Bengal. II. Genus *Lepadella* Bory St. Vincent, 1826. *Hydrobiologia*, 58: 83–86.

23. Sharma, B.K. (1979). Rotifers from West Bengal. III. Further studies on the Eurotatoria. *Hydrobiologia*, 64: 239–250.

24. Sharma, B.K. (1980). Contributions to the rotifer fauna of Punjab state, India I. Family Brachionidae. *Hydrobiologia*, 76: 249–253.

25. Sharma, B.K. (1987). Rotifera : Eurotatoria : Monogononta (Freshwater). In: *Fauna of Orissa Series*, 1(1): 323–340. Zoological Survey of India Publ.

26. Sharma, B.K. (1997). Biodiversity of freshwater rotifera in India: A status report. *Proc. Zool. Soc.*, 49(2): 73–85.

27. Sharma, B.K. (1998). Freshwater rotifers (Rotifera : Eurotatoria). In: *Fauna of West Bengal Part* II State Fauna Series, 3: 341–461.

28. Sharma, B.K. and Michael, R.G. (1980). Synopsis of taxonomic studies on Indian Rotatoria. *Hydrobiologia*, 73: 229–236.

29. Siddiqi, S.Z. and Chandrasekhar, S.V.A. (1995). New distributional records of *Trichotria tetractis* (Rotatoria/Trichotridae) and *Daphnia lumholtzi* (Branchipoda/ Cladocera/Daphniidae) from Kolleru lake, Andhra Pradesh with a notes on indicator value. *J. Bom. Nat. Hist. Soc.*, p. 309–310.

Biodiversity of Aquatic Reseources (2012)
Editors: **Mamta Rawat & Sumit Dookia**
Published by: **DAYA PUBLISHING HOUSE, NEW DELHI**

Pages **124-138**

Chapter 9

Fish Diversity in the Pairy River of Mahanadi River System in the Chhattisgarh

S. Singh*, Om Prakesh, H.K. Vardia and M.C. Chari

Department of Fisheries, Indira Gandhi Agricultural University,
Raipur – 492 006, Chhattisgarh

ABSTRACT

The river Pairy is the major tributary of the Mahanadi river system. The ichthyofauna of the catchment of Pairy river was recorded during the period January to July 2004. The study analyzed the fish diversity in the Pairy river of Mahanadi river system in the Chhattisgarh. Fish specimens were collected from pond, reservoir and rivers at Gariaband, Chhura blocks of South Raipur district. The results of empirical study revealed that the list of 39 species classified under 5 orders, 14 families and 26 genera. Analysis of the different species revealed that 76.92 per cent of them belong to a single order Cypriniformes and 6.67 per cent of the species belong to two orders Clupiformes and Mastacembeliformes. Order Beloniformes and Ophiocephaliformes have 3.33 per cent catch and rest 10 per cent species belong to Perciformes. Pairy River had 19 species common, 5 species dominant and 15 species rare. In the present study 39 species belonging to 5

* Corresponding Author E-mail: sattupoonia@yahoo.com

orders, 14 families and 26 genera were enlisted including 4 exotic fish species. 15 new species have been put on record and 20 species are not recorded in present study compared for periods 1940 through 2004.

Keywords: Ichthyofauna, Fish diversity, Pairy River, Mahanadi river, Chhattisgarh.

Introduction

Chhattisgarh is situated between 17°46' N–24°80 'N latitude and 80°15' E–84°24' E longitude. The state lies in catchment areas of the Mahanadi River. Chhattisgarh contributes 56 per cent of catchment area for Mahanadi. Pairy river is a tributary of Mahanadi originating from Bhatragadh hilly area of Raipur district, which is above 493 m msl. This joins Mahanadi at southern part of Raipur district. Its length in Raipur district is 90 km and covers 14 per cent of area of the Raipur district. The freshwater body of include river, pond and market. Presently, there are altogether tanks, ponds covering 1,22,550 ha water area under fish cultivation in the blocks (Bindra Nawagarh) Chhura, Gariaband) of this district. Knowledge of fish diversity of a particular region is essential not only for national management of ichthyogauna of that region but also for their conservation strategies. The works of Hora (1937, 1938, 1940, 1941, 1949) Hora and Mukherjee, 1936. Bio-diversity refers to the totality of genes, species and ecosystems of a region. Biological diversity means the variability among living organisms from all sources including *intera alia*, terrestrial, marine and other aquatic ecosystem and the ecological complexes of which they are part, this includes diversity within species, between species and of an ecosystem.

Biodiversity therefore continued improvement of life and hence the benefits will depend on new and enhanced resources from nature. At any given time, changes in bio-diversity, *i.e.* the increase or reduction or maintenance of the diversity of genes, species or ecosystems will depend largely on human activities. Access to these resources will therefore depend on scientific knowledge of these resources through the studies of bio-diversity to enable prediction of most promising species and choosing sites of prospective biological resources, which in turn will provide relevant information from the countless number of specie The concept of bio-diversity emphasizes the interrelatedness of biological world and the importance of these interrelationships in maintaining diversity. It covers the terrestrial, marine and other aquatic environment such as river, streams, wetland and groundwater systems. Bio-diversity of inland water is important to sustained health of the ecosystem (both 'natural' and 'managed' ecosystem). Bio-diversity of inland water is also important for its economic value as habitat for species of commercial value.

India is fortunate to be endowed with a bounty of natural habitats, including snow covered Himalayas, the Indo-Gangetic plains, the Deccan plateau, deserts of Rajasthan and coastal regions. Such areas support a broad extent of water resources including cold, warm, brackish and marine water inhabiting varied types of fishes. For the exploitation and scientific development of Aquaculture knowledge of existing fish fauna of the area of study is a pre-requisite. Although some workers have studied icthyofuna of lotic and lentic water of the various riverine systems of Chhattisgarh region long ago, a fresh review of the existing fauna was highly desired. The earlier works are very old and biodiversity studies need updating at frequent intervals to understand the changes in fish species and their habitat. The catchment areas of the rivers are widely used for various types of agricultural practices and this has eventually resulted in the reduction of river width. Pollution of river water due to the leaching and discharges of various chemicals by factories are also increasing.

The present study was undertaken to survey the existing water resources, water quality and fish fauna of Chhattisgarh state in order to explore the possibility of fisheries development in the State. A survey of literature revealed that published information is available on ichthyofauna of this region, but there are reports on the fishes of different places of Chhattisgarh such as headwater the Mahanadi river, Raipur (Hora, 1940), Mahanadi drainage system (Jayaram and Majumdar, 1976), Ravishankar Sagar reservoir (Desai *et al.*, 2004),on a collection of fish from Bastar district (Karamakar and Datta, 1988) and on a collection of fish from Bailadila range (Hora, 1938). Keeping in mind the paucity of information on consolidated account of fishes of this state the present study was undertaken which would be certainly helpful for fishery workers in planning the future strategies for fishery development of the region.

Review of Literature

The review of literature on work done in the past is essential to understand the problem in depth and provides necessary guidelines as well as feed back for the fulfillment of the study

Hora and Mukherjee (1936) noted the fish of the Eastern Doons, United Provinces and listed 20 species of fish from the Song and Suswa rivers and one from the Bhatta stream near Mussoorie. Hora (1937) studied comparision of the fish-fauna of the northern and southern faces of the great Himalayan range. He also reported distribution of Himalayan fishes and its bearing on certain palaeogeopgrahical problems.

Hora (1938) reported on a small collection of fish from the Bailadila range, Bastar state, Central Provinces. The Bailadila range is situated to the South of the Indravati river, a tributary of the Godawari, and runs through the centre of the Bastar state from North to South. This collection was made from four different streams and noted the ecological conditions of their waters. In this collection a total of 13 species of fish were obtained from different localities. The thirteen species of fish found in different locations are namely *Danio aequipinnatus* (McClelland), *Rasbora daniconius* (Hamilton), *Garra mullya* (Sykes), *Parapsilorhynchus tentaculatus* (Annandale) *Barbus ticto* (Hamilton), *Nemachilus dayi* (Hora), *Nemachilus evezardi* (Day), *Glyptothorax dekhanensis* (Gunther), *Ophiocephalus gachua* (Hamilton), *Mastacembelus armatus* (Lacepdi), *Barbus amphibius* (Cuvier and Valenciennes) *Barbus pinnauratus* (Day), *Nemachilus botia* var *aureus* (Day). Among the 13 species of fish obtained *Mastacembelus armatus, Danio aequipinnatus, Rasbora daniconius, Barbus ticto* and *Ophiocephalus gachua* are fairly widely distributed all over India. The remaining species are generally restricted to Peninsular India, but their occurrence in the Bastar state deserves special attention. He observed that the fish fauna of Bailadila range has very close affinity to the Satpuras and the Western Ghats.

Hora (1940) studied the collection of fish from the Head waters of the Mahanadi river, Raipur district of the Central Provinces and published a list of 43 species. Twenty three belong to the order Cyprinoidea (20 species in Cyprinidae and 3 species in Cobitidae), 10 species to the order Siluroidea (4 species in Bagridae, one species in Amblycepidae, 2 species in Sisoridae, one species in Schilbeidae, one species in Clariidae and one species in Heteropneustidae), while the remaining ten species are of the families Mastacembelidae (3 species), Belonidae (1 species), Ophicephalidae (2 species), Ambassidae (one species), Nandidae (1 speices), Pristolepidae (1 species) and Gobiidae (1 species). With the exception of few small species of carp-minnows all the others are fairly well known and do not call for any special comments from a systematic point of view. However, notes are given on the distribution of *Rhynchod della aculeata, Nemachilus denisonii* and *Amblyceps mangois*.

Hora (1941) also recorded the fishes of the Satpura range, Hoshangabad district, Central Provinces and listed 40 species, of these 26 belonging to the order Cyprinoidea (22 species Cyprinidae and 4 species Cobitidae), 5 to the order Siluridea (3 species Bagridae, 1 species Amblycepidae and 1 species Sisoridae), while the remaining species are distributed among the families Notopteridae (1 species) Belonidae (1 species), Mastacembelidae (2 species), Ophicephalidae (2 species), Nandidae (one species), Pristolepidae (1 species) and Gobiidae (one species). With the exception of a few species of carp-minnows, all others are fairly well known and do not call for any comments from a systematic point of view. However, a few remarks are necessary on *Danio aequipinnatus* (Mcclelland), *Barbus* (Puntius) *Chrysopoma* (Cuv and Val), *Nemachilus dayi* (Hora), *Amblyceps mangois* (Ham.) and *Laguvia ribeiroi* (Hora). It shows that the fish-fauna of the Satpuras are closely allied to the fauna of the Western ghats. Some of these species, such as *Parapsilorhynchus tentaculatus, Nemachilus evezardi* and *Rita pavimentata*, were not found in the Sihawa range, Raipur District, but the remaining species are widely distributed in India.

Hora (1949) further found 42 species of fish in the Rihand river above and below the pipri dam and power station project site and their occurrence was correlated with the physical features of the river. Of the 42 species, 14 are of special interest from a zoogeographical point and these are divided into 3 categories, *viz.* (1) species common to peninsular India and Ceylon, (ii) species widely distributed in peninsular India, (iii) species with restricted distribution in India. It is has been seen that out of 14 species of zoogeographical interest, two were found from Ceylon in the South to the Vindhyas in the North, four are found in peninsular India, out of the eight species referred to in the last category, show affinities to the fauna of the Eastern Himalayas, Assam hills and Farther East.

Jayaram and Majumdar (1976) obtained fish samples from the Mahanadi drainage system in Orissa and Madhya Pradesh in a stretch extending from Cuttuck to Seorinarayan, a distance of about 350 km. Fourty two species were recorded of which some continue to live above and below the Hirakund Dam. Taxonomic notes on a few interesting species are given. *Osteobrama cotio, Peninsularies silas* are recorded for the first time from this river system.

This fish collection was made in the head waters of Godawari and Mahanadi river systems, namely Indirawati river, Saberi river, Mahanadi, Dudhnadi and Kukri (Bhor) rivers. Collections were also made from lakes, tanks and other water mass.

Desai *et al.* (2004) reported the fish fauna of Ravishankar Sagar reservoir, The Ravishankar Sagar reservoir is a large reservoir in Chhattisgarh region of Madhya Pradesh, with water spread of 9,540 ha at full water level. It has catchment area of 3,670 sq km. It came into being in 1978 with damming of the river Mahanadi, originated from village Pharsia in South-eastern part of the district of Raipur in Madhya Pradesh. The reservoir is also known as Gangrel Dam. It was constructed during the period 1973 to 1978 for irrigation and supplying water to Bhilai Steel Plant and Raipur. The detailed studies of Ravishankar Sagar reservoir on its ecology and fishery were conducted by Central Inland Capture Fisheries Research Institute (CICFRI) from 1987 to 1993. The fish fauna of the reservoir recorded were 48 species belonging to 15 families and 32 genera. However the numbers reported by Hora (1940) and Jayaram and Majumdar (1976) were 19 and 23, respectively, while 25 species recorded by CICFRI from the reservoir were common to those reported before from the river, 22 species were encountered for the first time. The commercially important fish species like *Catla catla, Labeo rohita* and *Cirrhius mrigala* did not exist in Mahanadi during 1940. However, the dominant catfishes of the reservoir (*Aorichthys aor* and *Aorichthys seenghala*) existed even earlier in the river which got in strong foothold after the impoundment. The minor carp fishery of the reservoir was mainly represented by *Gudusia chapra* followed by *Osteorbrama cotio*. The species were also not present in the river but their abundance after impoundment provided forage base for the catfishes of the reservoir.

Result and Discussion

Fishes were collected during January to July 2004 from the river, Sikasaur reservoir and ponds by using cast and drag nets and also from the various local fish markets and fish landing centers of the region. Small fishes were directly preserved in 10 per cent formalin but large sized fishes were given an injection of 10 per cent formalin to prevent spoilage of their visceral organs before their preservation. Morphometric and meristic characters of each species were studied by using fish measuring board and compass for the purpose of their identification. The works of Day (1988), Datta and Srivastava (1988), Talwar and Jhingran, (1991) were referred.

The present study indicated that the freshwater bodies is rich in fish fauna, A total number of 39 species belonging to 5 orders, 14 families and 26 genera were recorded from this region. Their common and vernacular names, fin formula, distribution and relative abundance are given in Table 9.1. Besides native fishes, some exotic fishes are also available in the region. Silvar carp, *Hypopthalmichthys molitrix* and common carp, *cyprinus carpio* are thriving best among the exotic fishes introduced in the region for their cultivation. The relative abundance of the various species of ichthyofauna of the region has been categorized into dominant (I), common (II), rare (III) species. The categories I, II and III include 5, 19, 15 species respectively.

The analysis of the different species revealed that 76.92 per cent of the species belong to the order Cypriniformes, 6.67 per cent to orders Clupiformes and

Table 9.1: Details of Fish Fauna of Pairy River

Phylum–Vertebrata, Sub-phylum–Craniata, Super class—Gnathostomata, Series–Pisces, Class–Teleostomi, Sub-class–Actinopterygii

Sl.No.	Systematic Position and Scientific Name	Local Name	English Name	Fin Formula	Distribution	Relative Abundance
	Order–Clupeiformes					
	Family–Clupeidae					
	Genus–Gudusia					
1.	Gudusia chapra (Hamilton)	Chhuria	Ganger shed	D.14 (3/11); P.12; V.8; A.22 (2/20); C.17; L.l-80; L.tr. 30	India, Pakistan, Nepal, Bangladesh, Malaya	=
	Family–Notopteridae					
	Genus–Notopterus					
2.	Notopterus Chitala (Hamilton)	Patola	Humped feather	D.9 (1/8); P.15; V.6; A.110; C.12; L.l-160	Freshwater and brackish waters of India, Pakistan, Burma, Siam and Philippines	=
	Order–Cypriniformes					
	Family–Cyprinidae					
	Genus–Labeo					
3.	Labeo angra (Hamilton)	Gadela	Labeo angra*	D.13 (3/10); P.16; V.8; A.8(3/5); C.19; L.l-43; L.tr. 16 (8/8) : barbles 2 pairs	U.P. Bihar, Bengal, Orissa, East Pakistan and Burma	≡
4.	Labeo rohita (Ham–Buch)	Rohu	Rohu	D.15(2/13); P.17; V.9; A.7 (2/5); C.19; L.l-42; L.tr. 7, barbles one pairs	India, Pakistan and Burma	–
5.	Labeo calbasu (Ham–Buch)	Kalbaz	Calbasu Orange finlabeo	D.16; P.18; V.9; A.7(2/5); C.19; L.l-40; L.tr. 6/7, barbles two pairs	India, Pakistan and Burma	≡
6.	Labeo bata (Ham.)	Bata	Bata Labeo*	D.11; P.18; V.9; A.7(2/5); C.19; L.l-40; L.tr. 5/6, barbles one pairs	India, Bangladesh, Nepal and Pakistan	=

Contd...

Table 9.1.–Contd...

Sl.No.	Systematic Position and Scientific Name	Local Name	English Name	Fin Formula	Distribution	Relative Abundance
	Genus–Hypophthalmichthys					
7.	*Hypophthalmichthys molitrix*	Silvar carp	Silvar carp	D.10(3/7); V.8(1/7); A.14(2/12); L.l110	India, Pakistan, Burma	I
	Genus–Cyprinus					
8.	*Cyprinus carpio* (Linn)	Carpio	Common carp	D.19(1/18); P.16(1/15); V.8(1/7); A7(1/6); C.21; L.l.34; L.tr.13, barbels 2 pairs	India, Pakistan, Burma	I
	Genus–Puntius					
9.	*Puntius sarana* (Ham-Buch)	Kotra	Olive barb*	D.11(2/9); P. 15; V.9; A.8(3/5); C.19; L.l-32; L.tr. 6$^{1/2}$/5, barbles two pairs	India, Pakistan, Burma, Nepal, Bangladesh, Thailand, Sri Lanka and China	II
10.	*Puntius chola* (Ham-buch)	Kotri	Green barb	D.11(3/8); P.14; V.9; A.8(3/5); C.19; L.l-27; L.tr. 5$^{1/2}$/5	India, Pakistan and Burma	III
11.	*Puntius dorsalis* (Jerdon)	Kotri	Long Snouted barb*	D.11(3/8); P.15; V.9; A.8(3/5); C.19; L.l-24; L.tr. 4$^{1/2}$/4	India, Pakistan and Burma	III
12.	*Puntius ticto* (Ham)	Kotri	Ticto barb*	D.11(3/8); P.15; V.9; A.8(3/5); C.19; L.l-25; L.tr. 5-6$^{1/2}$/6-6$^{1/2}$	India, Nepal, Pakistan, Sri Lanka, Bangladesh, Burma, Thailand and Siam	II
	Genus–Barilius					
13.	*Barilius bendelisis* (Ham)	Jori	Hamilton's barila*	D.8(1/7); P.14; V.9; A.10(2/8); C.15; L.l. 38; L.tr. 11 (8/3), barbels two pairs	India, Pakistan, Nepal, Bangladesh, Sri-Lanka, Burma, Egypt, West Africa	III
	Genus–Catla					
14.	*Catla catla* (Ham.)	Katla	Catla	D.16; P.18; V.9; A.7; C.19; L.l-38; L.tr. 7$^{1/2}$/6	Northern India, Pakistan and Burma	I

Contd...

Table 9.1–Contd...

Sl.No.	Systematic Position and Scientific Name	Local Name	English Name	Fin Formula	Distribution	Relative Abundance
	Genus–Cirrhinus					
15.	Cirrhinus mrigala (Ham.)	Mrigal	Mrigal	D.16(3/3); P.18; V.9; A.8(2/6); C.15; L.l.-42; L.tr. 7; barbles one pair	Northern part of India, Pakistan, Nepal, Burma	I
16	Cirrhinus reba (Ham)	Borai	Reba carp*	D.10; P.15; V.9; A.6; C.17; L.l.-32; L.tr. 6/5; barbles one pair	India, Pakistan, Nepal, Bangladesh	III
	Genus–Salmostoma					
17.	Salmostoma bacaila (Ham.)	Sirangi	Large razor-belly minnow	D.9(2/7); P.12; V.8; A.14; C.17; L.l.-43; L.tr. 12/10	India except Malabar, Mysore and Madras	II
	Genus–Garra					
18.	Garra annandalei (Hora)	Gatuea	Annandale garra*	D.10(2/8); P.14; V.10; A.7(1+6); L.l.-30; L.tr. $4^{1/2}/3^{1/2}$ barbles two pairs	India, Pakistan, Nepal, Bangladesh, Burma	III
	Genus–Gonoproktopterus					
19.	Gonoproktopterus kolus (Sytes)	Gulti	Kolus*	D.10; P.15; V.9; C.17, L.l.-30, L.tr.3 ½/2 ½ Barbels one pair	India, Pakistan, Nepal, Bangladesh, Burma, Thailand, West Africa	III
	Genus–Noemacheilus					
20.	Noemacheilus bolia (Ham.)	Rudwa	Loach	D.11; P.13; V.8; A.7; C.18; L.l.-105; L.tr.28; Barbels three pairs	Northern India, Nepal, Burma, Pakistan, Bangladesh and Sri Lanka	II
	Family–Siluridae					
	Genus–Ompok					
21.	Ompok bimaculatus (Bloch)	Bolia	Butter cat fish*	D.4; P15(11/4)V.7; A60; C.17; Barbels two pairs	Freshwater of India, Nepal, Sri Lanka, Burma, Bangladesh, Thailand, Pakistan	I

Contd...

Table 9.1–Contd...

Sl.No.	Systematic Position and Scientiiic Name	Local Name	English Name	Fin Formula	Distribution	Relative Abundance
	Family–Bagridae					
	Genus–Mystus					
22.	Mystus bleekeri (Day)	Tengna	Day's Mystus*	D.8(1/7); P10(1/9); V.6; A.9(3/6); C.16: barbles 4 pairs	Northern India, Pakistan, Burma and Sumatra	=
23.	Mystus cavasius (Ham.)	Jaliya tenga	Dwarf Cat fish*	D.8(1/7); P.9(1/8).V.6; A.10(2/8); C.15; Barbles four pairs	Northern India, Pakistan and Burma	=
24.	Mystus tengara (Ham.)	Tengra	Tengara mystus*	D.8(1/7); P.8(1/7); V.6; A.10(2/8), C.15; barbles four pairs	North India and Pakistan	=
25.	Mystus (Aorichthys) aor (Ham.)	Singhar	Long whiskered cat fish*	D.8(1/7); P.10(1/9); V6; A.11(3/8) C.19; barbles four pair	South India Pakistan Bangladesh, Burma and China	III
26.	Mystus vittatus (Bloch)	Tengra	Striped dwarf cat fish*	D.8(1/7); P.10(1/9); V.7; A.10(2/8); C.17; Barbles four pair	India, Pakistan, Bangladesh, Burma	=
27.	Mystus (Aorichthys) seenghala (Sykes)	Tengra	Gaint river cat fish*	D.9(1/8); P.11(1/10); V.6; A.11(2/9); C.19; barbles 4 pairs	India, Pakistan and Burma	=
	Genus–Rita					
28.	Rita rita (Ham.)	Bhunda	Rita	D.7(1/6); P.8 (1/7); V.7; A.10(2/8) C.20, Barbles four pairs	India-freshwaters of Uttar Pradesh, Bihar, West Bengal, Pakistan	=
	Family–Schilbeidae					
	Genus–Eutropiichthys					
29.	Eutropiichthys vacha (Ham-Buch)	Bachra	Vacha	D.8(1/7); P.16(1/15); V.6; A.45 (3/42); C.17; Barbles four pairs	Northern India, Orissa, Bengal and Assam	III
	Genus–Pseudeutropius					
30.	Pseudeutropius atherinoides (Bloch)	Ketia-kuli	Indian potasi*	D.6(1/5); P.7(1/6); A.32(2/30); C.15; Barbels four pairs	India, Pakistan, Malaya, and China	III

Contd...

Table 9.1–Contd...

Sl.No.	Systematic Position and Scientific Name	Local Name	English Name	Fin Formula	Distribution	Relative Abundance
	Family–Saccobranchidae or Heteropneustidae					
	Genus–Heteropneustes					
31.	Heteropneutes fossilis (Bloch)	Singhi	Stinging cat fish	D.6; P8(1/7); V.6; A.63; C.18 barbles 4 pairs	India, Pakistan, Ceylon, Burma and China	III
	Family–Clariidae					
	Genus–Clarias					
32.	Clarias batrachus (Linn.)	Mangur	Air breathing cat fish	D.65; P.19(1/18); V.6; A.52; C.16; barbles 4 pairs	India, Pakistan, Ceylon, Burma and Malaya	II
	Order–Beloniformes					
	Family–Belonidae					
	Genus–Xenentodon					
33.	Xenentodon cancila (Hamilton)	Sodiya	Freshwater gar fish	D.16; P.11; V.6; A.17; C.15	India, Pakistan, Ceylon and Burma	III
	Order–Ophiocephaliformes					
	Family–Ophiocephalidae					
	Genus–Channa					
34.	Channa marulius (Ham.)	Khokshi	Giant snake-head murrel	D.51; P.18; V.5; A.32; C.15; L.l.-65; L.tr. 5$^{1/2}$/12	India, Pakistan, Ceylon and China	II
	Order–Perciformes					
	Family–Centropomidae					
	Genus–Chanda					
35.	Chanda nama (Ham.)	Chan-deni	Glassy perchlet	D.22(1+6/1+14); P.13; V.6(1/5); A19(3/16); C.17; L.tr. Scales deciduous	India, Pakistan and Burma	II

Contd...

Table 9.1.–Contd...

Sl.No.	Systematic Position and Scientific Name	Local Name	English Name	Fin Formula	Distribution	Relative Abundance
	Family–Anabantidae					
	Genus–Anabas					
36.	Anabas testudineus (Bloch)	Kimi	Climbing perch*	D.26(18+9); P.15; V.6(1/5); A.18; C.16; L.l.29; L.tr. 4/8	India, Pakistan and Burma, Bangladesh	=
37.	Anabas oligolepis (Bleeker) or Anabas cobojius*	Kinni	Gangetic koi*	D.25(16+9); P.15; V.6 (1/5); A20(10/10); C. 17; L.l.–29; L.tr. 4/8	Bhimavaram, West Godavari district, Andhra Pradesh, Orissa, West Bengal	≡
	Order–Mastacembeliformes					
	Family–Mastacembelidae					
	Genus–Mastacembelus					
38.	Mastacembelus armatus (Lacepede)	Bambi	Sny eel	D.37/73; P.22; V. abst; A.3/75; C.20 (confluent with dorsal and anal)	India, Pakistan, Ceylon, Nepal, Burma and Malaya	=
	Genus–Macrognathus					
39.	Macrognathus aculeatus (Bloch)	Bambi	Lesser spiny eel	D.19+44; P19 V.abst. A.3/42; C15; vert 15/20	India, Thailand, Pakistan, Burma and Malaya	≡

*: Fish Base 2000.

D: Dorsal fin; P: Pectoral fin; V: Ventral fin or pelvic fin; L.l.: Lateral line of perforated scales; L.tr.: Lateral transverse row of scales; B: Barbels.

Mastacembeliformes. Order Beloniformes and Ophiocephaliformes have 3.33 per cent catch and rest 10 per cent species belong to Perciformes.

The fish fauna of Mahanadi systems as reported by present study from Hora, Jayaram and Majumdar are compared with the Pairy river and are depicted in Table 9.2. Perusal of these data shows that of 39 species reported in the present study, 10 species were found common with the earlier two reports and thus 15 species have been placed and reported as new in this report which have not been reported (Hora, 1940, Jayaram and Majumdar 1976). The present study compared with Hora (1940) shows that the common species (7) are *Puntius chola, Puntius dorsalis, Mystus aor, Heteropneustes fossilis, Pseudeutropius atherinoides, Noemacheilus botia* and *Clarias batrachus*. Similarly when compared with Jayaram and Majumdar (1976) the common species are *Ompok bimaculatus, Mystus seenghala, Chanda nama, Labeo rohita, Catla catla* and *Labeo bata*.

Table 9.2: Comparison of Ichthyofauna as Reported by Different Workers in the Mahanadi Basin

Sl.No.	Fish Species	Pairy River (2004)	Hora (1940)	Jayram and Majumdar (1976)
1.	*Gudusia chapra* Hamilton	N	–	–
2.	*Notopterus chitala* Ham.	N	–	–
3.	*Barilius bendelisis* Ham.-Buch	+	+	+
4.	*Puntius sarana* Ham.-Buch	+	+	+
5.	*Puntius chola* Ham.-Buch	+	+	–
6.	*Puntius dorsalis* Jerdon	+	+	–
7.	*Puntius ticto* Ham.	+	+	+
8.	*Labeo angra* Ham.-Buch	N	–	–
9.	*Gonoproktopterus kolus* Sykes	N	–	–
10.	*Salmostoma bacaila* Ham.–Buch	+	+	+
11.	*Garra annandalei* Hora	N	–	–
12.	*Ompok bimaculatus* Bloch	+	–	+
13.	*Mystus (Aorichthys) aor* (Ham)	+	+	–
14.	*Mystus (Aorichthys) seenghala* Sykes	+	–	+
15.	*Mystus bleekeri* Day	N	–	–
16.	*Mystus tengara* (Ham)	+	+	+
17.	*Mystus vittatus* Bloch	+	+	+
18.	*Rita rita* Ham.	N	–	–
19.	*Eutropiichthys vacha* Ham.-Buch	N	–	–
20.	*Pseudeutropius atherinoides* Bloch	+	+	–
21.	*Heteropneustes fossilis* Bloch	+	+	–
22.	*Xenentodon cancila*	+	+	+
23.	*Channa marulius* Ham.	N	–	–

Contd...

Table 9.2–Contd...

Sl.No.	Fish Species	Pairy River (2004)	Hora (1940)	Jayram and Majumdar (1976)
24.	*Chanda nama* Ham.	+	–	+
25.	*Anabas testudineus* Bloch	N	–	–
26.	*Anabas oligolepis* Bleeker	N	–	–
27.	*Mastacembelus armatus* Lacepede	+	+	+
28.	*Macrognathus aculeatus* Bloch	+	+	+
29.	*Laubuca laubuca* Hamilation	–	+	–
30.	*Barilius barna* Ham.	–	+	+
31.	*Brachydanio rerio*	–	+	+
32.	*Danio aequipinnatus*	–	+	–
33.	*Esomus danricus*	–	+	+
34.	*Aspidoparia morar*	–	+	+
35.	*Barbus (Puntius) gelius*	–	+	+
36.	*Barbus (Puntius) guganio*	–	+	–
37.	*Garra mullya*	–	+	–
38.	*Oreichthys cosuatus*	–	+	–
39.	*Nemachilus denisonii* Day	–	+	–
40.	*Amblyceps mangois*	–	+	–
41.	*Erethister hara*	–	+	–
42.	*Bagarius bagarius*	–	+	+
43.	*Badis badis*	–	+	+
44.	*Labeo rohita*	+	–	+
45.	*Hypophthalmichthys molitrix val.*	N	–	–
46.	*Cyprinus carpio* Linn.	N	–	–
47.	*Catla catla*	+	–	+
48.	*Noemacheilus botia*	+	+	–
49.	*Mystus cavasius* Ham.-Buch	+	+	+
50.	*Clarias batrachus* Linn.	+	+	–
51.	*Barilius vagra*	–	–	+
52.	*Labeo bata*	+	–	+
53.	*Rita chrysea*	–	–	+
54.	*Gagata cenia*	–	–	+
55.	*Ailia coila*	–	–	+
56.	*Clupisoma garua*	–	–	+
57.	*Cirrhinus mrigala*	N	–	–
58.	*Labeo calbasu*	N	–	–

Recorded: +: Not recorded; –: New record = (N).

Tallying with Hora (1940), Jayaram and Majumdar (1976) in the present study, the following 20 species were not available: *Laubuca laubuca, Barilius barna, Brachydanio rerio, Danio aequipinnatus, Esomus dansicus, Barilius vagra, Rita chrysea, Gagata cenia, Ailia coila, Clupisoma garua. Aspidoparia morar, Barbus (Puntius) gelius, Barbus (Puntius) guganio, Garra mullya, Oreichthys cosuatus, Nemachilus denisonii, Amblyceps mangois, Erethistes bara, Bagarius bagarius* and *Badis badis.*

In place of these fishes the fifteen new species recorded are *Gudusia chapra* (Ham.), *Notopterus chitala* (Ham.), *Labeo angra* (Ham.-Buch), *Gonoproktopterus kolus* (Sykes), *Garra annandalei* (Hora), *Mystus bleekeri* (Day), *Rita rita* (Ham.), *Eutropiichthys vacha* (Ham.-Buch), *Channa marulius* (Ham.), *Anabas testudineuds* (Bloch), *Anabas oligolepis* (Bleeker), *Hypophthalmichthys molitrix* (Val.),*Cyprinus carpio* (Linn.), *Cirrhinus mrigala* and *Labeo calbasu.*

Biodiversity threats in the form of diverse types of human interventions are the main reasons for the alarming variations of fish populations in most of the rivers. The great altitudinal differences coupled with varied physiography have contributed to great variations in the region, having definite pockets representing tropical and sub–tropical areas. The most common attribute would be to measure in a broad spatial scale is habitat. In conservation, it may be assumed that if habitat in the adjacent catchments is good, if riparian zones are intact, if structural attributes of the in-stream habitat are maintained and reasonable environmental flows are maintained, and then biodiversity values are likely to remain intact. Use of small meshed fishing gears boat, drag net, gill net, wallnet, cast net etc. The use of fishing gears with less than 5 mm mesh size is very rampant in the down streams of most of the river Pairy. Such practices for a short–term profit kills the fry and fingerlings of the fishes, thus leading to a drastic reduction and change in population.

The catchment areas of the rivers are widely used for various types of agricultural practices and this has eventually resulted in the reduction of river width. Pollution of river water due to the leaching and discharges of various agricultural chemicals are also increasing. During the transport of desired fish seed from outside the state (especially Kolkatta) some other unintentional and undesirable seed is also introduced where by the species diversity has increased and changed.

In the present study 39 species belonging to 5 orders, 14 families and 26 genera were enlisted including 4 exotic fish species. 15 new species have been put on record and 20 species are not recorded in present study compared for periods 1940 through 2004. Exotic species as *Cyprinus carpio* and *Hypophthalmichthys molitrix* are exclusively found in the pond culture system along with Indian Major Carps. *Tilapia* has also made introduce into the ponds affecting pond fish production. Also monsoonal flooding and unintention introduction have led to many Minor Carps establishing in ponds and reservoir. Such species changes are mainly due to human interventions rather than natural. It is likely that at some point time only uneconomical fishes may become dominant due to habitat changes.

Acknowledgements

With great reverence I express my warmest feelings with deep sense of gratitude to Dr. H.K.Vardia and Dr. M.S. Chari, Professor, Department of Fisheries, Indira Gandhi Agricultural University, Raipur.

References

1.   Datta Munshi, J.S. and Shrivastava, M.P. (1988). *Natural History of Fishes and Systematic of Freshwater Fishes of India*. Narendra Publishing House, Delhi, pp. 403.

2.   Day, F. (1988). *The Fishes of India: A Natural History of the Fishes Known to Inhabit the Seas and Freshwaters of India, Burma and Ceylon, Vols. I and II*. Today and Tomorrows Book Agency, New Delhi, 778 pp.

3.   Desai, V.R., Kumar, D. and Shrivastava, N.P. (2004). Fish fauna of Ravishankar Sagar Reservoir. *Journal Inland Fish Society of India*, 29(2): 54–57.

4.   Hora, S.L. (1937). Comparison of the fish-faunas of the northern and the southern faces of the Great Himalayan Range. *Records of the Indian Museum*, 39: 241–259.

5.   Hora, S.L. (1938). On a collection of fish from the Bailadila range, Bastar State, central provinces. *Records of the Indian Museum*, 40: 237–241.

6.   Hora, S.L. (1940). On a collection of fish from the headwaters of the Mahanadi river, Raipur district, central provinces. *Records of the Indian Museum*, 42(20): 365–374.

7.   Hora, S.L. (1941). Fishes of the Satpura range, Hoshangabad district, central provinces. *Records of the Indian Museum*, 43: 361–373.

8.   Hora, S.L. (1949). The fish fauna of the Rihand river and its zoogeographical significance. *Journal of the Zoological Society of India*, 1 1): 1–7.

9.   Hora, S.L. and Mukherji, D.D. (1936). Fish of the Eastern Doons, United Provinces. *Rec. Ind. Mus.*, 38: 139–368.

10. Jayaram, K.C. and Majumdar, N. (1976). On a collection of fish from the Mahanadi. *Records the Zoological Survey of India*, 69: 305–323.

11. Karmakar, A.K., and Datta, A.K. (1988). On a collection of fish from Bastar district, Madhya Pradesh. *Records of the Zoological Survey of India*. Occasional Paper No. 98: 1–50.

12. Talwar, P.K. and Jhingran, A.G. (1991). *Inland Fishes of India and Adjacent Countries*. Vols. I and II. Oxford and IBH Publishing Co., New Delhi, India, 1158 pp.

Biodiversity of Aquatic Reseources (2012)
Editors: **Mamta Rawat & Sumit Dookia**
Published by: **DAYA PUBLISHING HOUSE, NEW DELHI**

Pages **139-144**

Chapter 10

Icthyo-faunal Diversity in Hussainsagar Lake, Hyderabad, Andhra Pradesh, India

S. Anitha Kumari[1]* and N. Sree Ram Kumar[2]

[1]*Department of Zoology,*
Osmania University College for Women, Koti, Hyderabad, A.P.
[2]*Department of Zoology,*
Nizam College, Basheerbagh Hyderabad, A.P.

ABSTRACT

The present study deals with the icthyofaunal diversity and its abundance in Hussainsagar Lake, Hyderabad, Andhra Pradesh (A.P.) India. The fish diversity is correlated with various physico-chemical parameters of the water that regulate the distribution of different species of fishes. It is observed that the distribution of fish species in Hussainsagar Lake in the present condition is quite variable when compared to its past history harbouring only pollution resistant forms of fishes, *i.e., Channa punctatus, Channa striatus, Channa gachua, Heteropneustes fossilis* and *Clarias batrachus*. Of these, *Channa punctatus* was found to be the most abundant. It is thus indicated that the biodiversity in the pollution–stressed environment becomes poor.

Keywords: *Icthyo-faunal diversity, Hussainsagar lake, Physico-Chemical parameters, Pollution-resistant forms.*

* Corresponding Author E-mail: anitha_shinde2001@yahoo.com

Introduction

Inland fisheries in India have great potential of contributing to the food security of the country. Reservoirs and lakes are the main resources exploited for inland fisheries and understanding of fish faunal diversity is a major aspect for its development and the sustainability management. Lakes in India support rich variety of fish species, which in turn support the commercial exploitation of the fisheries potential (Krishna and Piska 2006).

Hussainsagar, the picturesque lake situated in between the twin cities of Hyderabad and Secunderabad, is an ecological and cultural landmark on the map of Hyderabad, the capital of Andhra Pradesh. It was excavated mainly to store drinking water brought from the river Moosi. However, over the last 50 years, the lake is subjected to gross pollution due to the constant disposal of industrial effluents and domestic sewage. The expanding slums along the western bank of the lake, traditional washing activities by the washermen community and the immersion of Lord Ganesh idols during *Ganesh Chaturthi* festival, every year further contribute to the lakes pollution. The ever increasing aquatic pollution and hyper-eutrophication of the lake and the resulting deterioration in aesthetic, physico-chemical and biological quality of lake-water has eventually led to frequent fish kills in the lake as evidenced by several reports on fish mortality (Khan and Hussain, 1976; Muley, 1987; Manjula Devi, 1988 and Siddiqui and Rao, 1991). As a result there is a gradual elimination of sensitive fish species by pollution tolerant forms. Hence, the present study was conducted to find out the past and the present status of ichthyo-faunal diversity and its abundance in Hussainsagar Lake.

Material and Methods

In the present study the fishes were collected from different stations of Hussainsagar Lake throughout the period of work covering all seasons by repeated netting and preserved in 4 per cent formalin and were identified upto the species level using standard literature (Beavan, 1990, Jhingran, 1991 and Jayaram, 2002). Physico-chemical parameters reflecting on water quality were also taken into consideration and the analysis of the physico-chemical parameters of the water samples were carried out as per the procedures described in the standard methods (APHA, 1998 and Trivedi *et al.*, 1998).

Result and Discussion

Water quality is an important criterion for fish habitat. Hence the important physico-chemical characters of water from the lake were analyzed and are presented in Table 10.1.

The water quality data revealed the water temperature range between 28 to 36°C. The pH ranged from 7.2 to 8.62 which are affected not only by the levels of Carbon di–oxide but also by other organic and inorganic components of water. The specific conductance ranged from 1517 to 2900 whereas turbidity showed fluctuation within a range of 22 to 45. The oxygen is one of the most important factor in any aquatic ecosystem. The main sources of dissolved oxygen are from the atmosphere and

Table 10.1: Physico-chemical Parameters of Hussainsagar Lake

Sl.No.	Parameter	Range	
		Min.	*Max.*
1.	Water temperature ºC	28	36
2.	pH	7.2	8.62
3.	Specific conductance µ mhos/cm³ at 20ºC	1517	2900
4.	Turbidity	22	45
5.	DO (Dissolved oxygen)	1.2	4.5
6.	CO_2 (Free carbon dioxide)	9	12.5
7.	Biochemical oxygen demand at 20ºC (BOD)	28	52
8.	Chemical oxygen demand (COD)	80	215
9.	Total Solids (TS)	1120	1880
10.	Total dissolved solids (TDS)	530	1540
11.	Total suspended solids (TSS)	254	480
12.	Total alkalinity	210	640
13.	Total hardness	320	440
14.	Calcium as Ca^{2+}	82	256
15.	Magnesium as Mg^{2+}	25	50.74
16	Chlorides	220	458
17.	Sulphates	200	305
18.	Nitrates	20.50	40
19	Phosphates	6.5	12.40
20.	Fluorides	0.5	1.08

All values are in mg/l except Temperature, pH, conductance and turbidity.

photosynthesis. In the present investigation the Dissolved Oxygen (DO) was found to be in the range of 1.2 to 4.5mg/l which is less than the permissible limit. The Free CO_2 content of the water depends upon the temperature, depth range of respiration, decomposition of organic matter, chemical nature of the bottom and geographical features, surrounding the water body. During the present study, free CO_2 was observed to be in the range of 9 to 12.5 mg/l. The Biological Oxygen Demand (BOD) ranged between 28 to 52 mg/l whereas Chemical Oxygen Demand (COD) ranged between 80 to 215 mg/l. The total solids were recorded to be in the range of 1120 to 1880 mg/l and the total dissolved solids were recorded in the range of 530 to 1540 mg/l. However the total suspended solids showed a range of 254 to 480mg/l. A high content of total solids elevate the density of water and such a medium increases osmoregulatory stress on aquatic biota as advocated by (Verma *et al.*, 1978). The total alkalinity is generally imparted by the salts of CO_3, HCO_3, phosphates, nitrates etc together with the hydroxyl ions in free state. The total alkalinity was recorded in the range of 210 to 640mg/l. Hardness is governed by the contents of calcium and magnesium salts largely combined with bicarbonates and carbonates giving temporary hardness and with sulphates, chlorides and other anions of mineral acids causing permanent hardness. In the present study the total hardness of water varied between 320 to 440 mg/l. the calcium was recorded in the range of 82 to 256 mg/l whereas magnesium in the range of 25 to 50.74 mg/l. The chlorides ranged between 220 to 458mg/l.The sulphates the recorded 200-305mg/l and the Nitrates to be in the range of 20.50 to 40mg/l. phosphates were recorded to be in the range of 6.5 to 12.40mg/l and fluorides in the range of 0.5 to 1.08mg/l.

Table 10.2: Fish Fauna of Hussainsagar Lake (Babu Rao and Siva Reddy, 1984)

Sl.No.	Species
1.	*Channa punctata* (Bloch)
2.	*Channa marulia* (Hamilton)
3.	*Channa striata* (Bloch)
4.	*Channa gachua* (Hamilton)
5.	*Heteropneustes fossilis* (Bloch)
6.	*Clarias batrachus* (Linnaeus)
7.	*Mystus vittatus* (Bloch)
8.	*Mystus bleekeri* (Day)
9.	*Wallagu attu* (Schneider)
10.	*Chela bacaila* (Hamilton)
11.	*Ambly pharyngodon moal* (Hamilton)
12.	*Rasbora daniconius* (Hamilton)
13.	*Puntius sophore* (Hamilton)
14.	*Puntius ticto* (Hamilton)
15.	*Puntius chola* (Hamilton)
16.	*Puntius amphibia* (Valenciennes)
17.	*Puntius sarana* (Hamilton)
18.	*Puntius filamentosus* (Valenciennes)
19.	*Labeo boggut* (Sykes)
20.	*Garra mullya* (Sykes)
21.	*Lepido cephalus (Lepido cephalichthys) guntea* (Ham)
22.	*Gambusia affinis patruelis* (Barid and Girard)
23.	*Poecilia reticulata* (Peters)
24.	*Etroplus maculatus* (Bloch)
25.	*Glossogobius giuris* (Hamilton)
26.	*Osphronemus goramy* (Lecepede)
27.	*Notopterus notopterus* (Pallas)

The alterations in various physico-chemical characteristics of the lake water clearly indicate the pollution load of Hussainsagar Lake and thus play an important role in the distribution of fish species. The fish fauna is an important aspect of fishery potential of a water body. It is observed that the distribution of fish species in Hussainsagar Lake in the present condition is quite variable when compared to its past history as the study revealed that out of the 27 hardy species of fish reported earlier by Babu Rao and Siva Reddy (1984) as shown in Table 10.2, only 5 species of fish were found inhabiting the lake waters whereas the other species totally disappeared from the lake. These include *Channa punctatus, Channa striatus, Channa gachua, Heteropneustes fossilis* and *Clarias batrachus*. All these fishes are useful as food fishes.

Jaya raju *et al.* (1994) has also studied fish diversity in collaboration with physico-chemical parameters from the river Munneru, a tributary of Krishna River, Andhra Pradesh. Rajaram *et al.* (2004) has also studied icthyofaunal diversity of great Nicobar Island whereas Pawar *et al.* (2006) studied the fish fauna of Pethwadas dam in Maharashtra. Battul *et al.* (2007) have also studied fish diversity from Ekrukh Lake. As per their studies, Apart from other physico-chemical parameters, temperature and DO were observed to be major controlling factors in the distribution of fishes. Our results are also in accordance with their findings.

Besides forming the simple diet for human beings, fishes are also very useful indicators of the real state of purity of water (Peter, 1987). The fish can also be utilized in monitoring the water quality for the toxic constituents (Seth *et al.,* 1967).

If is thus evident from the present study that Hussainsagar Lake once upon a time supported fairly rich fish diversity, However in the recent times due to the constant pollution-stress, the lake is facing major threat for its very existence and the fish diversity in it harbouring only pollution resistant forms thus indicating that the biodiversity in the pollution–stressed environment becomes poor.

References

1. APHA (1998). *Standard Methods for Examination of Water and Wastewaters* 20ᵗʰ Edition. American Public Health Association. Washington D.C.

2. Babu Rao, M. and Siva Reddy, Y. (1984). Fish fauna of Hussainsagar. *Jantu.* V–2: 1–6.

3. Battul, P.N., Rao, K.R., Navale, R.A., Bagale, M.B. and Shah, N.V. (2007). Fish diversity from Ekrukh Lake near Solapur, Maharashtra. *J. Aqua Biol.,* 22(2): 68–72.

4. Beavan, R. (1990). *Handbook of the Freshwater Fishes of the India.* A venture of Low Price Publications.

5. Jaya Raju, P.B., Prasad Rao, G.D.V. and Sharma, S.V., (1994). Seasonal variations in physico-chemical parameters and diversity in the flora and fauna of the river Munneru, a tributary to river Krishna, Andhra Pradesh, India. *J. Aqua. Biol.,* 9(1 and 2): 19–22.

6. Jayaram, K.C. (2002). *The Freshwater Fishes of the Indian Region.* Narendra Publishing House, Delhi.

7.   Jhingran, V.G. (1991). *Fish and Fisheries of India*, 3rd Edition. Hindustan Publishing Corporation, New Delhi, p. 252–257.

8.   Khan, M.A. and Hussain, A.M. (1976). Preliminary observations on pollution of Lake Hussainsagar caused by Industrial effluents. *Indian J. Environ. Health*, 3: 227–232.

9.   Krishna, M. and Piska, R.S., 2006. Ichthyofaunal diversity in secret Lake, Durgamcheruvu, Ranga Reddy District, Andhra Pradesh, India. *J. Aqua Biol.*, 21(1): 77–79.

10. Manjula Devi, P. (1988). Ecotoxicological studies on freshwater fish with special reference to pollution. *Ph.D. Thesis*, Osmania University, Hyderabad.

11. Muley, E.V. (1987). Preliminary report on the fish kills in lake Hussainsagar. Project sponsored by Municipal Corporation of Hyderabd and Andhra Pradesh Pollution Control Board, pp. 42.

12. Pawar, S. K., Mane, A.M. and Pulle, J.S. (2006). The fish fauna of Pethwadas Dam, Taluka Kandhar in Nanded district Maharashtra, India. *J. Aqua Biol.*, 21(2): 55–58.

13. Peter, T. (1987). Fish fisheries, Aquatic macrophytes and water quality in Inland waters. *Water Quality Bull.*, 12(3): 103–129.

14. Rajaram, R., Srinivasan, M., Ajmal Khan, S. and Kannan, L. (2004). Icthyofaunal diversity of Great Nicobar Islands, Bay of Bengal. *Journal of Indian Fisheries Association*, 31: 13–26.

15. Seth, A.K., Srivastava, S.K., George, N.G. and Bewtra, J.K. (1967). Monitoring of certain toxic constituents in water supply by fish. *Ind. J. Environ Hlth.*, 9 (1): 34–37.

16. Siddiqui, S.Z. and Rao, K.R.V. (1991). Limnologic investigation on a recent major fish kill (*Notopterus notopterus*) in Hussainsagar, Hyderabad, India. *Poll. Res.*, 10(4): 191–198.

17. Trivedi, R.K., Goel, P.K. and Trisal, C.L. (1998). *Practical Methods in Ecology and Environment Science*. Enviro Media Publications, Karad, India.

18. Verma, S.R., Tyagi, A.R. and Dalela, R.C. (1978). Physico-chemical and biological characteristic of Karadabad drain in U.P. *Indian J. Environmental Health*, 20(1): 1–13.

Biodiversity of Aquatic Reseources (2012)
Editors: **Mamta Rawat & Sumit Dookia**
Published by: **DAYA PUBLISHING HOUSE, NEW DELHI**

Pages **145-158**

Chapter 11

Avifauna of Kole Wetlands: Species Diversity and Abundance Distribution Patterns

C. Sivaperuman[1]* and E.A. Jayson[2]
[1]*Zoological Survey of India,*
Andaman and Nicobar Regional Centre, Hodda,
Port Blair – 744 102
[2]*Division of Forest Ecology and Biodiversity Conservation,*
Kerala Forest Research Institute, Peechi, Thrissur, Kerala – 680 653

ABSTRACT

Avifauna of Kole wetlands have been studied from November 1998 through October 2001. The name "Kole" refers to the peculiar type of paddy cultivation carried out from December to May and this Malayalam word indicates bumper yield of high returns in case floods do not damage the crops. Total count methods were employed to census the bird populations. The species richness and abundance of birds varied in different months. Similarly, diversity Index (H') also varied in different months and also in the intensive study areas. High similarity observed between Enamavu and Kanjany. During the migratory seasons the more number of shorebirds was recorded in the mudflats and these are of regional importance to shorebirds. The species richness and abundance of birds showed high value in

* Corresponding Author E-mail: c_sivaperuman@yahoo.co.in

the Kole wetlands. The Kole wetlands appear to serve as important intermediate stopover sites for migratory birds.

Keywords: Avifauna, Abundance, Diversity, Kerala, Kole-wetlands.

Introduction

Wetlands are fragile ecosystem and are susceptible to changes even with minor change in their composition of biotic and abiotic components. Wetland birds have been attracted the attention of the public and scientists because of their beauty, abundance, visibility, and social behavior, as well as for their recreational and economic importance. Recently, they have become of interest as indicators of wetland quality, and as parameters of restoration success and regional biodiversity. Wetlands in Kerala are distributed all along the coast and in the inlands. According to Gopalan (1991), as much as two-third area of Vembanad Lake has been either reclaimed as land or converted into fields for agricultural and fishery activities. Wetlands in Kerala are mainly used for agriculture, aquaculture, reclamation for housing and industrial purposes, disposing the waste materials, discharging the industrial effluents and municipal waste water, wood seasoning, feeding waters for ducks, dumping dredged soil, coir retting and for fishing (Balachandran *et al.*, 2002). Wetlands in Kerala come under Central Asian-Indian flyway (Anon., 1996). During the annual migration, water birds halt at sites for very short periods to rest and feed and these 'stepping stones' are essential for their survival.

The ornithology of Kerala wetlands received attention after Neelakantan's extensive explorations (Neelakantan, 1969 and 1970; Neelakantan *et al.*, 1981; Neelakantan and Sureshkumar, 1981). Uthaman and Namasivayam (1991) explored the birds of Kadalundy. No detailed information on the wetland birds of the Kole wetlands of Thrissur was reported earlier, except a few reports of occurrence of species (Ravindran, 1993, 1995, 1999 and 2001; Jayson and Sivaperuman, 1999). Therefore, an attempt has been made to study the species diversity and abundance distribution patterns in the Kole wetlands of Thrissur, which is part of Vembanad-Kole Ramsar site.

Study Area

The study area located in Thrissur and Malappuram District of Kerala State, (10° 20'–10° 40' N, 75° 58'–76° 11' E) with an extent of 13,632 ha and it extends from the northern banks of Chalakudy River in the South to the southern banks of Bharathapuzha River in the North. Eastern side of Kole wetlands is Thrissur town and western side extends up to Arabian Sea (Figure 11.1). This wetland are low lying tracts located 0.5 to 1 m below MSL and it remains submerged for about six months in a year.

The climate of the area is moderate and there are three different distinct seasons. The dry season (December to April), wet season-I (May to August) during the period of southwest monsoon and wet season-II (September to November) during northeast monsoon. The average annual rainfall is 3,200 mm (James, 1983) and there is a variation in the temporal distribution of rainfall. Maximum rainfall is received during

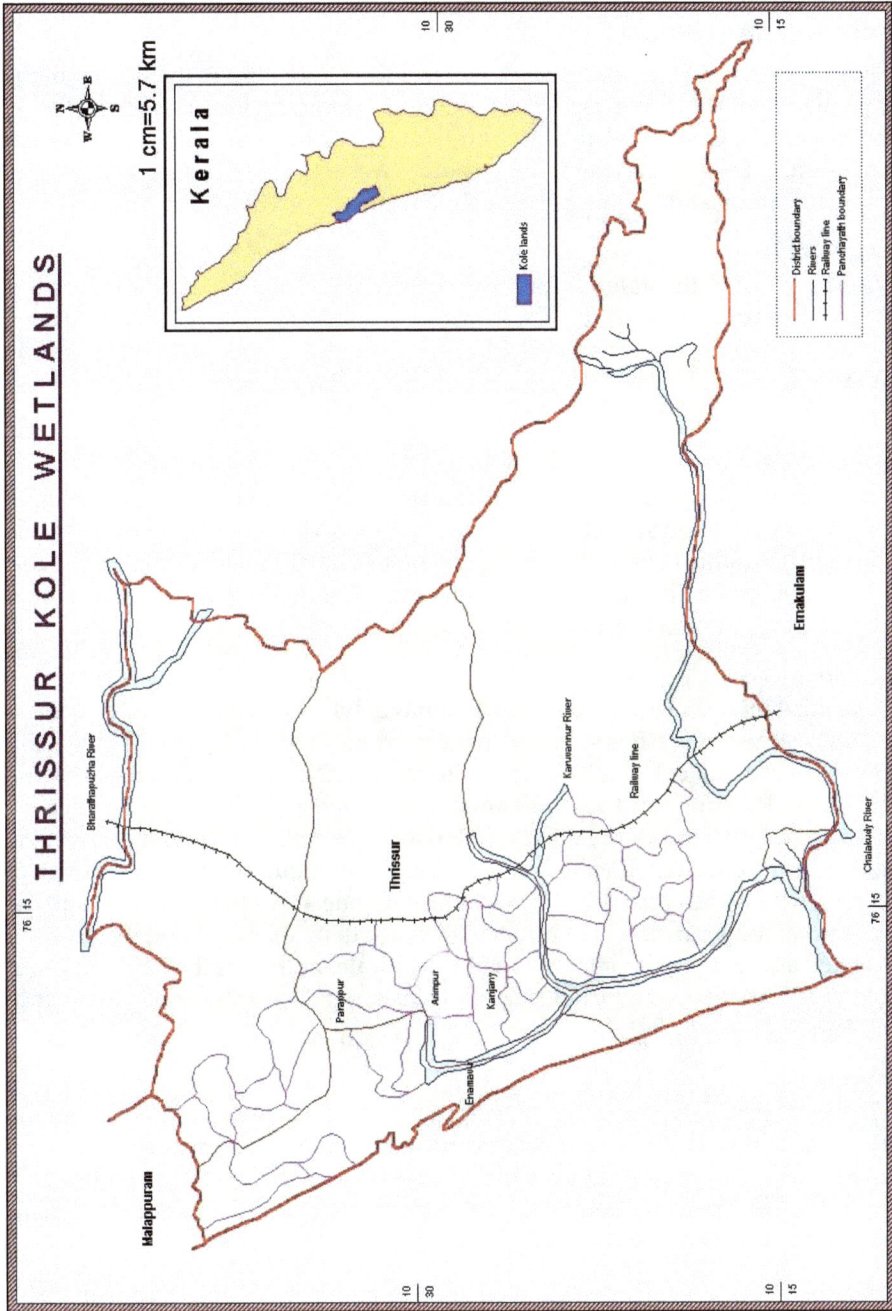

Figure 11.1: Kole Wetlands of Thrissur

the month of June followed by July. Extremes of heat or cold are not felt and the temperature varied from 28° C to 31.5° C. Atmosphere is always damp along the coastal belt due to high humidity.

Intensive Study Areas

The Kole wetlands are divided into many blocks for better management of the paddy cultivation. They are Adat, Aranattukara, Chettupuzha, Enamavu, Manakkodi, Mullurkayal, Palakkal, Pudukkad, Anthikad and Parappur. After a thorough reconnaissance survey, four intensive study areas were selected namely Chettupuzha, Kanjany, Enamavu and Parappur in the Kole wetlands of Thrissur.

Chettupuzha

Chettupuzha is the gateway of the Kole wetlands from Thrissur town and is located 10 km from West of the Thrissur town. Southern side of the wetlands is connected to the Kanjany Kole fields. Through the eastern side, the main irrigation canal is passing and the western side is adjoining with coconut plantations (Plate 11.1).

Kanjany

The second intensive study area at Kanjany is located in the central portion of Kole wetlands. Southern side of Kanjany extends up to Muriyad Kole of Irinjalakuda and the northern side is connected to the Enamavu. Eastern and western side of this Kole is adjoining with the human habitations and coconut plantations and one main canal is passing through the area (Plate 11.1).

Enamavu

The third intensive study area was Enamavu, which is located on the western end of the Kole wetlands. It is situated on the northern side of Kanjany and the main canal dividing this block finally empty into the sea. Northern side of this region extends up to Parappur and Adat. Enamakkal barrage was constructed about five decades ago to prevent salinity intrusion into the Kole wetlands from the Chettuva. Thrissur Kole wetlands used to have salinity intrusion through openings at Chettuva and Kottapuram. The regulator at Enamakkal and one at Kottenkottuvalavu in the lower reach of the Karuvannur River act both as a spill way for the flood waters from the wetland and as a regulator for preventing salt water entry into the Kole wetlands. Regulator is usually opened from June to August and closed during September to May (Plate 11.1).

Parappur

The fourth intensive study area, Parappur is located on the northern side of the Kole wetlands. It is continuous from Enamavu and is divided by Thrissur–Chavakkad main road and Thrissur–Guruvayur railway line. Northern side of the wetlands is connected to the Ponnani Kole. Small patch of coconut and rubber plantations is located in the middle of the wetlands (Plate 11.1).

Methods

The study was conducted from November 1998 to October 2001 and the bird population was estimated using the total count method (Hoves and Bakewell, 1989).

Kanjany

Chettupuzha

Parappur

Enamavu

Plate 11.1: Intensive Study Area in the Kole Wetlands of Thrissur

In this method, representative blocks were identified and birds in the blocks were counted using a telescope (15x–45x). The time of observation was from 07.00 h to 10.00 h. Community parameters like species richness, abundance, diversity indices, evenness indices, density of birds and similarity indices were calculated and presented for all the months. These parameters were also presented for each month and for each intensive study area by taking the mean values.

Collector's Curve

The collector's curve is used for assessing the sampling efficiency. This curve was drawn by plotting the cumulative number of species observed in each month against the month of observation (Pielou, 1975).

Species Richness and Abundance

Species richness indices like Margalef Index (R1) and Menhinick Index (R2) were calculated using the formula given by Magurran (1988).

Species Diversity Indices

Shannon-Weiner (H'), Simpson's (*l*), and Hill's diversity number N1 and N2 were calculated using the computer program SPDIVERS.BAS developed by Ludwig and Reynolds (1988).

Density

Density of birds in each month and individual density of selected species were calculated for the whole area and for the intensive study areas.

Similarity Indices

Similarity Indices between the intensive study areas were calculated using Jaccard Index, Sorenson Index (Magurran, 1988).

Distribution Models

Species-abundance model was constructed as explained in Magurran (1988). Species of birds were ranked in order of abundance, as represented by individuals seen for each species and this was plotted in decreasing order for all species against the number of individuals for the whole area.

Dominance Index

The dominance of the each bird species in the Kole wetlands was calculated using the dominance index.

Results

Collector's Curve

The collector's curve showed the sampling efficiency of the study. An increase in the number of species was recorded from the month of November 1998 to January 2001. After that, there was no change in the species richness, which indicated that the sampling was adequate (Figure 11.2).

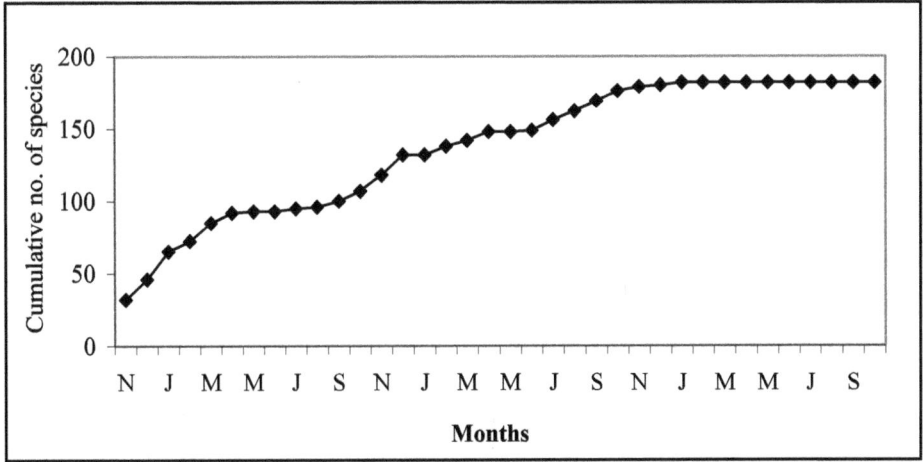

Figure 11.2: Collector's curve from the Kole Wetlands

Species Richness

The overall variation in the species richness during the study period is presented in Figure 11.3. Species richness increased during the migratory season (September–March) and decreased during the southwest monsoon (May–August).

Abundance of Birds

Total number of birds varied from 35 to 8,033 individuals in a month. The highest number of birds was observed during December and the lowest in July (Figure 11.4).

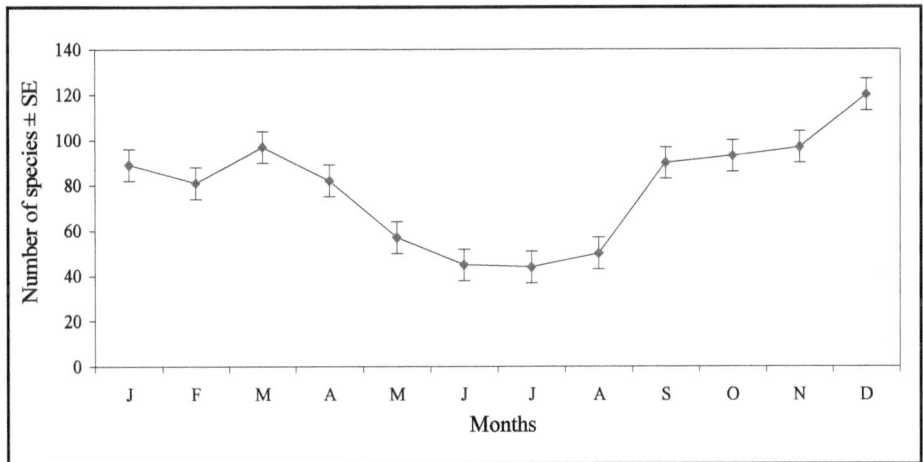

Figure 11.3: Species Richness of Birds in the Kole Wetlands (n = 36)

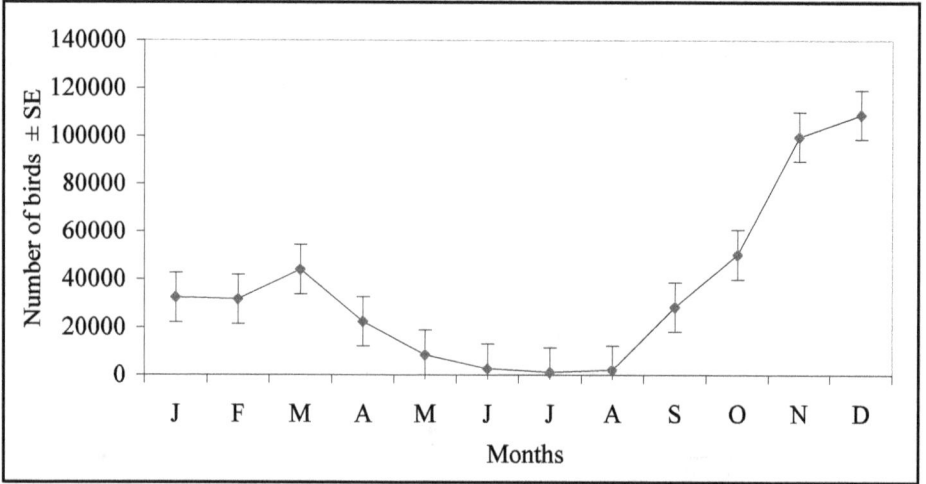

Figure 11.4: Total Number of Birds Recorded in the Kole Wetlands in Different Months (n = 36)

Diversity Indices

During the period of study, the highest diversity index (H') was recorded in the month of December (3.01), July (2.96) and the lowest H' value (2.11) was in October (Figure 11.5).

Density

Highest density of 29,158 birds/ha was recorded in December followed by November (24,373 birds/ha). Lowest density of birds was observed in July (272). Density of birds in different months and standard error is presented in Figure 11.6.

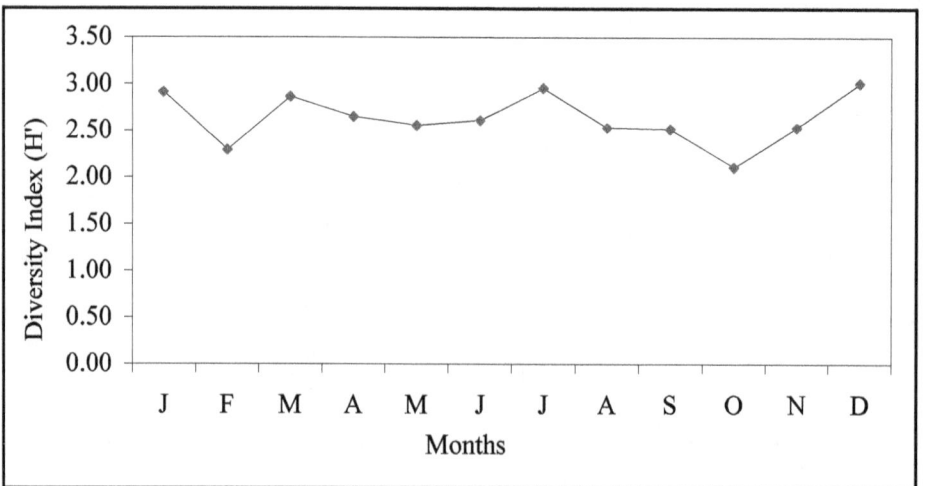

Figure 11.5: Diversity Index (H') in Different Months

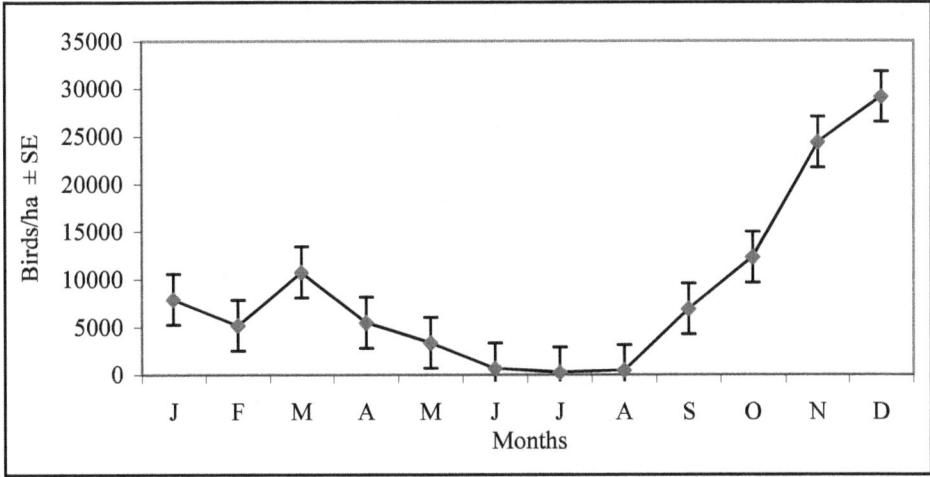

Figure 11.6: Density (Birds/ha) of Birds in Different Months in the Kole Wetlands (n = 36)

Species Richness in the Intensive Study Areas

Species richness of birds varied in different months in the intensive study areas. Highest number of species was recorded at Parappur during December (79) and lowest was at Chettupuzha during June (10). The number of species increased during the winter months and decreased during monsoon months (Table 11.1).

Table 11.1: Species Richness of Birds in the Intensive Study Areas

Intensive Study Areas	Months											
	J	F	M	A	M	J	J	A	S	O	N	D
Chettupuzha	30	28	20	25	17	10	17	21	22	28	39	30
Kanjany	58	58	63	51	28	17	22	27	36	45	65	76
Enamavu	32	21	37	25	29	14	21	20	46	36	30	27
Parappur	42	44	69	53	33	38	32	32	53	63	56	79

Abundance of Birds in the Intensive Study Areas

Total number of birds in the intensive study areas varied from 140 to 73,604 individuals in a month. Highest number of birds was recorded at Kanjany during December (73,604) and lowest was at Chettupuzha during July (140) (Table 11.2).

Diversity Indices (H') in the Intensive Study Areas

The highest diversity Index (H') was observed in June (2.90) at Parappur and the lowest diversity Index in February (1.53) at Parappur (Table 11.3). Similarly, the lowest and the highest diversity Index were recorded at Parappur.

Table 11.2: Abundance of Birds in the Intensive Study Areas

Intensive Study Areas	Months											
	J	F	M	A	M	J	J	A	S	O	N	D
Chettupuzha	1456	1680	982	753	460	208	140	301	733	544	9871	1230
Kanjany	25514	21501	20110	11012	2466	353	346	903	5881	11603	70683	73604
Enamavu	1994	707	4144	1531	4353	1453	156	408	20351	6864	4715	3649
Parappur	2912	7780	18799	9111	1274	739	486	425	1509	31146	13452	14380

Table 11.3: Diversity Indices (H') of Birds in the Intensive Study Areas

Intensive Study Areas	Months											
	J	F	M	A	M	J	J	A	S	O	N	D
Chettupuzha	2.33	2.20	2.02	2.37	2.03	1.84	2.08	2.14	2.00	2.61	2.15	2.43
Kanjany	2.81	2.51	2.37	2.02	2.36	2.39	2.64	1.90	2.40	1.87	2.28	2.84
Enamavu	2.31	2.18	2.41	1.99	2.04	1.62	2.51	2.18	2.22	2.16	2.04	1.98
Parappur	2.37	1.53	2.27	2.31	2.20	2.90	2.69	2.88	2.33	1.78	2.52	2.63

Similarity Indices between the Intensive Study Areas

Two similarity indices namely Jaccard Index and Sorenson Index were calculated. Similarity indices between the intensive study areas were computed using qualitative data (Table 11.4). Both indices showed highest similarity between Enamavu and Kanjany (0.62 and 0.76). Similarity between Chettupuzha, Kanjany and Parappur was less than fifty per cent.

Table 11.4: Jaccard Index and Sorenson Index Values (in parenthesis) for the Intensive Study Areas

Intensive Study Areas	Chettupuzha	Enamavu	Kanjany	Parappur
Chettupuzha	0 (0)	0.56 (0.72)	0.46 (0.62)	0.43 (0.60)
Enamavu		0 (0)	0.62 (0.76)	0.48 (0.52)
Kanjany			0 (0)	0.51 (0.67)
Parappur				0 (0)

Distribution Model

The distribution model indicates the absence of a single dominant species or group of species and the presence of long series of very rare species at Kole wetlands. The species, which is represented by less than 2 individuals, can be called as rare.

The observed and expected number of species was compared using the χ^2 goodness of fit test. The test showed that there is no significant difference between the observed and expected distribution ($\chi^2 = 18.31$; $P = 0.08$). Table 11.5 indicate that the bird community is following the truncated lognormal distribution pattern.

Table 11.5: Truncated Lognormal Distribution at Kole Wetlands (χ^2 test)

Class	Upper Boundary	Observed	Expected	χ^2
Behind veil line	0.5	–	7.21	–
1	2.5	7	14.92	4.20
2	4.5	16	9.15	5.13
3	8.5	14	12.51	0.18
4	16.5	18	15.09	0.56
5	32.5	13	18.02	1.40
6	64.5	19	20.34	0.09
7	128.5	16	20.01	0.80
8	256.5	15	20.08	1.29
9	512.5	14	17.49	0.70
10	1024.5	11	15.59	1.35
11	2048.5	9	12.04	0.77
12	4096.5	13	9.52	1.27
13	8192.5	5	6.74	0.45
14	∞	12	10.82	0.13
	Total	182	209.53	18.31

Overall bird community parameters are presented in Table 11.6. Diversity Index (H') was 3.11 and (λ) 0.08. Species Richness Index R1 was 13.96 and R2 was 0.28. Similarly, high values were obtained for Hill's number N1 and N2. Hill's number N1 was 22.38 and Hill's numbers N2 was 12.36. Evenness Index (E1) was 0.60 and (E2) 0.12.

Table 11.6: Bird Community Parameters in the Kole Wetlands

Species Richness Index		Shannon Index H'	Simpson's Index λ	Hill's Number		Evenness Index	
R1	R2			N1	N2	E1	E2
13.96	0.28	3.11	0.08	22.38	12.36	0.60	0.12

Discussion

Species richness and abundance of birds showed high values in the Kole wetlands, which is comparable to other wetland in India (Sampath, 1989; Nagarajan and Thiyagesan, 1996). The highest number of birds was recorded during November, which showed the influx of birds into the region due to the trans-continental migration. Population was low during June to July, when the migratory species were absent and the few resident species moved away to avoid the heavy rain. As the whole wetland lay inundated during this period, availability of food was also low. Only diving species like Little Cormorant, Darter and Little Grebe preferred the area during the months of southwest monsoon. Occurrence of 97 species of birds in a month is commendable, which showed the importance of the area for the migratory species. The increase in wetland species from September to March implies the presence of their preferred microhabitat and higher production of benthic and macro fauna. As reported earlier from the Western Ghats, highest number of birds was recorded during the winter and there was a reduction in population size during the monsoon (Daniels, 1989).

Species richness in an area is dependent on the availability of food, climate evolutionary history and predation pressure. Diversity indices are dependent on two factors, species richness and evenness. It is directly correlated with the stability of the ecosystem. It will be higher in the biologically controlled systems and low in disturbed ecosystem. In the Kole wetlands, diversity indices were higher. As the evenness measures also showed high values, it could be concluded that species of individuals uniformly presented and this indicated the conservation value of the wetlands.

Among the four intensive study sites, species richness and diversity was highest at Parappur and abundance and density of birds was highest at Kanjany. The geographic position of Kanjany is in the middle of the Kole wetlands. Many models are available for describing the species-abundance distribution. Preston (1948) introduced the lognormal distribution to explain the species-abundance data. Usually in ecological work, distribution of species is always truncated at the left side (Preston, 1962). Geometric series patterns are usually found in species poor or harsh environments. Log series patterns are usually observed where one or a few factors dominate the ecology of a community. Lognormal distribution is found in most biological populations. The Broken-stick model distribution shows the maximum equitable distribution of available resources. Species-abundance distribution at Kole wetlands followed the truncated lognormal model, which indicated the presence of natural bird community in the area.

A number of hypotheses have been forwarded to explain the characteristic diversity profiles of different habitats. Habitat heterogeneity, in addition to the area, is an important determinant of species richness. Shannon Index obtained for the area is comparable with other wetlands. Even though the total number of birds and species richness reduced during the southwest monsoon, it was not reflected in the diversity indices. All the study years showed high diversity index values. Evenness indices indicated high values during June to July, when the abundance of birds was lowest. High similarity was observed between Kanjany and Enamavu. This is because these two intensive study sites are continuous stretch of the Kole wetlands. High abundance of insectivores in this wetland showed the importance of wetlands to the insectivorous bird species. During the migratory seasons the more number of shorebirds was recorded in the mudflats and these are of regional importance to shorebirds in the Kole wetlands. The species richness and abundance of birds showed high value in the Kole wetlands. The Kole wetlands appear to serve as an important intermediate stopover sites for migratory birds.

References

1. Anon. (1996). Asia-Pacific Water bird Conservation Strategy, 1996–2000. Wetlands International Asia–Pacific, Kula Lumpur, Publication No. 117, and International Waterfowl and Wetlands Research Bureau–Japan Committee, Tokyo.

2. Balachandran, P.V., Mathew, G. and Peter, K.V. (2002). Wetland agriculture problems and prospects. In: *Wetland Conservation and Management in Kerala*, (Ed.) Jayakumar. State Committee on Science Technology and Environment, Thiruvananthapuram, pp. 50–68.

3. Daniels, R.J.R. (1989). A conservation strategy for the birds of the Uttara Kannada District. *Ph.D. Thesis*, Indian Institute of Science, Bangalore.

4. Hoves, J.G. and Bakewell, D. (1989). *Shore Bird Studies Manual*. AWB Publications No. 55 Kula Lumpur, 362 p.

5. James, E.J. (1983). Establishing a hydro-metrological data bank at CWRDS. In: *Proc. National Workshop on Scientific Methods to Collection and Documentation of Hydrologic Data*. CWRDM, Kozhikode, Kerala, pp. 29–35.

6. Jayson, E.A. and Sivaperuman, C. (1999). Kole lands of Thrissur: A threatened wetland ecosystem. *Evergreen*, 43: 10–11.

7. Ludwig, J.A. and Reynolds, J.F. (1988). *Statistical Ecology: A Premier on Methods and Computing*. A Wiley-Interscience Publication, 337 p.

8. Magurran, A.E. (1988). *Ecological Diversity and its Measurement*. Croom Helm Ltd., London, 179 p.

9. Nagarajan, R. and Thiyagesan, K., 1996. Waterbirds and substrate quality of the Pichavaram wetlands, Southern India. *Ibis*, 138: 710–721.

10. Neelakantan, K.K. and Sureshkumar, V.K. (1981). Occurrence of the Blackwinged Stilt *Himantopus himantopus* in Kerala. *J. Bombay Nat. Hist. Soc.*, 77(3): 510.

11. Neelakantan, K.K., Sreenivasan, K.V. and Sureshkumar, V.K. (1981). The Crab Plover *Dromas ardeola* in Kerala. *J. Bombay Nat. Hist. Soc.*, 77(3): 508.

12. Pielou, E.C. (1975). *Ecological Diversity*. A Wiley-Interscience Publication, John Wiley and Sons, New York.

13. Preston, F.W. (1948). The commonness and rarity of species. *Ecology*, 29: 254–283.

14. Preston, F.W. (1962). The canonical distribution of commonness and rarity. *Ecology*, 43: 185–215, 410–432.

15. Ravindran, P.K. (1993). Occurrence of the Glossy Ibis in Kole wetland, Thrissur District Kerala. *Newsl. Birdwatchers*, 33(6): 109.

16. Ravindran, P.K. (1994). Gadwall at Kadalundy Estuary, Kerala. *Newsl. Birdwatchers*, 34(2): 33.

17. Ravindran, P.K. (1995). The Kole Wetlands: An avian paradise in Kerala. *Newsl. Birdwatchers*, 35(1): 2–5.

18. Ravindran, P.K. (1999). Whitenecked stork in Kole wetlands. *Newsl. Birdwatchers*, 39(3): 51.

19. Ravindran, P.K. (2001). Occurrence of the White-winged Black Tern *Chlidonias leucopterus* in Kerala. *J. Bombay Nat. Hist. Soc.*, 98(1): 112–113.

20. Sampath, K. (1989). Studies on the ecology of Shorebirds (Aves: Charadriiformes) of the Great Vedaranyam Swamp and the Pichavaram Mangroves of India. *Ph.D. Dissertation*, Annamalai University, Tamil Nadu, 202 p.

21. Uthaman, P.K. and Namasivayam, L. (1991). The birdlife of Kadalundi Estuary. *Blackbuck*, 7(1): 3–11.

Biodiversity of Aquatic Reseources (2012)
Editors: **Mamta Rawat & Sumit Dookia**
Published by: **DAYA PUBLISHING HOUSE, NEW DELHI**

Pages 159–183

Chapter 12

Importance of Aquatic Avifauna in Southern Rajasthan, India

Sarita Mehra[1], Satya Prakash Mehra[1]*
and Krishan Kumar Sharma[2]

[1]*Rajputana Society of Natural History,*
Kesar Bhawan, 16/747, P.No.90, B/d Saraswati Hospital,
Ganeshnagar, Pahada, Udaipur – 313 001 Rajasthan
[2]*Laboratory of Biodiversity and Molecular Development,*
Department of Zoology, Maharshi Dayanand Saraswati University,
Ajmer – 305 009 Rajasthan

ABSTRACT

Rajasthan is the largest state of India area wise. Approximately sixty per cent area of Rajasthan is under desert ecosystem due to which the importance of water bodies is well known to the general public. World's oldest hill range, Aravalli, delimits the desert expansion on the eastern side. Amidst of Aravalli, southern parts of the state hold many water bodies of significance either in form of lakes, dams and rivers. Many aquatic lives are associated with these water bodies. Aquatic avifauna is one of the most important and eye-catching organisms among them. Southern Rajasthan has the key sites marked on the world tourism map especially due to lakes. Chittorgarh-Udaipur-Mt Abu forms a major point of attraction.

The paper documents the birds of the southern region of Rajasthan which are directly or indirectly dependent on the wetlands during the period of nine

* Corresponding Author E-mail: spmehra@yahoo.com

years (1999-2008). The importance of this part of state is tried to highlight through the presence of avifauna. Further, birding could be used as a tool for employment generation as well as conservation of wetlands in this belt of the State. Birdwatching could be encouraged in this area to uplift the socio-economic conditions of the tribal population.

The aquatic habitats of southern region of state are home for 111 species which represents 27 families of birds. 13 species of global interest are found in the wetlands of the study area.

Monitoring of water birds can provide valuable information on the status of wetlands and can be a key tool for increasing the awareness of importance of wetlands and conservation values.

Keywords: Wetland, Conservation, Rajasthan, Udaipur, Banas.

Introduction

Waterbirds play an important role in several spheres of human interest: culturally, socially, scientifically and as food resource (Pandey 1993, Manihar and Trisal 2001). They represent important components of our wetland ecosystems as they form important links in the food web and nutrient cycles (Chen and Zhang 1998). Many wetland species also play a role in the control of agricultural pests, whilst some species are themselves considered as pests of certain crops. After fish, birds are probably the most important faunal group that attracts people to wetlands. Many waterbirds are migratory, undertaking annual migrations along different flyways spanning the length and breadth of the globe between their breeding and non-breeding grounds (Ali 1959, Alerstam 1990).

According to the "Ramsar Convention", the waterbirds are broadly defined as "the birds depend on the wetlands for their living", including what we used to say: ducks and geese, shorebirds and waders and some other species depending on wetlands, such as kingfishers, raptors and some passerines. However, only 20 families of birds are included in the Species List of "Ramsar Convention". Except Glareolidae, they are all natatorial birds, wading birds and shorebirds. In these 20 families, there are at least 404 species occurring in Asia-Pacific region and 243 species of them are migratory birds (Anon., 2001).

Out of total 510 avian species of state, 155 species comprising approximately 30 per cent of the total number of birds found in Rajasthan depend on the wetlands (Ali and Ripley, 1968-1999; Grimmett *et al.*, 1999). Most are collectively called waterbirds and include the grebes (Podicipediformes); pelicans and cormorants (Pelecaniformes); herons, ibis, spoonbills and bitterns (Ciconiiformes); ducks, geese (Anseriformes); cranes, gallinules (Gruiformes); and waders, guls and terns (Charadriiformes). Species such as fish-eagles, osprey, harrier not usually called waterbirds, also depend on wetlands. There are also some waterbirds which have virtually lost any association with wetlands (plovers, pratincoles) (Ali and Ripley, 1968-1999; Grimmett *et al.*, 1999).

Alike all over the world aquatic birds attracted the attention of ornithologists, specialists on hunting management and hunters from the very past time in the princely state of Rajasthan (Adam 1873, Barnes 1891, Oates, 1899; Messurier, 1904; Impey, 1909; Whistler, 1938; Prakash, 1960; Kushlan, 1986). Keoladeo is a well known man-made wetland from the State. It was estimated that roughly 40118.4 sq. km is under 872 wetlands in India (Anon. 1990). The State of Rajasthan is one of the driest state of the country and the total surface water resources in the State is only about 1 per cent of the total surface water resources of the country. Nevertheless there are thousands of temporary freshwater and salt aquatic bodies in the region, varying enormously in size. 52 wetlands, including three natural, have been identified in state which expands in approx. 34 per cent of the geographic area of state (Anon., 1990). The rivers of the state are rainfed and identified by 14 major basins divided into 59 subbasins (Rajasthan Irrigation Department 2007).

Wetland is a complex ecosystem intermediate between purely aquatic and terrestrial systems where water is commonly accumulated to induce characteristic hydrophytes and anaerobic humic soils. IUCN (1971) defines wetlands as "area of submerged or water saturated lands, both natural or artificial, permanent or temporary, with water that is static or flowing, fresh, brackish or salty including area of marine water, the depth of which at low tide does not exceed 6m". Ecologically, wetlands may be viewed as complex hydrological and biogeochemical systems, endowed with specific structural and functional attributes (Gopal, 1973). There are many wetlands in the state which are well recognized. The different origin and ecological peculiarities of wetlands make up the typology of wetlands which are the main habitats of aquatic birds. The surface water play major role in providing the ground to aquatic birds along with its characteristics with respect to the food availability and protection.

The surface water resources in Rajasthan are mainly confined to south and southeastern part of the State. The paper deals with the southern part of the state with respect to the avifauna.

Study Area

The southern part of the State under study comprises of six districts of Udaipur region namely, Banswara, Chittorgarh, Dungarpur, Pratapgarh, Rajsamand and Udaipur and one district from Jodhpur region, namely, Sirohi (Figure 12.1). Pratapgarh was formed in 2008 and was taken along with Chittorgarh for the collection of observations.

Aquatic Bodies of Study Area

The study area forms 12.3 per cent total geographic area of the state. This part of the state has four major river basins, *viz.*, Mahi, Sabarmati, Banas and West Banas. Jakham Reservoir, Jaisamand Lake, Udaipur Lake Complex and Gambhiri are large sized (> 1000 ha) man-made wetlands of the area. Orai and Bhupalsagar are medium sized (500–999 ha) man-made wetlands of the area. Bhatewar, Kapasan, Murlia, Dingoli, Bankia, Borda, Soniyna, Saropa, Gadola and Baghela are small sized (100–499 ha) man-made wetlands of the area.

Besides above mentioned wetlands, there are many small less known lakes and dams: Rajsamand, Nandsamand in Rajsamand District; Udaisagar, Bari, Sei Dam,

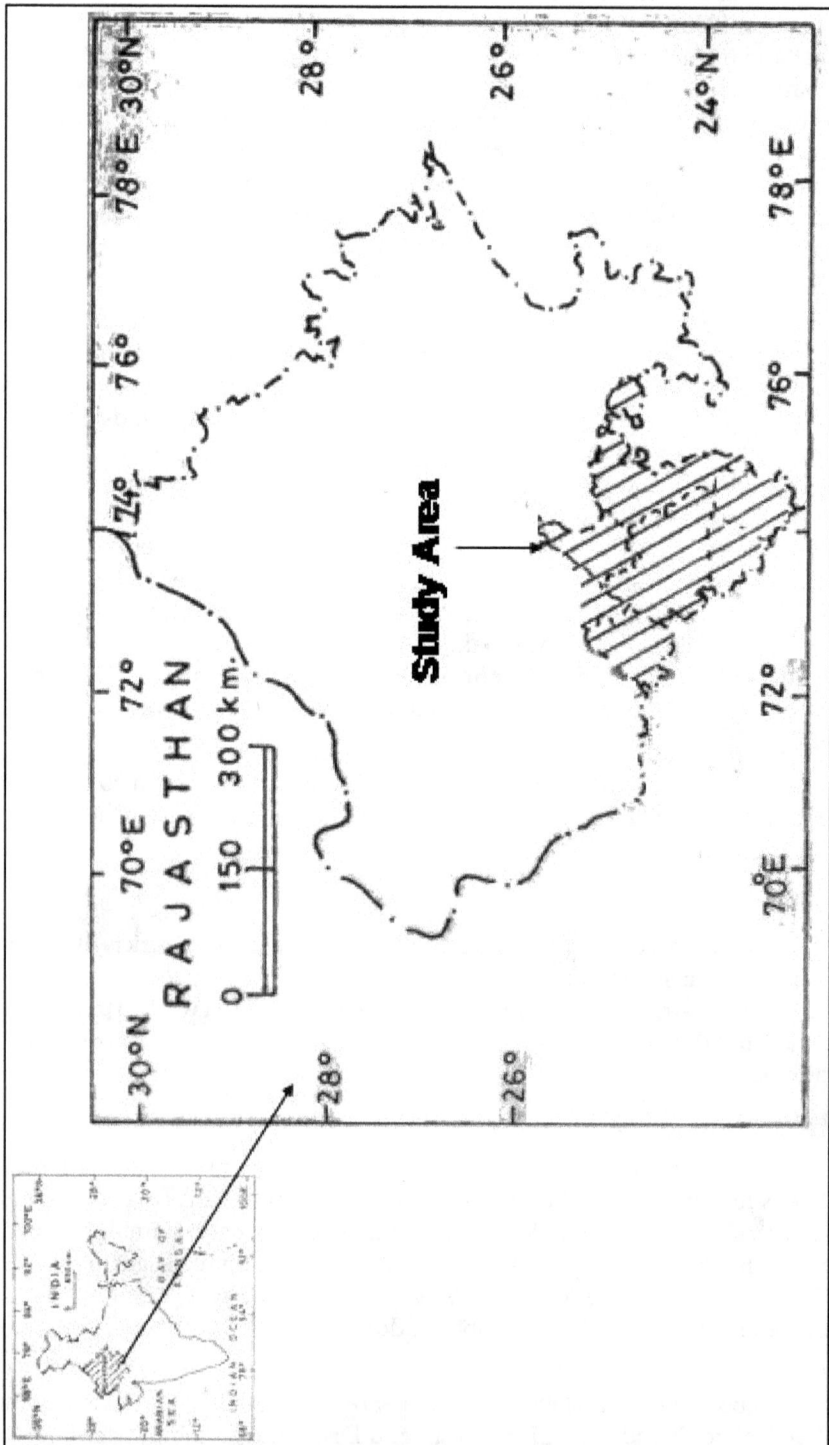

Figure 12.1: Location Map of Study Area

Jhadol Dam, Vallabhnagar Dam, Salumbar Dam, Bhatt *Talao*, Manpur, Tuta Madha Kothar, Jhighni, Gopalpur, Badora, Kejad, Obra Chalka, Badora Dam in Udaipur District; Gap Sagar and more than 70 small waterbodies in Dungarpur; wetlands created by back water of Mahi in Banswara; West Banas Dam, Ora Dam, Kalkaji *Talao*, Posindra Dam, Kadambari Dam, Tokra Dam, Kameri Dam etc. in Sirohi and small dams in Chittorgarh.

Demographic Status of Study Area

Total population of the state is 44,005,990 (State Census 2001). The study area is included in the Tribal Sub Plan (TSP) area. The Scheduled Tribe population of the state is 5,474,881 and its percentage to total population is 12.4 per cent. The Scheduled Tribe population in the Tribal sub plan area is 2401711 and its percentage to total population of TSP areas is 68.24 per cent.

Methodology

Assessment of Avifauna

The water birds of the study area were documented from 1999 to 2005 in winters and summer. Periodical studies, along with the involvement of volunteer local trained team members, were made to know the status and distribution of the water birds at the respective sites. After 2005, systematic sampling and observations were recorded for the major sites of interest. The surveys still continued in the whole of the area for the long term assessment. During the winter seasons, the frequency of visits by the team members was weekly whereas for other season the visits were monthly or bimonthly depending on the water availability in the aquatic bodies.

Taxonomic and ecological data on avifaunal species in the area were gathered systematically from July 2006 to March 2010. Different field methods were adapted as according to the species in question (Bibby *et al.*, 2000; Javed and Kaul, 2002; Urfi *et al.*, 2005). Total count method for selected wetland habitats were used to assess the bird species from 2006 onwards.

Waterbirds were identified with the help of field guides Grimmett *et al.* (1999), Kazmierczak (2000) and Grimmett *et al.* (2004). The scientific names and classification were used as according to Manakadan and Pittie (2001).

The local members were trained by the specialists to identify the waterbirds. The waterbird census was started from 2005-06 to monitor their population. The total count method is used for the census. The census was taken thrice in one season for selected wetlands.

The terms used for the status were as according:

Very common (VC): when the species has high frequency of sighting (>80 per cent occurrence and/or sighting during the total surveys).

Common (C): when the species has moderate frequency of sighting (>50 per cent but <80 per cent occurrence and/or sighting during the total surveys).

Less Common (LC): refers when the species has low frequency of sighting (>10 per cent but <50 per cent occurrence and/or sighting during the total surveys).

Rare (R): when the species has very low instances of sighting (<10 per cent occurrence and/or sighting during the total surveys)

Assessment of Characteristics of Selected Birding Sites

The characteristics of the selected wetlands were assessed under following heads to rate its importance for the local residents in terms of socio-economic upliftment.

Selected aquatic habitats were considered for the assessment which includes:

Banswara–Mahi Bajaj Sagar, discrete wetlands; Chittorgarh–Bassi, Ghosunda, Sitamata, discrete wetlands; Dungarpur–Urban wetlands which include Gap Sagar and Sabela, discrete wetlands; Rajsamand–Rajsamand, discrete wetlands; Sirohi–West Banas, Ora, discrete wetlands; Udaipur–Udaipur Lake Complex which include Pichola, Fatehsagar and linked canals, Udaisagar, discrete wetlands.

Accessibility Level

The accessibility level of human denotes the approach of site for a common man. The assessment included three points–(a) Nearness from urban residential area (0-10 kms), (b) Frequency of use by local residents (>5 times a week); and (c) Ownership–public property and/or permitted site for common man. Based on these three criteria, rating on human accessibility was given:

1. Sites fulfilling all the three points–Rating 1
2. Sites fulfilling either points 'a' and 'c' or 'b' and 'c'–Rating 2
3. Site does not fulfilling point 'c' and/or all the three points–Rating 3.

Importance Level

The assessment of the importance of the site was made through interactions with at least 50 visitors (local and/or tourist) per season found at particular site on the direct or indirect benefits for respondent:

Based on the Use of Local Community

1. Frequently used–Rating 1
2. Occasionally used–Rating 2
3. Rarely used–Rating 3

Based on the Use of Global Community

1. Frequently used–Rating 1
2. Occasionally used–Rating 2
3. Rarely used–Rating 3

Based on the Potential of Birding

1. Could be hotspot for birding–Rating 1
2. Could be used alternative site for birding–Rating 2
3. Least important for birding–Rating 3

Results

Status and Occurrence of Birds

About 111 bird species representing 27 families were recorded in different aquatic habitats of southern Rajasthan, with one doubtful record (Tables 12.1A and 12.1B). Out of total aquatic species, 90 species were wetland species and 21 species were wetland dependent species. More than 100 species were recorded from aquatic habitats of Chittorgarh, Sirohi and Udaipur whereas near about 100 species were recorded from aquatic habitats of Banswara, Dungarpur and Rajsamand.

Eleven species namely, Little Grebe, Little Egret, Indian Pond-Heron, White-breasted Waterhen, Red-wattled Lapwing, Black-winged Stilt, River Tern, White-breasted Kingfisher, Wire-tailed Swallow, Red-rumped Swallow and Streak-throated Swallow were among the most common species recorded from all over the study area in all surveys whereas 58 species were commoners with limited sightings. 34 species were less commonly distributed and six species were rarely showed their presence in aquatic habitats of study area. One species, namely, Yellow-legged Gull were doubtful sighting in 2005 from West Banas (Sirohi) in 2005.

Four aquatic bird species *viz.* Spot-billed Pelican from Udaipur Lake Complex, Falcated Duck from waterbodies of Dungarpur, Eurasian Woodcock from Mount Abu and Indian Skimmer from Mount Abu and Udaipur were either reported in the past and or reported by other workers from the study area (Table 12.1C).

Fifteen species were breeding resident of the area, 25 species were breeding resident showing local movement and one species breeding resident showed its winter migration. Nine species were non-breeding residents showing local movement, six species non-breeding resident showed winter migration and one species was non-breeding resident showing both winter migration as well as local movement. 54 aquatic species recorded from the area were winter migrants.

Globally Threatened Species

Thirteen aquatic species of global importance were recorded from the habitats of study area (Table 12.2). Out of these thirteen, three species *viz.* Dalmatian Pelican, Lesser White-fronted Goose and Sarus Crane are listed as vulnerable whereas remaining ten species, namely Darter, Painted Stork, Black-necked Stork, Oriental White Ibis, Lesser Flamingo, Ferruginous Pochard, Lesser Grey-headed Fish-Eagle, Black-tailed Godwit, Eurasian Curlew and Black-bellied Tern are listed in near threatened category.

Eight globally threatened species including two vulnerable species–Dalmatian Pelican, Sarus Crane and six near threatened species–Darter Painted Stork, Black-necked Stork, Oriental White Ibis, Ferruginous Pochard and Black-tailed Godwit showed their presence in all the districts. Lesser White-fronted Goose was sighted only in year 2007 from the small wetland of Deograh in Rajsamand whereas Lesser Grey-headed Fish Eagle was sighted in Sitamata Forests in Chittorgrah since 2005. Lesser Flamingo was recorded from Mangalwar *talao* and Ghosunda Dam of Chittorgarh; wetlands of Dungarpur; West Banas and Ora dams of Sirohi; and

Table 12.1A: Aquatic (Wetland) Bird Species Recorded from Districts of Southern Rajasthan

Sl.No.	Common Name Scientific Name	Status with Notes on Sighting Records	Distribution Records (During study period)					
			BAN	CTR	DGR	RJS	SIR	UDR
A.	**Grebes Podicipedidae**							
1.	Little Grebe (5) *Tachybaptus ruficollis* (Pallas, 1764)	VC, BR; all districts	x	x	x	x	x	x
2.	Great Crested Grebe (3) *Podiceps cristatus* (Linnaeus, 1758)	LC, WM; recorded only at Rajsamand Lake (RJS) and Baghdarra (UDR) (31/12/2006)	–	–	–	x	–	x
B.	**Pelicans Pelecanidae**							
3.	Great White Pelican (20) *Pelecanus onocrotalus* (Linnaeus, 1758)	C, WM; all districts	x	x	x	x	x	x
4.	Dalmatian Pelican (22) *Pelecanus crispus* (Bruch, 1832)	LC, WM, all districts	x	x	x	x	x	x
C.	**Cormorants/Shags Phalacrocoracidae**							
5.	Little Cormorant (28) *Phalacrocorax niger* (Vieillot, 1817)	C, BR/LM; all districts with restricted breeding sites (BAN, DGR)	x	x	x	x	x	x
6.	Indian Shag (27) *Phalacrocorax fuscicollis* (Stephens, 1826)	LC, NBR/LM; all districts with restricted breeding sites (BAN, DGR)	x	x	x	x	x	x
7.	Great Cormorant (26) *Phalacrocorax carbo* (Linnaeus, 1758)	C, BR/LM; all districts with restricted breeding sites (BAN, DGR, SIR)	x	x	x	x	x	x
D.	**Darters Anhingidae**							
8.	Darter (29) *Anhinga melanogaster* (Pennant, 1769)	LC, BR/LM; all districts with restricted breeding sites (BAN, DGR)	x	x	x	x	x	x
E.	**Herons, Egrets and Bitterns Ardeidae**							
9.	Little Egret (49) *Egretta garzetta* (Linnaeus, 1766)	VC, BR; all districts	x	x	x	x	x	x
10.	Grey Heron (35-36) *Ardea cinerea* (Linnaeus, 1758)	C, NBR/WM; all districts	x	x	x	x	x	x
11.	Purple Heron (37-37a) *Ardea purpurea* (Linnaeus, 1766)	C, BR/LM; all districts	x	x	x	x	x	x

Contd...

Table 12.1A—Contd...

Sl.No.	Common Name Scientific Name	Status with Notes on Sighting Records	Distribution Records (During study period)					
			BAN	CTR	DGR	RJS	SIR	UDR
12.	Large Egret (45-46) *Casmerodius albus* (Linnaeus, 1758)	C, BR/LM; all districts	x	x	x	x	x	x
13.	Median Egret (47, 48) *Mesophoyx intermedia* (Wagler, 1829)	C, BR/LM; all districts	x	x	x	x	x	x
14.	Cattle Egret (44) *Bubulcus ibis* (Linnaeus, 1758)	C, BR; all districts	x	x	x	x	x	x
15.	Indian Pond-Heron (42-42a) *Ardeola grayii* (Sykes, 1832)	VC, BR/LM; all districts	x	x	x	x	x	x
16.	Little Green Heron (38-41) *Butorides striatus* (Linnaeus, 1758)	LC, BR/LM; all districts	x	x	x	x	x	x
17.	Black-crowned Night-Heron (52) *Nycticorax nycticorax* (Linnaeus, 1758)	LC, BR/LM; recorded from all the districts	x	x	x	x	x	x
F.	**Storks Ciconiidae**							
18	Painted Stork (60) *Mycteria leucocephala* (Pennant, 1769)	C, BR/LM; all districts with restricted breeding sites (BAN, CTR, DGR)	x	x	x	x	x	x
19.	Asian Openbill-Stork (61) *Anastomus oscitans* (Boddaert, 1783)	C, BR/LM; recorded from all the districts with restricted breeding sites (BAN, CTR, DGR, UDR)	x	x	x	x	x	x
20.	Black Stork (65) *Ciconia nigra* (Linnaeus, 1758)	R, WM; West Banas (SIR) (19/01/2008), Bassi (CTR) and different sites in BAN	x	x	–	–	x	–
21.	White-necked Stork (62) *Ciconia episcopus* (Boddaert, 1783)	C, BR/LM; all districts with restricted breeding sites (BAN, CTR, DGR, SIR)	x	x	x	x	x	x
22.	Black-necked Stork (66) *Ephippiorhynchus asiaticus* (Latham, 1790)	LC, NBR/LM; restricted sites in all districts	x	x	x	x	x	x

Contd...

Table 12.1A–Contd...

Sl.No.	Common Name Scientific Name	Status with Notes on Sighting Records	Distribution Records (During study period)					
			BAN	CTR	DGR	RJS	SIR	UDR
G.	**Ibises and Spoonbills Threskiornithidae**							
23.	Glossy Ibis (71) *Plegadis falcinellus* (Linnaeus, 1766)	LC, NBR/WM/LM; restricted sites in all districts	×	×	×	×	×	×
24.	Oriental White Ibis (69) *Threskiornis melanocephalus* (Latham, 1790)	C, BR/LM; all districts with restricted breeding sites (BAN, CTR, DGR, UDR)	×	×	×	×	×	×
25.	Black Ibis (70) *Pseudibis papillosa* (Temminck, 1824)	C, BR/LM; all districts	×	×	×	×	×	×
26.	Eurasian Spoonbill (72) *Platalea leucorodia* (Linnaeus, 1758)	C, BR/LM; all districts with restricted breeding sites (BAN, DGR)	×	×	×	×	×	×
H.	**Flamingos Phoenicopteridae**							
27.	Greater Flamingo (73) *Phoenicopterus ruber* Linnaeus, 1758	C, NBR/WM; all districts	×	×	×	×	×	×
28.	Lesser Flamingo (74) *Phoenicopterus minor* (Geoffroy, 1798)	R, WM; recorded from different sites in DGR, CTR, West Banas (SIR) and Udaisagar (UDR)	–	×	×	–	×	×
I.	**Geese and Ducks Anatidae**							
29.	Lesser Whistling-Duck (88) *Dendrocygna javanica* (Horsfield, 1821)	C, NBR/LM; all districts	×	×	×	×	×	×
30.	Lesser White-fronted Goose (80) *Anser erythropus* (Linnaeus, 1758)	R, WM; small wetland Deogarh (RJS) in 2007	–	–	–	×	–	–
31.	Greylag Goose (81) *Anser anser* (Linnaeus, 1758)	C, WM; restricted to wetlands of BAN, CTR, DGR and UDR	×	×	×	–	–	×
32.	Bar-headed Goose (82) *Anser indicus* (Latham, 1790)	C, WM; all districts	×	×	×	×	×	×
33.	Brahminy Shelduck (90) *Tadorna ferruginea* (Pallas, 1764)	C, WM; all districts	×	×	×	×	×	×

Contd...

Table 12.1A–Contd...

Sl.No.	Common Name Scientific Name	Status with Notes on Sighting Records	Distribution Records (During study period)					
			BAN	CTR	DGR	RJS	SIR	UDR
34.	Comb Duck (115) Sarkidiornis melanotos (Pennant, 1769)	C, BR/LM; all districts	x	x	x	x	x	x
35.	Cotton Teal (114) Nettapus coromandelianus (Gmelin, 1789)	C, NBR/LM; all districts	x	x	x	x	x	x
36.	Gadwall (101) Anas strepera (Linnaeus, 1758)	C, WM; all districts	x	x	x	x	x	x
37.	Eurasian Wigeon (103) Anas penelope (Linnaeus, 1758)	C, WM; all districts	x	x	x	x	x	x
38.	Mallard (100) Anas platyrhynchos (Linnaeus, 1758)	LC, WM; all districts	x	x	x	x	x	x
39.	Spot-billed Duck (97-99) Anas poecilorhyncha (J.R. Forester, 1781)	C, BR/LM; all districts	x	x	x	x	x	x
40.	Northern Shoveller (105) Anas clypeata (Linnaeus, 1758)	C, WM; all districts	x	x	x	x	x	x
41.	Northern Pintail (93) Anas acuta (Linnaeus, 1758)	C, WM; all districts	x	x	x	x	x	x
42.	Garganey (104) Anas querquedula (Linnaeus, 1758)	LC, WM; all districts	x	x	x	x	x	x
43.	Common Teal (94) Anas crecca (Linnaeus, 1758)	C, WM; all districts	x	x	x	x	x	x
44.	Red-crested Pochard (107) Rhodonessa rufina (Pallas, 1773)	LC, WM; except CTR all districts	x	–	x	x	x	x
45.	Common Pochard (108) Aythya ferina (Linnaeus, 1758)	C, WM; all districts	x	x	x	x	x	x
46.	Ferruginous Pochard (109) Aythya nyroca (Guldenstadt, 1770)	C, WM; all districts	x	x	x	x	x	x
47.	Tufted Pochard (111) Aythya fuligula (Linnaeus, 1758)	LC, WM; all districts	x	x	x	x	x	x
J.	**Cranes Gruidae**							
48.	Sarus Crane (323-324) Grus antigone (Linnaeus, 1758)	C, BR/LM; all districts	x	x	x	x	x	x
49.	Demoiselle Crane (326) Grus virgo (Linnaeus, 1758)	LC, WM; few sites, large congregation at Ora village (SIR)	x	x	–	–	x	x

Contd...

Table 12.1A–Contd...

Sl.No.	Common Name Scientific Name	Status with Notes on Sighting Records	Distribution Records (During study period)					
			BAN	CTR	DGR	RJS	SIR	UDR
50.	Common Crane (320) *Grus grus* (Linnaeus, 1758)	LC, WM; few sites	×	×	×	–	×	×
K.	**Rails, Crakes, Moorhens, Coots Rallidae**							
51.	Brown Crake (342) *Amaurornis akool* (Sykes, 1832)	LC, NBR/LM; few sites	×	×	×	×	×	×
52.	White-breasted Waterhen (343-345) *Amaurornis phoenicurus* (Pennant, 1769)	VC, BR; all districts	×	×	×	×	×	×
53.	Purple Moorhen (348-349) *Porphyrio porphyrio* (Linnaeus, 1758)	C, BR/LM; all districts	×	×	×	×	×	×
54.	Common Moorhen (347-347a) *Gallinula chloropus* (Linnaeus, 1758)	C, BR/LM; all districts	×	×	×	×	×	×
55.	Common Coot (350) *Fulica atra* (Linnaeus, 1758)	C, NBR/WM; all districts	×	×	×	×	×	×
L.	**Jacanas Jacanidae**							
56.	Pheasant-tailed Jacana (358) *Hydrophasianus chirurgus* (Scopoli, 1786)	C, BR/LM; all districts	×	×	×	×	×	×
57.	Bronze-winged Jacana (359) *Metopidius indicus* (Latham, 1790)	C, BR/LM; all districts	×	×	×	×	×	×
M.	**Painted-Snipes Rostratulidae**							
58.	Greater Painted-Snipe (429) *Rostratula benghalensis* (Linnaeus, 1758)	LC, NBR/LM; all districts	×	×	×	×	×	×
N.	**Plovers, Lapwings Charadriidae**							
59.	Little Ringed Plover (379-380) *Charadrius dubius* (Scopoli, 1786)	C, BR/WM; all districts	×	×	×	×	×	×
60.	Kentish Plover (381-382) *Charadrius alexandrinus* (Linnaeus, 1758)	LC, WM; few sites	–	×	–	–	×	×

Contd...

Table 12.1A–Contd...

Sl.No.	Common Name Scientific Name	Status with Notes on Sighting Records	Distribution Records (During study period)					
			BAN	CTR	DGR	RJS	SIR	UDR
61.	Yellow-wattled Lapwing (370) *Vanellus malabaricus* (Boddaert, 1783)	C, BR/LM; all districts	x	x	x	x	x	x
62.	Red-wattled Lapwing (366-368) *Vanellus indicus* (Boddaert, 1783)	VC, BR; all districts	x	x	x	x	x	x
63.	White-tailed Lapwing (362) *Vanellus leucurus* (Lichtenstein, 1823)	LC, WM; few sites	x	x	x	–	x	x
O.	**Sandpipers, Stints, Snipes, Godwits and Curlews Scolopacidae**							
64.	Pintail Snipe (406) *Gallinago stenura* (Bonaparte, 1830)	LC, WM; all districts	x	x	x	x	x	x
65.	Common Snipe (409) *Gallinago gallinago* (Linnaeus, 1758)	C, NBR/WM; restricted sites of all districts	x	x	x	x	x	x
66.	Jack Snipe (410) *Lymnocryptes minimus* (Brünnich, 1764)	LC, WM; restricted sites of all districts	x	x	x	x	x	x
67.	Black-tailed Godwit (389-390) *Limosa limosa* (Linnaeus, 1758)	C, WM; all districts	x	x	x	x	x	x
68.	Bar-tailed Godwit (391-391a) *Limosa lapponica* (Linnaeus, 1758)	LC, WM; only from CTR, SIR, UDR	–	x	–	–	x	x
69.	Eurasian Curlew (387-388) *Numenius arquata* (Linnaeus, 1758)	LC, WM; only from restricted sites of CTR, DGR, SIR	–	x	x	–	x	–
70.	Spotted Redshank (392) *Tringa erythropus* (Pallas, 1764)	C, WM; all districts	x	x	x	x	x	x
71.	Common Redshank (393, 394) *Tringa totanus* (Linnaeus, 1758)	C, WM; all districts	x	x	x	x	x	x
72.	Marsh Sandpiper (395) *Tringa stagnatilis* (Bechstein, 1803)	C, WM; all districts	x	x	x	x	x	x

Contd...

Table 12.1A–Contd...

Sl.No.	Common Name Scientific Name	Status with Notes on Sighting Records	Distribution Records (During study period)					
			BAN	CTR	DGR	RJS	SIR	UDR
73.	Common Greenshank (396) *Tringa nebularia* (Gunner, 1767)	LC, WM; all districts	x	x	x	x	x	x
74.	Green Sandpiper (397) *Tringa ochropus* (Linnaeus, 1758)	C, WM; all districts	x	x	x	x	x	x
75.	Wood Sandpiper (398) *Tringa glareola* (Linnaeus, 1758)	C, WM; all districts	x	x	x	x	x	x
76.	Common Sandpiper (401) *Actitis hypoleucos* (Linnaeus, 1758)	C, NBR/WM; all districts	x	x	x	x	x	x
77.	Little Stint (416) *Calidris minuta* (Leisler, 1812)	C, WM; all districts	x	x	x	x	x	x
78.	Temminck's Stint (417) *Calidris temminckii* (Leisler, 1812)	LC, WM; all districts	x	x	x	x	x	x
79.	Ruff (426) *Philomachus pugnax* (Linnaeus, 1758)	C, WM; all districts	x	x	x	x	x	x
P.	**Avocets and Stilts Recurvirostridae**							
80.	Black-winged Stilt (430-431) *Himantopus himantopus* (Linnaeus, 1758)	VC, BR; all districts	x	x	x	x	x	x
81.	Pied Avocet (432) *Recurvirostra avosetta* (Linnaeus, 1758)	LC, WM; all districts	x	x	x	x	x	x
Q.	**Stone-Curlew and Stone-Plovers/Thick-knees Burhinidae**							
82.	Great Stone-Plover (437) *Esacus recurvirostris* (Cuvier, 1829)	C, BR/LM; all districts	x	x	x	x	x	x
R.	**Coursers and Pratincoles Glareolidae**							
83.	Small Pratincole (444) *Glareola lactea* (Temminck, 1820)	LC, BR/LM; all districts	x	x	x	x	x	x
S.	**Gulls, Terns Laridae**							
84.	Yellow-legged Gull (451) *Larus cachinnans* (Pallas, 1811)	WM; Doubtful record from West Banas (SIR) (Dec 2005)	–	–	–	–	x	–

Contd...

Table 12.1A—Contd...

Sl.No.	Common Name Scientific Name	Status with Notes on Sighting Records	Distribution Records (During study period)					
			BAN	CTR	DGR	RJS	SIR	UDR
85.	Brown-headed Gull (454) *Larus brunnicephalus* (Jerdon, 1840)	LC, WM; restricted to few sites of CTR (Ghosunda), RJS (Nand-samand), SIR (West Banas)	–	×	–	×	×	–
86.	Black-headed Gull (455) *Larus ridibundus* (Linnaeus, 1766)	LC, WM; restricted to few sites of RJS (Nandsamand), SIR (West Banas), Pichola (UDR)	–	–	–	×	×	×
87.	Gull-billed Tern (460-461) *Gelochelidon nilotica* (Gmelin, 1789)	LC, WM; restricted to few sites	×	×	×	–	×	×
88.	River Tern (463) *Sterna aurantia* (J.E. Gray, 1831)	VC, BR; all districts	×	×	×	×	×	×
89.	Black-bellied Tern (470) *Sterna acuticauda* J.E. (Gray, 1831)	LC, WM; restricted to few sites	×	×	×	–	×	×
90.	Whiskered Tern (458) *Chlidonias hybridus* (Pallas, 1811)	LC, WM; restricted to few sites	×	×	×	–	×	×
	Total		81	85	81	77	87	85

Table 12.1B: Aquatic (Wetland Dependent) Bird Species Recorded from Districts of Southern Rajasthan

SI.No.	Common Name Scientific Name	Status with Notes on Sighting Records	Distribution Records (During study period)					
			BAN	CTR	DGR	RJS	SIR	UDR
A.	**Hawks, Eagles, Buzzards, Old World Vultures, Kites, Harriers Accipitridae**							
1.	Brahminy Kite (135) *Haliastur indus* (Boddaert, 1783)	LC, NBR/LM; all districts	×	×	×	×	×	×
2.	Lesser Grey-headed Fish-Eagle *Ichthyophaga humilis* (S. Muller and Schlegel, 1841)	R, BR; only record from Sitamata Wildlife Sanctuary (CTR)	–	×	–	–	–	–
3.	Western Marsh-Harrier (193) *Circus aeruginosus* (Linnaeus, 1758)	C, WM; all districts	×	×	×	×	×	×
4.	Steppe Eagle (169) *Aquila nipalensis* Hodgson, 1833	C, WM; all districts	×	×	×	×	×	×
B.	**Osprey Pandionidae**							
5.	Osprey (203) *Pandion haliaetus* (Linnaeus, 1758)	C, WM; all districts	×	×	×	×	×	×
C.	**Falcons Falconidae**							
6.	Peregrine Falcon (209-211) *Falco peregrinus* Tunstall, 1771	R, NBR/LM; sighted only in Ghosunda (CTR), West Banas 9SIR), Lake Pichola (UDR, Dec 2005)	–	×	–	–	×	×
D.	**Owls Strigidae**							
7.	Brown Fish-Owl (631-632) *Ketupa zeylonensis* (Gmelin, 1788)	LC, NBR/LM; only in Ghosunda (CTR), West Banas (SIR), Udaisagar (UDR)	–	×	–	–	×	×
E.	**Kingfishers Alcedinidae**							
8.	Small Blue Kingfisher (722-724) *Alcedo atthis* (Linnaeus, 1758)	C, BR; all districts	×	×	×	×	×	×
9.	Stork-billed Kingfisher (730-732) *Halcyon capensis* (Linnaeus, 1766)	R, BR; only in restricted sites	×	×	×	–	–	–
10.	White-breasted Kingfisher (735-738) *Halcyon smyrnensis* (Linnaeus, 1758)	VC, BR; all districts	×	×	×	×	×	×
11.	Lesser Pied Kingfisher (719-720) *Ceryle rudis* (Linnaeus, 1758)	C, BR; all districts	×	×	×	×	×	×

Contd...

Table 12.1B–Contd...

Sl.No.	Common Name Scientific Name	Status with Notes on Sighting Records	Distribution Records (During study period)					
			BAN	CTR	DGR	RJS	SIR	UDR
F.	**Bee-eaters Meropidae**							
12.	Blue-tailed Bee-eater (748) *Merops philippinus* (Linnaeus, 1766)	LC, WM; all districts	×		×	×	×	×
G.	**Swallows and Martins Hirundinidae**							
13.	Common Swallow (916-918) *Hirundo rustica* (Linnaeus, 1758)	LC, WM; all districts	×	×	×	×	×	×
14.	Wire-tailed Swallow (921) *Hirundo smithii* (Leach, 1818)	VC, BR; all districts	×	×	×	×	×	×
15.	Red-rumped Swallow (923-928) *Hirundo daurica* (Linnaeus, 1771)	VC, BR; all districts	×	×	×	×	×	×
16.	Streak-throated Swallow (922) *Hirundo fluvicola* (Blyth, 1855)	VC, BR; all districts	×	×	×	×	×	×
H.	**Wagtails and Pipits Motacillidae**							
17.	White Wagtail (1885-1890) *Motacilla alba* (Linnaeus, 1758)	C, WM; all districts	×	×	×	×	×	×
18.	Large Pied Wagtail (1891) *Motacilla maderaspatensis* (Gmelin, 1789)	C, NBR/WM; all districts	×	×	×	×	×	×
19.	Citrine Wagtail (1881-1883) *Motacilla citreola* (Pallas, 1776)	C, WM; all districts	×	×	×	×	×	×
20.	Yellow Wagtail (1875-1880) *Motacilla flava* (Linnaeus, 1758)	C, WM; all districts	×	×	×	×	×	×
21.	Grey Wagtail (1884) *Motacilla cinerea* (Tunstall, 1771)	C, WM; all districts	×	×	×	×	×	×
	Total		18	21	18	17	19	19

(225-256): Numbers within brackets after the common names are the numbers given to species in Ripley's (1982) Synopsis, which was also followed in Ali and Ripley's Handbook

BAN: Banswara; CTR: Chittorgarh (including Pratapgarh); DGR: Dungarpur; RJS: Rajsamand; SIR: Sirohi; UDR: Udaipur.

VC: Very Common; C: Common; LC: Less Common; R: Rare; BR: Breeding Resident; NBR: Non-Breeding Resident; LM: Local Movement; WM: Winter Migrant; PM: Passage Migration; SM: Summer Migrant.

Table 12.1C: Aquatic Bird Species with Past or Doubtful Records from Districts of Southern Rajasthan

Sl.No.	Common Name Scientific Name	Status with Notes on Sighting Records	Distribution Records (During study period)					
			BAN	CTR	DGR	RJS	SIR	UDR
A.	**Pelicans Pelecanidae**							
1	Spot-billed Pelican (21) *Pelecanus philippensis* (Gmelin, 1789)	?, WM, reported from Udaipur Lake Complex before 1996 (R. Tehsin, *pers. comm.,* 2001)	–	–	–	–	–	x
B.	**Geese and Ducks Anatidae**							
2.	Falcated Duck (102) *Anas falcata* (Georgi, 1775)	?, ?; presence was documented in historical literature of Dungarpur	–	–	x	–	–	–
C.	**Sandpipers, Stints, Snipes, Godwits and Curlews Scolopacidae**							
3.	Eurasian Woodcock (411) *Scolopax rusticola* (Linnaeus, 1758)	?, ?; past record from Mount Abu (Shivrajkumar, 1949)	–	–	–	–	x	–
D.	**Skimmers Rynchopidae**							
4.	Indian Skimmer (484) *Rynchops albicollis* (Swainson, 1838)	?, ?; past record from Mount Abu and recently from Udaipur	–	–	–	–	x	x

(225-256): Numbers within brackets after the common names are the numbers given to species in Ripley's (1982) Synopsis, which was also followed in Ali and Ripley's Handbook

BAN: Banswara; CTR: Chittorgarh (including Pratapgarh); DGR: Dungarpur; RJS: Rajsamand; SIR: Sirohi; UDR: Udaipur.

Udaisagar and Bhatwariya *talao* of Udaipur. Single or two individuals of Eurasian Curlew was sighted once in Mahi backwater (Banswara) in December 2007 whereas it was frequently recorded from Ghosunda (Chittorgarh); different small wetlands (*talao*) of Dungarpur; and West Bans and Ora dams in Sirohi.

Table 12.2: Bird Species of Global Importance from Districts of Southern Rajasthan

Sl.No.	Common Name Scientific Name	Distribution Record
A.	**Vulnerable**	
1.	Dalmatian Pelican (22) *Pelecanus crispus* Bruch, 1832	All districts
2.	Lesser White-fronted Goose (80) *Anser erythropus* (Linnaeus, 1758)	Rajsamand
3.	Sarus Crane *Grus antigone*	All districts
B.	**Near Threatened**	
4.	Darter *Anhinga melanogaster* Pennant, 1769	All districts
5.	Painted Stork *Mycteria leucocephala* (Pennant, 1769)	All districts
6.	Black-necked Stork *Ephippiorhynchus asiaticus* (Latham, 1790)	All districts
7.	Oriental White Ibis *Threskiornis melanocephalus* (Latham, 1790)	All districts
8.	Lesser Flamingo *Phoenicopterus minor* (Geoffroy, 1798)	All districts except Banswara and Rajsamand
9.	Ferruginous Pochard *Aythya nyroca* (Guldenstadt, 1770)	All districts
10.	Lesser Grey-headed Fish-Eagle *Ichthyophaga humilis* (S. Muller and Schlegel, 1841)	Chittorgarh
11.	Black-tailed Godwit (389-390) *Limosa limosa* (Linnaeus, 1758)	All districts
12.	Eurasian Curlew (387-388) *Numenius arquata* (Linnaeus, 1758)	Banswara, Chittorgarh, Dungarpur, Sirohi
13.	Black-bellied Tern (470) *Sterna acuticauda* J.E. Gray, 1831	All districts except Rajsamand

Source: IUCN 2009.

Assessment of Birding Sites

While assessing the characteristics of the sites for birding, it was found that wetlands within urban settlements in form of lakes were most accessible (Table 12.3). These include urban wetlands of Dungarpur (Gap Sagar and Sabela), Rajsamand (Rajsamand Lake) and Udaipur (Urban Lake Complex). The forest wetlands *viz*. Bassi and Sitamata of Chittorgarh were the sites which were not accessible for the local urban residents for their routine recreational activities. Rest all the sites were accessible for the nearby urban residents.

It was further observed that the accessibility level affected the importance level of the particular site for the residential urban population. The most accessible sites, namely lakes of Dungarpur, Rajsamand and Udaipur within urban limits were of great importance for the urban residents (Table 12.3). The discrete wetlands sites in all the districts were of least importance for the urban residents except in Dungarpur (Table 12.3). Globally almost all the sites except few wetland sites in districts of

Table 12.3: Assessment and Rating of Birding Sites (Aquatic Habitats) of Southern Rajasthan

Characteristics / Districts →		BAN		CTR				DGR		RJS		SIR				UDR		
	Sites →	MBS	BDW	B	G	S	CDW	UW	DDW	RS	RDW	WB	O	SDW	ULC	U	J	UDW
Accessibility Level		2	2	3	2	3	2	1	2	1	2	2	2	2	1	2	2	2
Importance Level	Local	2	3	2	2	2	3	1	2	1	3	2	2	1	1	2	2	3
	Global	3	3	3	3	2	3	2	3	2	3	3	3	1	1	2	1	3
Potential and Scope		2	2	2	2	2	1	1	2	1	2	1	1	2	1	1	2	2

Banswara (BAN)–Mahi Bajaj Sagar (MBS), discrete wetlands (BDW); Chittorgarh (CTR)–Bassi (B), Ghosunda (G), Sitamata (S), discrete wetlands (CDW); Dungarpur (DGR)–Urban wetlands (UW) which include Gap Sagar and Sabela, discrete wetlands (DDW); Rajsamand (RJS)–Rajsamand (RJ), discrete wetlands (RDW); Sirohi (SIR)–West Banas (WB), Ora (O), discrete wetlands (SDW); Udaipur (UDR)–Udaipur Lake Complex (ULC) which include Pichola, Fatehsagar and linked canals, Udaisagar (U), Jaisamand (J), discrete wetlands (UDW).

Dungarpur, Rajsamand and Udaipur, were of least importance. Due to importance of lakes for tourism, Udaipur Lake Complex and Jaismand were the only sites of global importance (Table 12.3).

Thus, based on the above characteristics through questionnaire surveys, it was found that almost all the sites have the potential of developing birding areas but the main site include wetlands of Chittorgarh, Dungarpur, Rajsamand, Sirohi and Udaipur (Table 12.3).

Discussion

Rajasthan State has the long history of ornithological studies. The major studies were confined to the north-eastern and western part of the state. Keoladeo National Park, Bharatpur is well known world heritage site of birds especially for Siberian Cranes *Grus leucogeranus*. Western desert region has its own importance in the avifaunal composition and could be well studied due to Great Indian Bustard *Ardeotis nigriceps*. If considering the southern region of the state, the main sites which had been studied from the very past are Abu Hills (Sirohi), and Lakes of Udaipur. Butler (1875-1877) documented the avifauna of Abu (Sirohi) including few of the waterbirds. Hume (1878) pioneered the aquatic avifaunal explorations in the southern Rajasthan. He documented important species of waterbirds from Jaisamand (Dhebar) Lake and Kankroli *Talao* from Udaipur. Since then no major documentation was done by any of the workers. Although few records of the waterbirds were found in some books related to hunting of animals by the princely family members (*e.g.* Tanwar, 1956). Such type of documentation only presents the group of waterbirds–not the species so it is hard to assess the bird species of the respective period. After a long gap, scientific documentation was done by Tehsin (1989) in which he report 66 wetland birds from Udaipur Lake Complex. Sharma (1998) documented some of the wetlands birds around Sajjangarh Wildlife Sanctuary. Sharma and Tehsin (1994) first documented the avifaunal checklist from the southern Rajasthan. With these detail listings the reporting of individual species related to wetlands from different parts were also continued (Tehsin 1987, 1997, 1999). Sharma (2002), Mehra *et al.* (2010) attempted to document almost all the waterbirds of southern Rajasthan whereas Islam and Rahmani (2004) made an attempt to document all the sites of southern Rajasthan which are important with respect to birds of global interest listed by Birdlife International (2001). Udaipur Lake Complex, Sei Dam, Jaisamand Lake and Baghdarrah are the important sites with respect to birds of international concern and were identified as Important Bird Areas (IBAs) (Islam and Rahmani, 2004).

Wetlands were the first major ecosystem to be protected by an international treaty to stop the decline of waterfowl populations which was then linked to habitat loss (Amezaga *et al.*, 2002). Wetlands around the world that support high concentrations of waterbirds are well known (Alerstam, 1990; Finlayson and Moser, 1991; Perrennou and Mundkur, 1991). The number of waterbirds using a particular habitat is related to types and quality of habitats, abundance and availability of food, and level of disturbance (Mukherjee, 1969–1976; Krishnan, 1978; Pandit, 1982; Gopakumar, 1990; Green, 1996; Hafner, 1997).

Wetlands often support many different species of water birds. Anatomical adaptations, bill and leg shape of the waterbirds ensure that a wide variety of different food types are accessible in wetlands (Perrins, 1990). Significance of sustainable use has been emphasized in the global initiative, the convention on Biological Diversity 1992, for the protection of the planet and its biota, of which wetland organisms have not the least share (Baldassarre and Bolen, 1994) and not to mention the resource of great demand, the waterbirds. Monitoring of waterbirds can provide valuable information on the status of wetlands (Custer *et al.*, 1991; Kushlan, 1993), and can be a key tool for increasing the awareness of importance of wetlands and conservation values. There is growing concern of the need to conserve waterbirds and wetlands and recognition that birds can serve as indicators of the health of our surroundings (Anon. 2001).

Conclusions

Wetland supports variety of bird species with a wide variety of feeding and breeding habits. Among 110 species of waterbirds recorded from southern Rajasthan, 19 species were from family Anatidae, It is the group which is largely attracting the humans to observe aquatic avifauna. The wetlands of Chittorgarh, Dungarpur, Rajsamand, Sirohi and Udaipur were most important sites having the potential of birding which could be developed as an alternative source of income for locals involving the tribal youth.

Conservation through Community Participation

There is always a conflict between protection of habitats and human involvement. Those tribal belts of study areas which are living from past many decades evolved parallal with wild life and found natural means of their conservation. But uncontrolled urbanization has forced both tribal habitats as well as biodiversity in a situation that both the components are struggling for their existence. There is need to bring the concept of conserving wild life as well as local residents so that they live together symbiotically. This would be a community based nature conservation that is coming up very successful in many parts of the globe.

Acknowledgement

We are thankful to Laboratory of Biodiversity and Rajputana Academy of Natural History for their help and support provided in the work. We acknowledge all those people who directly and indirectly helped us in providing information on the presence of avifauna of interest in their nearby waterbodies. Special thanks to Mr. G. K. Mehra for their comments and inputs in the write up.

References

1. Adam, R.M. (1873). Notes on birds of Sambhar lake and its vicinity. *Stray Feathers*, 1: 361–404.

2. Alerstam, T. (1990). *Bird Migration*. Cambridge University Press, Cambridge, U.K., 420 pp.

3. Ali, S. (1959). Local movements of resident waterbirds. *J. Bombay Nat. Hist. Soc.,* 56(2): 346–347

4. Ali, S. and Ripley, S.D. (1968–1999). *Handbook of the Birds of India and Pakistan,* 10 Volume. Oxford University Press, Mumbai.

5. Amezaga, J.M., Santamaria, L. and Green, A.J. (2002). Biotic wetland connectivity: Supporting a new approach for wetland policy. *Acta Oecologica,* 23: 213–222.

6. Anonymous (1990). *Wetlands of India: A Directory.* Ministry of Environment and Forests, Govt. of India, New Delhi.

7. Anonymous (2001). *Asia-Pacific Migratory Waterbird Conservation Strategy: 2001–2005.* Asia-Pacific Migratory Waterbird Conservation Committee-2001. Wetlands International-Asia Pacific, Kuala Lumpur, Malaysia, 67pp.

8. Baldassarre, C.A. and Bolen, E.G. (1994). *Waterfowl Ecology and Wildlife Management.* University of Minnesota Press, Minneapolis.

9. Barnes, H.E. (1891). Nesting in Western India. *J. Bombay Nat. Hist. Soc.,* 6(3): 285–317.

10. Bibby, C., Burgess, N.D., Hill, D.A. and Mustoe, S.H. (2000). *Bird Census Techniques,* 2nd Edition. Academic Press, London, xvii+302 pp.

11. BirdLife International (2001). *Threatened Birds of Asia: The BirdLife International Red Data Book.* BirdLife International, Cambridge, UK.

12. Butler, E.A. (1875–1877). Notes on the avifauna of Mount Aboo and Northern Guzerat. *Stray Feathers,* 3: 437–500; 4: 1–41; 5: 207–235.

13. Chen Kelin and Zhang Guixin (Eds.) (1998). Wetland and waterbird conservation. In: *Proceedings of an International Workshop on Wetland and Waterbird Conservation in North East Asia.* Wetland International-China Programme. China Forestry Publishing House, Beijing, China, 294pp.

14. Clement, A., Tisdell, A. and Zhu Xiang (1996). Tourism development and nature conservation in Xishuanbanna, Yunnan: A case study. *Tiger Paper,* 23(2): 20–28.

15. Custer, T.W., Rattner, B.A., Ohlendorf, H.M. and Melancon, M.J. (1991). Herons and egrets proposed as indicators of estuarine contamination in the United States. In: *Proceedings of the International Ornithological Congress,* 20: 2474–2479.

16. Finlayson, C.M. and Moser, M. (1991). *Wetlands.* International Waterfowl and Wetlands Research Bureau, Simbridge. Toucan Books, London, 224 pp.

17. Gopakumar, G. (1990). Habitat utilization pattern of waterbirds in artificial wetlands of an arid to semi-arid region. *Indian J. For.,* 13(2): 85–91.

18. Gopal, B. (1973). A survey of the Indian studies on ecology and production of wetlands and shallowwater communities. *Pol. Arch. Hydrobiol.,* 2: 21–30.

19. Green, A.J. (1996). Analyses of globally threatened anatidae in relation to threats, distribution, migration patterns and habitat use. *Conserv. Biol.,* 10: 1435–1445.

20. Grimmett, R., Inskipp, C. and Inskipp, T. (1999). *Birds of the Indian Subcontinent.* Oxford University Press, New Delhi.

21. Grimmett, R., Inskipp, T. and Mehra, S.P. (2004). *Uttar Bharat Ke Pakshi* (in Hindi). Bombay Natural Histroy Society, Mumbai.

22. Hafner, H. (1997). Ecology of wading birds. *Colonial Waterbirds*, 20(1): 115–120.

23. Hume, A.O. (1878). A lake in Oodeypore. *Stray Feathers*, 7(1–2): 95–99.

24. Impey, L. (1909). Duck shooting in Rajputana. *J. Bombay Nat. Hist. Soc.*, 19(3): 750–751.

25. Islam, M.Z. and Rahmani, A.R. (2004). *Important Bird Areas in India: Priority Sites for Conservation*. Indian Bird Conservation Network: Bombay Natural History Society and BirdLife International (UK), xviii+1133.

26. IUCN (2009). *IUCN Red List of Threatened Species*. Version 2009.1. <www.iucnredlist.org>. Accessed on 22 July 2009.

27. Javed, S. and Kaul, R. (2002). *Field Methods for Bird Surveys*. Bombay Natural History Society; Department of Wildlife Sciences, Aligarh Muslim University, Aligarh and World Pheasant Association, South Asia Regional Office (SARO), New Delhi, India.

28. Kazmierczak, K. and van Perlo, B. (2000). *A Field Guide to the Birds of India*. Om Book Service, New Delhi.

29. Krishnan, M. (1978). The availability of nesting materials and nesting sites as vital factors in the gregarious breeding of Indian waterbirds. *J. Bombay Nat. Hist. Soc.*, 75(Supp): 1143–1152.

30. Kushlan, J.A. (1993). The management of wetlands for aquatic birds. *Colonial Waterbirds*, 9(2): 246–248.

31. Kushlan, J.A. (1993). Colonial waterbirds as bioindicators of environmental change. *Colonial Waterbirds*, 16: 223–251.

32. Manakadan, R. and Pittie, A. (2001). Standardised common and scientific names of the birds of the Indian subcontinent. *Buceros*, 6(1): ix+37.

33. Manihar, T. and Trisal, C.L. (2001). Sustainable development and water resources management of Loktak lake. In: *Asian Wetland Symposium 2001: Bringing Partnerships into Good Wetland Practices*, 27–30 August, Penang, Malaysia, 27pp.

34. Mehra, S., Mehra, S.P. and Sharma, K.K. (2010). Aquatic avifauna: Its importance for wetland conservation in Rajasthan, India. In: *Proceedings of Conservation of Lakes and Water Resources: Management Strategies*, February 19–20, 202–214 pp.

35. Messurier, A. le (1904). *Game, Shore, and Water Birds of India*. W. Thacker and Co., London.

36. Mukherjee, A.K. (1969–1976). Food habits of waterbirds of the Sundarban, 24-Parganas district, West Bengal, India, I–VI. *J. Bombay Nat. Hist. Soc.*, 66(2): 345–360, 68(1): 37–64, 68(3): 691–716, 71(2): 188–200, 72(2): 422–447, 73(3): 482–486.

37. Oates, E.W. (1899). *A Manual of the Game Birds of India, Vol. 2: Water Birds*, 2 Vols. Messrs. A.J. Combridge and Co., Bombay, 506 pp.

38. Pandey, S. (1993). *The Importance of a Man-made Reservoir in India for Conserving Waterbird Diversity.* Wetland and Waterfowl Conservation in South and West Asia, IWRB Spec. Publ. No. 25, AWB Publ. No. 85.

39. Pandit, A.K. (1982). Feeding ecology of breeding birds in five wetlands of Kashmir. *Indian J. Ecol.,* 9: 181–190.

40. Perrins, C.M. (1990). *The Illustrated Encyclopedia of Birds.* Marshall Editions Development, Ltd., London.

41. Perrennou, C. and Mundkur, T. (1991). *Asian Waterfowl Census 1991.* International Waterfowl and Wetlands Research Bureau, Simbridge, U.K., 84pp.

42. Prakash, I. (1960). Shikar in Rajasthan. *Cheetal,* 2(2): 68–72.

43. Rajasthan Irrigation Department (2007). State water policy derived from http: //www.rajirrigation.gov.in/5need.htm on 25 April 2007.

44. Sharma, S.K. (1998). Avian fauna of Sajjangarh wildlife sanctuary. *Newsletter for Birdwatchers,* 38(2): 25–27.

45. Sharma, S.K. (2002). Preliminary biodiversity survey of protected areas of southern Rajasthan, p. 1–23. Unpublished Report.

46. Sharma, S.K. and Tehsin, R. (1994). Birds of southern Rajasthan. *Newsletter for Birdwatchers,* 34(5): 109–113.

47. Shivrajkumar, Y.S. (1949). Occurrence of the Woodcock (*Scolopax rusticola* L.) at Mount Abu. *J. Bombay Nat. Hist. Soc.,* 48(3): 585.

48. Tanwar, D.T. (1956). *Shikari aur Shikar* (in Hindi).Udaipur, Mewar.

49. Tehsin, R. (1987). Migrating Demoiselle Cranes *Anthropoides virgo. Tiger Paper* 14(4): 26.

50. Tehsin, R. (1989). Faunal history of Fatehsagar lake. In: *Wetland Conservation,* (Eds.) L.N. Vyas and R.K. Garg. Environment Community Centre, Udaipur, pp. 109–117.

51. Tehsin, R.H. (1997). Little Green Heron *Butorides striatus* and White-eared Bulbul *Pycnonotus leucogenys* sighted in Southern Rajasthan. *Newsletter for Birdwatchers,* 37(5): 91.

52. Tehsin, R.H. (1999). Threetoed Kingfisher *Ceyx erithacus* Sighted at Panarwa. *J. Bombay Nat. Hist. Soc.,* 96(1): 142–143.

53. Urfi, A.J., Sen, M., Kalam, A. and Meganathan, T., (2005). Counting birds in India: Methodologies and trends. *Current Science,* 89(12): 1997–2003.

Part II

Pollution:
Monitoring and Management

Biodiversity of Aquatic Reseources (2012)
Editors: **Mamta Rawat & Sumit Dookia**
Published by: **DAYA PUBLISHING HOUSE, NEW DELHI**

Pages **187-192**

Chapter 13

Monitoring of Yamuna River Water for Assessing its Fitness in Supporting Aquatic Biota

Shalu Mittal*, Sangeet Prabha and P.K. Agarwal**

Department of Zoology,
B.S.A. (P.G.) College, Mathura, U.P.

ABSTRACT

Analysis of important Physico-chemical parameters was carried out to investigate the state of pollution in the holy river Yamuna at Mathura. The duration of study was from January 2008 to December 2008.Two major sites were selected for the sampling. The higher values of BOD (182 mg/l) and ammonia (22.02 mg/l) but a very low value of DO (1.2 mg/l) indicates a deplorable condition of the river which is unsafe for aquatic biota and human use. Findings suggest some major remedial steps should be taken immediately to maintain the water quality of the holy river.

Keywords: Pollutants, pH, DO, BOD, Sewage.

Introduction

Being the birth place of lord Krishna, Mathura (Uttar Pradesh) finds a religious mention in the Hindu Mythology. Yamuna River flows through the centre of this city.

E-mail: *shalumittalgarg@gmail.com, **dragarwalpk@gmail.com

Due to its religious importance, a large number of pilgrims from all over the country visit Mathura every year and take bath in this holy river. Due to mass bathing on important festivals, the problem of sewage and garbage disposal becomes uncontrolled and deplorable. Moreover, Mathura is a growing industrial city, so a large numbers of small and large industries (particularly saree printing and silver metallurgy ones) are developed in its vicinity. The harmful and non-biodegradable chemicals used by these industries are later discharged as effluents into the river Yamuna through dozens of wide drains. The municipal sewage mixed with industrial effluents is muddy-blackish in color and has pungent and irritating smell. Such a waste, when discharged into the river, greatly alters the normal physico-chemical characteristics of the water, which in turn adversely affects the aquatic life and the river ecosystem. Hence, the purpose of this work is to study the river water to asses and analyze its pollution status and to suggest some remedial measures.

Materials and Methods

To analyze the water of river Yamuna, two sampling sites A and B were identified.

1. Site A–near AIR (All India Radio station.)
2. Site B–At Swami Ghat.

The Water sampling was done, by Immerson type of sampling method, at an interval of 30 days (on every first Sunday morning 08.00 AM). The temperature was measured with a digital thermometer on the spot. On the spot, colour estimation was done on site whereas hydrogen ion concentration (pH) was observed with the help of portable digital pH meter. Other parameters like Dissolved Oxygen, turbidity, total hardness, ammonia, nitrate and chloride were done in the laboratory according to standard methods prescribed by APHA (1989). For Biological Oxygen Demand (BOD), the samples were taken in BOD bottles.

Observation and Discussion

The monthly data for various physico-chemical parameters at sampling sites A and B is given in Tables 13.1 and 13.2 respectively.

Temperature

It is one of the most important physical factors, which regulates the natural processes in the environment. It was found in accordance with seasonal changes. It was higher, ranging between 31°C to 40°C in the summers and lower during winters, ranging between 13°C to 14.5°C.

Turbidity

Turbidity is one of the common ways to measure the extent of pollution. It is generally caused by untreated sewage and industrial waste. It was very high during period of Hindu festivals because of mass bathing by a large crowd from all over the country. It ranged between 80.1 to 103.2 NTU.

Physico-chemical Examination of River Yamuna at Mathura (U.P.)

Table 13.1: Physico-chemical Parameters at Site-A

Parameters	Units	Jan.	Feb.	March	April	May	June	July	Aug.	Sep.	Oct.	Nov.	Dec.	Min.	Max.
Temperature	°C	13	16	17	23	32	40	38	33	28	25	18	15	13	40
Turbidity	NTU	80.4	83.6	81.6	86.4	85.2	95.4	91.6	98.4	92.7	103.2	84.6	81.0	80.4	98.4
pH	-	8.2	8.6	8.3	8.4	8.5	8.6	8.45	7.5	8.0	8.5	8.1	7.6	7.50	8.60
DO	mg/l	2.37	2.16	2.32	2.07	1.82	1.29	1.39	1.73	1.54	1.79	2.4	2.29	1.29	2.37
BOD	mg/l	98	116	101	126	148	182	165	146	155	132	114	99	98	182
Total hardness	mg/l	421	415	740	497	610	709	607	563	494	510	472	520	415	740
Ammonia	mg/l	12.12	16.21	15.74	18.10	22.02	16.21	19.24	21.12	13.19	14.10	15.40	5.92	5.92	22.02
Nitrates	mg/l	18.24	20.34	16.21	12.35	10.08	18.21	16.24	20.06	16.18	16.30	19.65	15.08	10.08	20.34
Chlorides	mg/l	550	515	560	618	670	712	682	650	570	515	576	482	482	712

Table 13.2: Physico-chemical Parameters at Site-B

Parameters	Units	Jan.	Feb.	March	April	May	June	July	Aug.	Sep.	Oct.	Nov.	Dec.	Min.	Max.
Temp	°C	15	18	21	25	29	39	38.6	31	26	27	19	14.5	14.5	39.0
Turbidity	NTU	80.1	83.4	81.3	86.3	85.2	95.0	92.4	99.6	92.4	99.6	91.7	103.2	80.1	99.6
pH	-	8.3	8.1	8.4	7.6	8.2	8.7	8.5	8.6	8.1	8.5	8.2	8.0	7.6	8.7
DO	mg/l	2.41	2.26	2.06	2.7	1.72	1.2	1.05	1.17	1.92	1.96	1.72	2.31	1.2	2.7
BOD	mg/l	105	111	129	107	135	175	163	151	125	132	140	120	105	175
Total hardness	mg/l	473	532	512	578	625	690	625	578	560	601	538	495	473	690
Ammonia	mg/l	15.25	16.27	18.01	12.41	18.04	20.54	19.41	18.04	19.08	16.09	13.17	11.65	12.4	20.5
Nitrates	mg/l	18.21	16.92	14.04	18.41	11.40	10.5	12.43	13.02	11.78	14.24	18.23	19.8	10.5	19.8
Chloride	mg/l	555	482	662	525	639	705	670	575	485	592	528	580	482	705

Hydrogen Ion Concentration (pH)

It is an important indicator that shows the acidic or alkaline nature of water. The water of Yamuna River was slightly alkaline, ranging from 7.6 to 8.7. It showed positive correlation with all the parameters except the D.O. (Dakshini, 1979)

Dissolved Oxygen

Oxygen is the important factor that supports the aquatic life and self purification capacity of water bodies as it is essential for the decomposition (oxidation) of chemical waste and organic matter. In the present study, dissolved oxygen was observed very less during summer season ranging from 1.05mg/l to 1.2 mg/l. The most probable reason for this observance was due to the low solubility of gases at high temperature (Klein, 1973).

BOD

BOD is the direct indication of the pollution in the water body. High BOD values indicate low availability of oxygen to the aquatic fauna since it is utilized in the biological oxidation of waste by microbes. The BOD was very high, 175 mg/l (site B) and 182 mg/l (site A) in summers than in winters.

Total Hardness

The water of Yamuna was found to be very hard with respect to total hardness. Calcium hardness was much higher than the magnesium hardness, possibly on account of waste water from the small scale industries. Furthermore, the groundwater in Mathura is itself very hard (Agarwal *et al.*, 2000). At site A it was 415mg/l to 740 mg/l, and at site B it ranged 473mg/l to 690mg/l.

Ammonia

Ammonical contents are another important indicator of pollution, especially of organic pollutants like sewage and garbage. Ammonia is developed from non oxidized animal waste (NEERI, 1984). The higher values (20.5 mg/l–22.02mg/l) were observed at both sites.

Nitrate

Analysis of nitrates in the water is essential to assess the state of pollution, as it provides a picture of easily decomposable organic matter (Klein, 1973). Analysis of nitrates revealed the incomplete oxidation of organic matter, as it was present in very less amount (Hynes, 1978). The nitrates showed a trend similar to that of dissolved oxygen. At site A and B it ranged 10.08 to 20.34 and 10.5 to 19.8 mg/l respectively.

Chloride

It is also one of the important factors, which directly indicate the extent of pollution. The main sources of chloride contents are the salts present in water and also the organic and animal waste. The presence of high level of chlorides also suggests that, the river is receiving a large amount of animal excreta (Kanta and Shah, 2004). The higher values of chlorides were 712 mg/l (site A) and 705mg/l (site B).

Conclusion

From the above discussion and analysis, it was concluded that the sites of Yamuna river were badly polluted due to intense human activities and influx of untreated sewage and industrial effluents. Following important reasons were observed for their deplorable condition:

1. Immersion of religious waste into the river by the local people due to their religious feelings for the sacred river.
2. The absence of proper sanitation facilities lead to open defecation near around this holy pilgrim, while visiting it.

3. Poor drainage system of the city.

4. Large number of unorganized small scale industries, involved in saree printing and silver metallurgy discards their harmful effluents into river due to absence of proper treatment plants and proper infrastructure.
5. Carelessness of local administration, pollution control board and municipality. Deterioration of the river water quality is causing many adverse effects on the livestock and humans.

Following remedial measures are required to sustain the good quality of water and to save the life.

1. Small scale industries owners should be legally forced to have water treatment plant.
2. Pilgrims must be given proper sanitation facilities, more public toilets and bathrooms should be constructed and drainage system must be strengthened.
3. People must be given knowledge of proper waste treatment methods with the help of television, radio, newspaper and NGOs.
4. Pollution control board should take effective step to regularly check the water quality and take action against the culprit.

Acknowledgement

The authors are thankful to the Principal, B.S.A. (P.G.) College, Mathura for providing the facilities required.

Glossary

Aquatic–Which lives in water for all of its life.

Biodegradable–Material can be break down by natural process into their natural components.

Biota–Total collection of organisms of a geographical region or a time period.

BOD–(Biological Oxygen Demand) Uptake rate of dissolved oxygen by the biological organisms in a body of water.

Decomposition–Process by which tissues of a dead organism break down into simple forms of matter.

DO–(Dissolved Oxygen) Amount of oxygen that is dissolved in a given medium.

Excreta–Waste matter.

Fauna–Animal life of any particular region.

Garbage–Waste, an unwanted or undesired material or substance.

Hardness–Mineral content.

Non-biodegradable–The material is totally immune from attack by any natural element.

Pollutants–The elements of pollution.

Sewage–Liquid and solid waste carried off in sewers or drains

References

1. Agarwal, P.K., Prabha, S. and Sharma, H.B. (2000). Water quality of sewage drains entering river Yamuna at Mathura. *J. Env. Biol.*, 21(4): 375–378.

2. APHA (1989). *Standard Methods for the Examination of Water and Wastewater*, 17th Edition. Washington D.C., U.S.A.

3. Dakshini, K.M.M. and Soni, J.K. (1979). Water quality of sewage drains entering Yamuna at Delhi. *Ind. J. Env. Hlth.*, 21(4): 354–360.

4. Hynes, H.B.N. (1978). *The Biology of Polluted Waters*. Liverpool University Press, Liverpool, 200–204 pp.

5. Kanta, S.A. and Shah, B.A. (2004). Risk of arsenic contamination in groundwater affecting the Ganga Alluvial Plain, India. *Environ. Health Perspect.* January, 112(1): A19–A21.

6. Klein, L. (1973). *River Pollution, II: Causes and Effect*, 5th Edition. Butterworth and Co. Ltd.

7. *Water and Wastewater Analysis* (1984). NEERI, Nagpur (India).

Biodiversity of Aquatic Reseources (2012)
Editors: **Mamta Rawat & Sumit Dookia**
Published by: **DAYA PUBLISHING HOUSE, NEW DELHI**

Pages **193–202**

Chapter 14

Health Hazards of Contaminated Underground Drinking Water at Moradabad, India: A Survey Work

D.K. Sinha[1] and Navneet Kumar[2]

[1]*Reader and Head, Department of Chemistry, K.G.K. (P.G.) College,*
Moradabad – 244 001, U.P.
E-mail: dkskgk@rediffmail.com
[2]*Assistant Professor, Department of Applied Science and Humanities,*
College of Engineering, Teerthankar Mahaveer University,
Moradabad – 244 001, U.P.
E-mail: navkchem@gmail.com

ABSTRACT

To study the health hazards of IM2 hand pump underground drinking water, an extensive 'Survey Work' using an exhaustive 'Questionnaire' was carried during 2006 over the residents of Moradabad who were totally dependent on the source sites of study. For this eight different sites and fifty residents of each site were selected. The selected ratio of men, women and children was 2:2:1 respectively. It was found that underground drinking water is severely contaminated and the entire population is at the risk for health problems due to intake of contaminated drinking water in the area of study. The residents of locality should be educated for health hazards of contaminated drinking water and water quality management.

Keywords: Health hazard, Drinking water contamination, Survey work, Questionnaire.

Introduction

It is well known fact that clean water is absolutely essential for healthy living. Adequate supply of fresh and clean drinking water is a basic need for all human beings on the earth, yet it has been observed that millions of people worldwide are deprived of this. Freshwater resources are threatened not only by over exploitation and poor management but also by ecological degradation (Biswas and Boruah, 2000) and Godwin, 2004). The main source of freshwater contamination can be attributed to discharge of untreated wastes, dumping of industrial effluents and run-off from agricultural fields. Industrial growth, urbanization and the increasing use of synthetic organic substances have serious and adverse impacts on freshwater bodies (Khan *et al.*, 2005, Madhuri *et al.*, 2004 and Tiwari *et al.*, 2003). Waterborne diseases and water-caused health problems are mostly due to inadequate and incompetent management of water resources Gupta, 1999 and Sharma *et al.*, 2001). Safe water for all can only be assured when access, sustainability and equity can be guaranteed. Chemicals in water can be both naturally occurring and induced by human interference and can have serious effects (Daniel, 1974, Park and Park, 1977 and Inforn, 1978). Chronic exposure to chemical contaminants can cause an array of health effects including cancers, neurological effects, reproductive and developmental outcomes, rashes, heart disease, diabetes and immune problems (De, 2005 and Haynes, 1982).

There are four different methods of collecting the data [Bailey and Cummings (1985)] *viz.*: (1) Library method, (2) Experimental method, (3) Observation, and (4) Questionnaire method. Questionnaire method is followed in the present study. The work of collecting facts is undertaken in a planned manner. Without proper planning, the facts collected may not be suitable for the purpose and a lot of time and money may be wasted. Human mind has its limitations; therefore, some process of condensation must take place. Condensation implies the organisation, classification, tabulation, and presentation of data in a suitable form. The 'Questionnaire' is designed in such a way that the questions have qualities: (1) clarity, (2) certain types of questions which are likely not to be answered in an honest way are avoided, (3) definition of answers, and (4) consistent number of questions. Direct personal observation method in which investigator obtains the data by a personal observation or interview is followed. This procedure is adopted because the field of inquiry is small and there is a desire for a greater accuracy. The work of editing primary data requires skill and scientific impartiality of a high degree. There are four types of editing the data: (1) editing for consistency, (2) uniformity, (3) completeness, and (4) accuracy. The process of statistical analysis is a method of abstracting significant facts from the collected mass of numerical data. The interpretation of the various statistical constants obtained through a process of statistical analysis is the final phase or the finishing process of the statistical technique. No pain is spared to see that the collected data are accurate, reliable and thorough (Snedecor and Cohran, 1967 and Daniel and Wood, 1980). Some approximations are definitely made without affecting the sanctity of degree of accuracy.

Materials and Method

To study the health hazards of contaminated IM2 hand pump underground drinking water on residents, eight different sampling sites are chosen at Moradabad. Fifty residents of each site are chosen who are totally dependent on the source sites of study. The ratio of men, women and children surveyed is 2:2:1 respectively. An extensive survey work using 'Questionnaire' is carried out in the area of study. For this, the questionnaires are completely filled, sampling sites and residents are thoroughly examined and the user also duly signed the questionnaire. A brief description of sampling sites is presented in Table 14.1.

Water at all sampling site was observed to have bad odour and taste also the colour of drinking water turned to yellow or yellowish–brown on standing. Occasionally, rusty water was also noticed at site no. IV and VII. A distinct oily layer was formed over the surface of water on standing at site no. V and VI.

Results and Discussion

Site-wise number and percentage of residents dependent on drinking water having different health problems at Moradabad, India is presented in Table 14.2. Site-wise and health problem-wise percentage of men, women and children is given in Table 14.3. A critical analysis of data reveals following meaningful facts regarding the health hazards of intake of contaminated drinking water on the residents of Moradabad district:

Site no. I is characterized by a very high percentage of people suffering from stomach problems, dental problems and skin problems. The percentage of people reported to have stomach ache, constipation and gastric problems are 56, 64 and 78 respectively. 82 per cent residents of this site are facing the problems of dental caries or tooth decay. Skin and hair problems are also prevalent in this area. At site no. II 24 per cent residents are hypertensive and 10 per cent are known diabetics. A fairly high percentage of people reported skin as well as hair problems. 56 per cent people are suffering from constipation, whereas, 72 per cent people are noticed to have dental caries or tooth decay. 17 and 27 out of 50 people are suffering from stomach ache and constipation at site no. III. The number of people with gastric problems is also very high and it is 66 per cent. Dental and hair problems are very prominent at this site. 50 per cent residents are having patches or worts on different parts of their body. Reported health hazards at site no. IV are less as compared to previous three sites. However, 72 per cent residents have dental caries or tooth decay. More than 50 per cent people are facing different kinds of hair problems.

Site no. V is characterised by highest percentage of hypertension and known diabetes and their percentage are 40 and 12 respectively. A very high number of people are suffering from gastric problems, stomach ache and frequent diarrhoea. 88 per cent people have dental caries or tooth decay. 33 people out of 50 face the problem to skin dryness after taking bath. About 60 per cent residents are suffering from different types of hair problems. At site no. VI, 92 per cent people have dental caries or tooth decay. Percentage of people with hair and skin problems can not be over looked. The occurrence of health problems is lower at this site. At site no. VIII the percentage

Table 14.1: A Brief Description of Sampling Sites

Sl.No.	Site No. and Name	Location of Site	Type of Source	Apparent Water Quality
1	I, Sri Gurdwara, Chandra Nagar	2 km West to collectorate, bored at a depth of 80 feet	IM2 hand pump, only source of drinking water	Objectionable odour, turns brownish-yellow on standing
2	II, Singh Mandap, Chandra Nagar	500 meter East to site no.I, bored at a depth of 80 feet	IM2 hand pump, complementary source	Fishy smell, turns yellowish on standing
3	III, Patel Nagar Mohallah	1.5 km West to site no.II, bored at a depth of 90 feet	IM2 hand pump, only source of water	Fowl smelling, turbid, turns yellowish on standing
4	IV, Shiv Shakti Vatika, Locoshed	500 meter East to site no.II, bored at a depth of about 80 feet	IM2 hand pump, complementary source	Objectionable odour, occasionally rusty colour
5	V, Chau Ki Basti, Linepar	Adjacent to Railway Station, bored at a depth of 85 feet	IM2 hand pump, only source of water	Fishy smell, turns yellowish brown and a district oily layer on standing
6	VI, Mata Mandir, Linepar	300 meter South to Railway Station, bored at a depth of 80 feet	IM2 hand pump, complementary source	Objectionable odour, oily layer over the surface on standing
7	VII, Prakashnagar Chauara	200 meter South to site no. VI, dored at a depth of about 80 feet	IM2 hand pump, only source of drinking water	Fishy smell, occasionally colour
8	VIII, Linepar Police Station	1 km South to s te no. VII, bored at a depth of 90 feet	IM2 hand pump, only source of water	Objectionable odour, yellow to yellowish-brown on standing

Table 14.2: Site-wise and Health Problem-wise Number and Percentage of Residents at Moradabad, India

Sl.No.	Health Problems	Site No. I		Site No. II		Site No. III		Site No. IV		Site No. V		Site No. VI		Site No. VII		Site No. VIII	
		Tallies	%	Tallies	%	Tallies	%	Tallies	%	Tallies	%	Tallies	%	Tallies	%	Tallies	%
1.	Hypertension		22		24		22		24		40		26		14		4
2.	Diabetes		2		10		2		10		12		6		–		–
3.	Stomach Problems																
	a) Stomach ache		56		34		56		34		56		42		18		22
	b) Constipation		64		56		64		56		68		38		56		34
	c) Frequent diarrhoea		22		–		22		–		–		–		–		–
	d) Dysentry/Colitis		4		8		4		8		18		10		4		6
	e) Worms		18		24		18		24		24		18		24		14
	f) Ulcer		4		2		4		2		2		–		–		–
	g) Gastric		78		44		78		44		80		62		66		54
4.	Hepatitis (Jaundice)		16		6		16		6		14		8		6		–
5.	Calcination (Stones)																
	a) Gall bladder stones		4		4		4		4		–		–		2		2
	b) Kidney stones		6		–		6		–		4		–		2		2

Contd...

Table 14.2–Contd...

Sl.No.	Health Problems	Site No. I %	Site No. II %	Site No. III %	Site No. IV %	Site No. V %	Site No. VI %	Site No. VII %	Site No. VIII %
6.	Thyroid/Goitre	–	–	–	–	–	–	–	–
7.	Dental Problems								
	a) Caries/Tooth decay	82	72	82	72	88	92	84	80
	b) Rotten Gum	36	22	36	22	16	34	32	3
8.	Skin Problems								
	a) Dryness/Scaly	86	58	86	58	66	46	36	34
	b) Itching	54	22	54	22	32	14	10	8
	c) Patches/Worts	30	6	30	6	18	6	8	4
9.	Hair Problems								
	a) Early greying	42	54	42	54	58	56	50	56
	b) Falling	72	76	72	76	66	36	38	42
	c) Dandruff	84	48	84	48	64	44	50	36
	D) Itching	64	34	26	26	56	20	26	20

Table 14.3: Site-wise and Health Problem-wise Percentage of Men, Women and Children at Moradabad, India

Sl.No.	Health Problems	Site No. I			Site No. II			Site No. III			Site No. IV			Site No. V			Site No. VI			Site No. VII			Site No. VIII		
		M	W	C	M	W	C	M	W	C	M	W	C	M	W	C	M	W	C	M	W	C	M	W	C
1.	Hypertension	6	16	–	6	18	–	8	10	–	–	14	–	14	24	2	8	18	–	4	10	–	2	2	–
2.	Diabetes	–	2	–	2	8	–	4	4	–	–	–	–	8	4	–	–	6	–	–	–	–	–	–	–
3.	Stomach Problems																								
	a) Stomachache	18	22	16	6	20	8	10	14	10	10	20	6	18	22	16	12	20	10	4	4	10	6	10	6
	b) Constipation	28	28	8	28	22	6	26	24	4	16	16	4	26	36	6	8	30	–	24	22	10	14	16	4
	c) Frequent diarrhoea	8	8	6	–	–	–	6	4	6	–	2	–	–	–	–	–	–	–	–	2	–	–	–	6
	d) Dysentry/Colitis	4	–	–	2	2	4	6	14	8	2	–	4	4	2	12	–	8	2	2	2	–	–	4	4
	e) Worms	4	4	10	8	4	12	–	2	10	2	–	4	2	10	12	6	2	10	4	4	16	6	4	4
	f) Ulcer	2	2	–	2	–	–	–	–	–	–	–	–	2	–	–	–	–	–	–	–	–	–	–	–
	g) Gastric	32	36	10	18	22	4	26	34	6	14	22	4	36	36	8	18	36	8	32	30	4	30	22	2
4.	Hepatitis (Jaundice)	8	2	6	–	4	2	8	4	–	–	–	–	6	4	4	–	4	4	–	4	2	2	–	–
5.	Calcination (Stones)	4	4	2	–	4	–	–	4	–	4	4	–	4	4	–	–	–	–	–	4	–	2	2	–
	a) Gall bladder stones	–	4	–	–	4	–	–	4	–	–	–	–	–	–	–	–	–	–	–	2	–	–	2	–
	b) Kidney stones	4	–	2	–	–	–	–	–	–	4	4	–	4	–	–	–	–	–	–	2	–	2	–	–
6.	Thyroid/Goitre	–	–	–	–	–	–	–	–	–	–	–	–	–	–	–	–	–	–	–	–	–	–	–	–
7.	Dental Problems																								
	a) Caries/Tooth decay	30	34	18	26	34	12	34	38	16	32	28	12	36	36	16	38	38	16	36	28	20	32	30	18
	b) Rotten Gum	14	14	8	4	16	2	26	36	16	4	4	–	6	6	4	14	16	4	12	12	8	10	14	8

Contd...

Table 14.3–Contd...

Sl.No.	Health Problems	Site No. I			Site No. II			Site No. III			Site No. IV			Site No. V			Site No. VI			Site No. VII			Site No. VIII		
		M	W	C	M	W	C	M	W	C	M	W	C	M	W	C	M	W	C	M	W	C	M	W	C
8.	Skin Problems																								
	a) Dryness/Scaly	32	34	20	16	30	12	10	22	2	14	10	6	28	28	10	14	24	8	4	16	16	12	16	6
	b) Itching	20	20	14	8	10	4	10	16	4	6	4	–	10	16	6	2	8	4	4	4	2	2	6	–
	c) Patches/Worts	12	12	6	–	6	–	20	12	18	–	2	–	6	6	6	4	2	–	–	–	8	–	2	2
9.A	Hair Problems																								
	a) Early greying	16	24	2	18	32	4	38	24	4	22	30	4	30	26	2	24	32	–	28	22	–	34	20	2
	b) Falling	30	32	10	26	38	12	26	36	2	8	34	8	26	36	4	6	30	–	10	22	6	14	26	2
	c) Dandruff	30	36	18	14	24	10	26	28	10	18	18	10	20	40	4	16	18	10	20	22	8	8	24	4
	d) Itching	28	18	18	6	22	6	8	12	6	4	16	6	16	36	4	6	14	–	10	12	4	6	14	–

of people with constipation and gastric problem is fairly high and it is 56 and 66respectively. 84 per cent residents are reported to have dental caries or tooth decay. Site no. VIII is characterized by 80 per cent people having dental caries, 54 per cent people with gastric problems and 56 per cent people suffering with early greying of hair. There is no known diabetic and the percentage of hypertensive people is lowest at this site. No case of Thyroid or Goitre is reported in the catchment area of study. The percentage of calcination is low at all the sites except site no. I where it is still on lower side and it is 10. This might be because of some other reasons.

A critical analysis of Table 14.2 reveals that in general, the percentage of men and women with health problems of different types is always more than that of children at all the sites. However, percentage of children facing health hazards in the area of study can not be ignored. The percentage of children with worms in their stomach, dental problems and skin problems is fairly high and it is quite alarming too. The percentage of women with different health problems surveyed is higher than that of men at almost all the sites. Surprisingly, women are suffering with hypertension in high percentage.

Conclusion

On the basis of above discussions it may be concluded that the underground drinking water at Moradabad, India in the catchment area of study is contaminated and is causing a number of diseases in the residents dependent on the water of sampling sites. The percentage of people suffering from dental, skin and hair problems is very very high. A considerable number of people are living with different kinds of stomach problems. Percentage of hypertensive persons is fairly good and it is highest at site no. V. Number of people suffering from hepatitis and calcination can not be over looked.

The percentage of women with different health problems is higher than that of men at almost all the sites of study. The percentage of children suffering from different health problems is low, however, the hazards can not be ignored.

While the entire population is at the risk for health problems due to drinking water contamination, various factors influence what specific effect that a person might experience. Pregnant women, infants and children, the frail elderly and persons with compromised immune system are more susceptible to the effect from exposure. Waterborne epidemics and health hazards in the aquatic environment are mainly due to improper management of water resources. Proper management of resources has become the need of the hour. People should take adequate precautions. Regular necessary steps should be taken to disinfect the water and water pipes should regularly be checked for leaks and cracks. The residents of locality should be educated for uncontaminated drinking water and water quality management.

References

1. Bailey, B. and Cummings (1985). *Introduction of Economics Statistics*. Patton and Tebbut, London.

2. Biswas, S.P. and Boruah, S. (2000). Ecology of river Dolphin (*Platamista gangetica*) in the upper Brahmaputra. *Hydrobiologia*, 430: 97–101.

3. Daniel, C. and Wood, F.S. (1980). *Fitting Equations of Data,* 2nd Edition. E.L.B.S., New York.

4. Daniel, W.W. (1974). *Biostatistics: A Foundation for Analysis in the Health Science.* John Wiley and Sons, New York Inc.

5. De, A.K. (2005). *Environmental Chemistry,* 5th Edition. Wiley Eastern Limited, New Delhi.

6. Godwin, W.S. (2004). Bio-accumulation of heavy metals by industrial molluscs of kanyakumari water. *Poll. Res.,* 23(1): 37–40.

7. Gupta, S.C. (1999). Chemical Character of groundwaters in Nagpur district, Rajasthan. *Indian J. Environ. Hlth.,* 33(3): 341–349.

8. Haynes, R. (1982). *Environmental Science Methods.* Chapman and Hall, London.

9. Inforn, S.L. (1978). *Quality Assurance Practices for Health Laboratories.* American Public Health Association, Washington, D.C.

10. Khan, N., Mathur, A. and Mathur, R. (2005). A study on drinking water quality in Lashkar (Gwalior). *Indian J. Env. Prot.,* 25(3): 222–224.

11. Madhuri, U., Srinivas, T. and Sirresha, K. (2004): A Study on groundwater quality in commercial area of Visakhapatnam. *Poll. Res.,* 23(3): 565–568.

12. Park, J.E. and Park, K. (1977). *Textbook of Preventive and Social Medicine.* Banarasidas Bhanot Publishers, New Delhi.

13. Sharma, B.C., Mishra, A.K. and Bhattacharya, K.G. (2001). Metals in drinking water in a predominantly rural area. *Indian J. Env. Prot.,* 21(4): 315–323.

14. Snedecor, G.W. and Cohran, W.O. (1967). *Statistical Method,* 6th Edition. The Iowa State University Press, Ames.

15. Tiwari, A.K., Dikshit, R.P., Tripathi, I.P. and Chaturvedi, S.K. (2003). Fluoride content in drinking water and groundwater quality in rural area of tahsil Mau, district Chitrakoot. *Indian J. Environmental Protection,* 23(9): 1045–1050.

Biodiversity of Aquatic Reseources (2012)
Editors: **Mamta Rawat & Sumit Dookia**
Published by: **DAYA PUBLISHING HOUSE, NEW DELHI**

Pages **203-216**

Chapter 15

Microbiological Study of Lonar Lake, Maharashtra, India

Pawar Vijaykumar Bhikusing*

*Dnyanopasak College of Arts, Commerce and Science,
Parbhani – 431 401, Maharashtra*

ABSTRACT

The impact origin of the Lonar crater has been well established based on the evidence of shock metamorphosed material. All the physico-chemical parameters of lake water including pH, Chloride (Cl), Total Hardness, Total Dissolved Solids, Ca, Mg, Na, K, CO_3, HCO_3, SO_4 etc. are higher and very greater than the BIS (1991) maximum permissible limits for drinking water, while the spring water is potable. The presence of bacteria related to water borne diseases was also found higher indicating the non-potable nature of the lake water but the stream water is normal and free from bacteria. Occurrence of few species of algae and fungi indicate the characteristics nature of bio-flora, which needs further investigation and interpretation.

Keywords: *Crater rim, MPN, SPC, IMVIC etc.*

Introduction

Lonar Lake (19°58' N; 76°31' E) in Buldhana district, Maharashtra, India is a circular lake occupied by saline water (Figure 15.1). It is formed by hypervelocity

* E-mail: vbpawar08@rediffmail.com

1: Site 'A' water from Ramgaya (East); 2: Site 'B' water from Lord Shiva Temple (West); 3: Site 'C' water from Shani Temple (South); 4: Site 'D' water from Devi Temple (North)

Figure 15.1: Lonar Lake View with Sampling Site Location

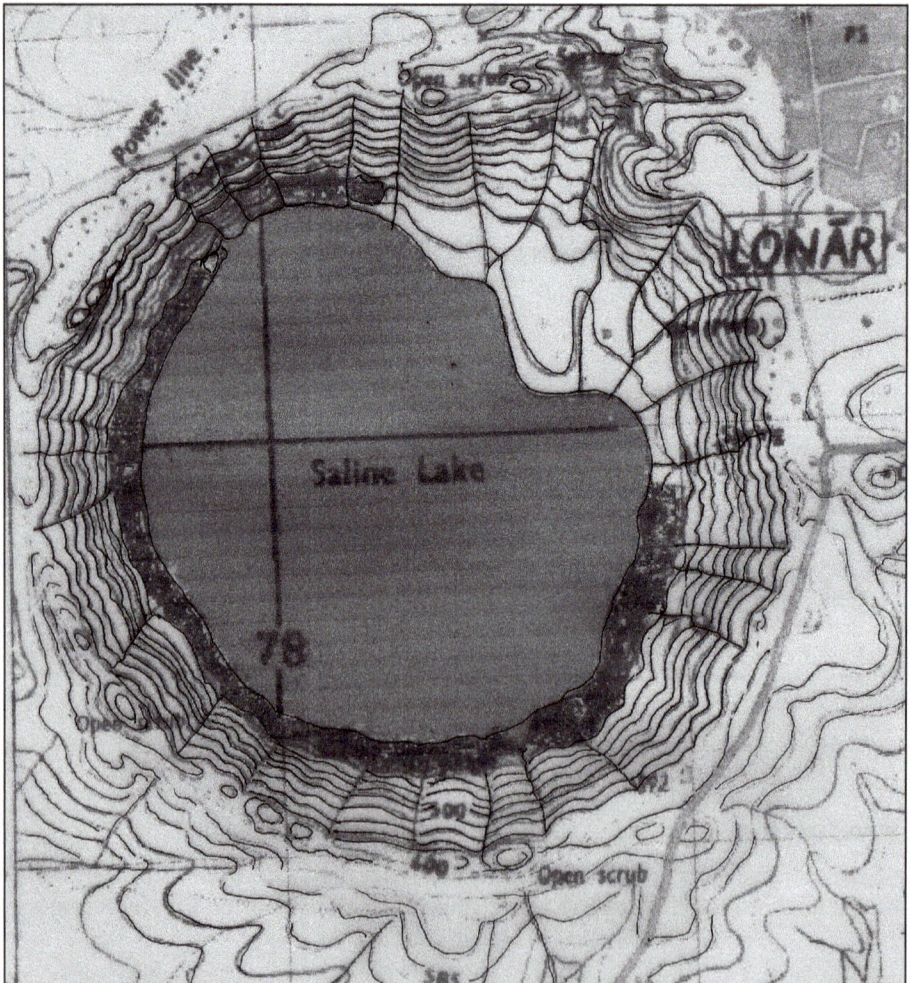

Figure 15.2: Counter Map of Lonar Crater Rim

meteoritic impact in basalt rock of the Deccan Traps near about 50 thousands years ago. It's diameter about 1830 m. and depth is 135 m. The lake is confined from all sides by crater rim of the crater a not a single channel of water for drainage out. The water is stagnant for thousands years. The water is salty, alkaline and the lake is rich with various biotic and abiotic assemblages. It offers unique opportunities for ecological investigation.

This holistic ecological assessment is necessary to implement conservation measures for long term sustainability of the unique lake ecosystem Lonar Crater Lake having a unique ecological status as only saline Crater Lake in Asia and an ecological wonder (Figure 15.2).

Material and Methods

Water samples were collected from four different sampling stations in airtight and opaque polythene container. These four stations are located in reservoir such as

1. Sampling station 'A' water from Ramgaya (East),
2. Sampling station 'B' water from Lord Shiva Temple (West),
3. Sampling station 'C' water from Shani Temple (South) and
4. Sampling station 'D' water from Devi Temple (North) was established along the periphery of lake basin for one year limnological study programme.

Standard analytical methods were employed to analyse water samples of Lonar lake (APHA, 1989) Microbiological examinations were carried out using the standard procedures given by (APHA, 1989, Trivedy and Goel, 1986) and the result were compared with maximum permissible was carried out by conducting MPN, SPC and IMVIC tests.

Methods of Sampling of Microorganisms

1 *Collection*: Algal blooms and water samples were collected in and around four different spot in vicinity of Lonar Lake.

2. *Culture media used*: The media employed for isolation of *Spirulina* sp. in Zarrouk media with pH 10.5 Zarrouk medium (Tables 15.1 and 15.2).

3. *Enrichment of samples*: The water samples are labeled as A, B, C and D and are enriched with both media in 500 ml flasks containing 250 ml media. The flasks were inoculated at room temperature $25° \pm 2° C$ with continuous illumination of 1500 lux provided with white fluorescent lamps for two weeks.

4. *Isolation*: After second enrichment the 5 per cent inoculum size was spread on solid media containing 2 per cent agar. The plates were incubated at room temp 25° C with continuous illumination of 1500 lux provided with white fluorescent lamp. The flasks were hand shaken twice daily. The isolates were chosen according to difference in their morphological nature with in microscopic view. This procedure is repeated thrice to bring unialgal culture and can be obtained using single cell isolation.

Table 15.1: Composition in g/l of Culture Media Used for Analysis

Composition	g/l	Composition	g/l
NaHCO$_3$	16.8	CaCl$_2$ 2H$_2$O	0.04
K$_2$HPO$_4$	0.5	FeSO$_4$ 7H$_2$O	0.01
NaNO$_3$	2.5	EDTA	0.08
K$_2$SO$_4$	1.0	Solution A5	1 ml.
NaCl	1.0	Solution B6	1 ml.
MgSO$_4$ 7H$_2$O	0.2	–	–
		*Solution A5	
H$_3$BO$_3$	2.86	CuSO$_4$ 5H$_2$O	0.079
MnCl$_2$ 4H$_2$O	1.81	MoO$_3$	0.015
ZnSO$_4$ 7H$_2$O	0.22	–	–
		*Solution B6	
NH$_4$VO$_3$	22.96	Na$_2$WO$_4$ 2H$_2$O	17.94
KCr(SO$_4$)$_4$12H$_2$O	192.0	TiOSO$_4$H$_2$SO$_4$ 8H$_2$O	61.6
NiSO$_4$ 6H$_2$O	44.8	CO (NO$_3$) 6H$_2$O	43.98

Table 15.2: Protocol for Maintaining Stock Cultures of *Spirulina* on Agar Slants

Parameters	Procedure
Nutrient medium	Zarrouk's medium
Preparation	a) 16.8 g of NaHCo$_3$ dissolved in 500 ml of water
	b) Other nutrients as in Table 15.1, Dissolved in 500 ml. water and added 20g (2 per cent) agar
	A and B were sterilized separately in an autoclave at 15lb pressure for 15 min. cooled to 50 °C and mixed in equal proportion the pH was 9-10 Distributed into sterile culture test tubes and slants made
Method if inoculation	a) 1 drop of *Spirulina* culture suspension was added directly onto agar slants and spread aseptically.
	b) Culture from another agar slant was aseptically streaked on the medium in a freshly prepared slant
Culture maintenance	Inoculated tests tubes were inoculated at room temperature and illuminated with 40W cool white fluorescence lamps at an irradiance of 3 to 5 klux
Interval for sub-culturing	20 days interval

5. *Pure Culture and Maintenance of Lonar Isolates*: Pure cultures are obtained using Single Cell Isolation Technique. Cultures were routinely maintained in sterilizing cotton plugged conical flasks in a culture room maintained at 25° C and illuminated under a fluorescent light intensity weekly to fresh media to keep the algae in logarithmic growth phase for use in the experiments.

6. *Studies on Lonar lake water by addition of N, P, K and rock-phosphate in growth medium on Spirulina isolates*: The water collected from Lonar lake and water sterilized by autoclaving at 15 lbs for 20 min. Media with various combination of Lonar water, distilled water, N, P, K and rock-phosphate were prepared and sterilized by autoclaving at 15 lbs for 20 min. five per cent inoculum of *S. fusiformis* and *S. patensis* were inoculated into 500ml sterilized autoclavable PVC conical flasks containing 250 ml media and grown at room temperature at 25° C with illuminating under a fluorescent light intensity of 1500 lux for 14 hrs/day. Flasks were hand shaken twice daily. The pH of media is same *i.e.* 10.5 adjusted. The all media are in duplicate. Growth was followed at 3 days intervals by measuring the optical density at 640 nm. Every time 50 ml of culture was removed from the each media and used for analysis of pigments and growth study.

Result and Discussions

The present paper deals with water quality parameters and micro-ecological aspects of Lonar Lake. The result of four sample stations and comparative data of other researchers are given in Table 15.3. Temperature of lake water ranged from 23.6 to 24.4, which varies with respect to depth season and environment. The depth of water varies at different places.

The colour of lake water is yellowish green to dark green which was result of dense algal population. The pH of water sample was found to be alkaline which ranged between 7.2 to 11.2, the higher values of pH was may be due to the increased primary production in aquatic ecosystem of lake (Zafar,1996 and Mohd. Mussaddiq *et al.*, 2001)

Low rate of primary production in aquatic ecosystem of lake is also indicate by the low values BOD and COD that ranged from 0.1 to 0.3 mg/lt and 0.01 to 0.04 mg/lt respectively.

The BIS (1991) Maximum permissible limit of Total Dissolved Solid (TDS) is 1500 mg/liter for drinking water. The TDS in study area ranged from 8900 mg/lt to 21800 mg/lt which is very high and above the permissible limit of BIS, 1991.

The chloride concentration in the lake varies from 55.5 mg/lt to 2816 mg/lt which is above the permissible limit of 250 mg/liter as given by BIS (1991) this means that the water is pollution due to organic matter and the other waste in the water.

Since the values of calcium (1285 mg/lt to 1420 mg/lt) and magnesium (658 mg/lt to 1043 mg/lt) were found to be very high, the water was very hard and not free from pollutants in it. The lake water characteristics very high chloride, calcium and magnesium so the total hardness of any water is dependent on these factors (Jain *et al.*, 1997)

The sulphate content in present study area is less that 45 to 54 mg/lt (Table 3) in lake water. Sulphate contributed from the rock weathering and in addition to the domestic waste, sewage, house hold effluents and human faeces.

Table 15.3: Physico-chemical Parameters of Lonar Lake Water Samples

Sl.No.	Parameters	BIS Max. Perm. Limit	Sample 'A'	Sample 'B'	Sample 'C'	Sample 'D'
	I. Physical Analysis					
1.	Odour	–	Highly Objectionable	Highly Objectionable	Highly Objectionable	Highly Objectionable
2.	Colour	–	Yellow-green	Dark-green	Dark-green	Dark-green
3.	Temperature (ºC)	–	23.6 ºC	24 ºC	24.2 ºC	24.4 ºC
4.	Total Solids (mg/lt)	–	25600	16800	9100	5400
5.	Total Suspended Solids (TSS) (mg/lt)	–	4780	4240	16500	5300
6.	Total Dissolved Solids (TDS) (mg/lt)	–	21800	12460	10400	8900
7.	D.O. (Dissolved Solids) (mg/lt)	–	0.03	0.07	0.02	0.08
8.	Biological Oxygen Demand (B.O.D.) (mg/lt)	–	0.1	0.3	0.2	0.1
9.	Chemical Oxygen Demand (C.O.D.) (mg/lt)	–	0.04	0.01	0.02	0.01
	II. Chemical Analysis					
10.	pH	6.2-9.2	10.2	11.2	10.5	7.2
11.	Total Alkalinity (mg/lt)	200	4213	5786	4390	370
12.	Calcium (mg/lt)	200	1285	1649	1420	32
13.	Magnesium (mg/lt)	100	658	1043	64.19	31
14.	Chloride (mg/lt)	1000	2414	2816	1442.5	55.5
15.	Fluoride (mg/lt)	–	3	2	2	2
16.	Ammonia (mg/lt)	–	2	0.5	2	0.5
17.	Sulphate (mg/lt)	400	45	40	54	67
18.	Phosphate (mg/lt)	45	2	6	3	4
	III. Bacteriological Analysis					
19.	MPN (Most Probable Number)		1800	1600	1800	900
20.	SPC (Standard Plate Count/100 ml.)		Highly Dense colonies observed at all stations Faecal coliforms were present at all stations			
21.	IMVIC test					

All parameters in mg/lt except temperature, pH and MPN

The naturally determined environmental conditions and persistent human intervention have caused eutrophication and have lead to senescence of the lake bringing it to the brink of a death (Khobragade, 2003)

Figure 15.3: **A–F**: Microscopic View of *Spirulina platensis;* **G**: Microscopic View of *Spirulina fusiformis* and **H**: Microscopic View of *Spirulina subsalsa*

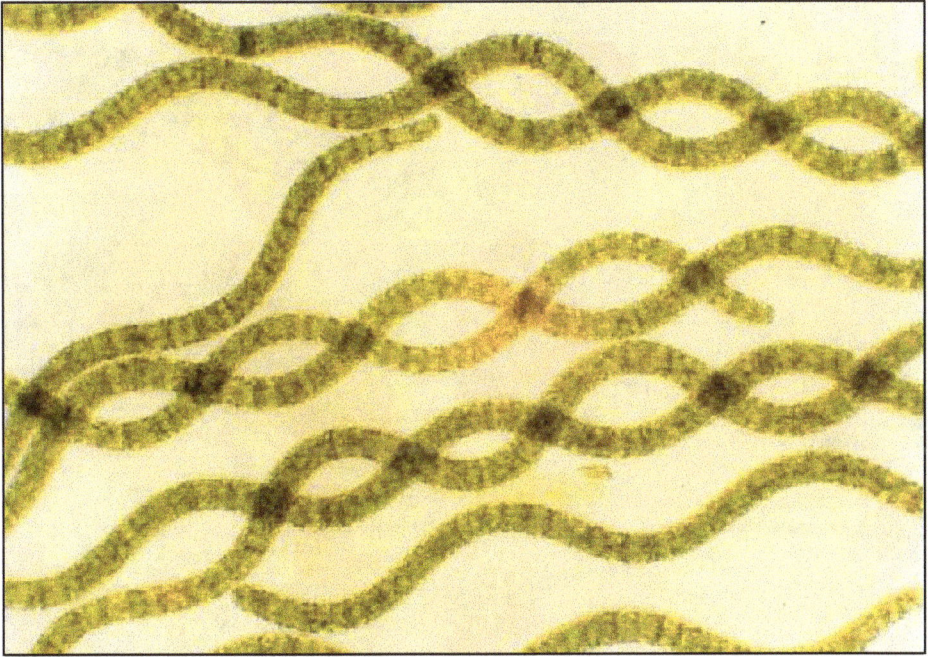

A

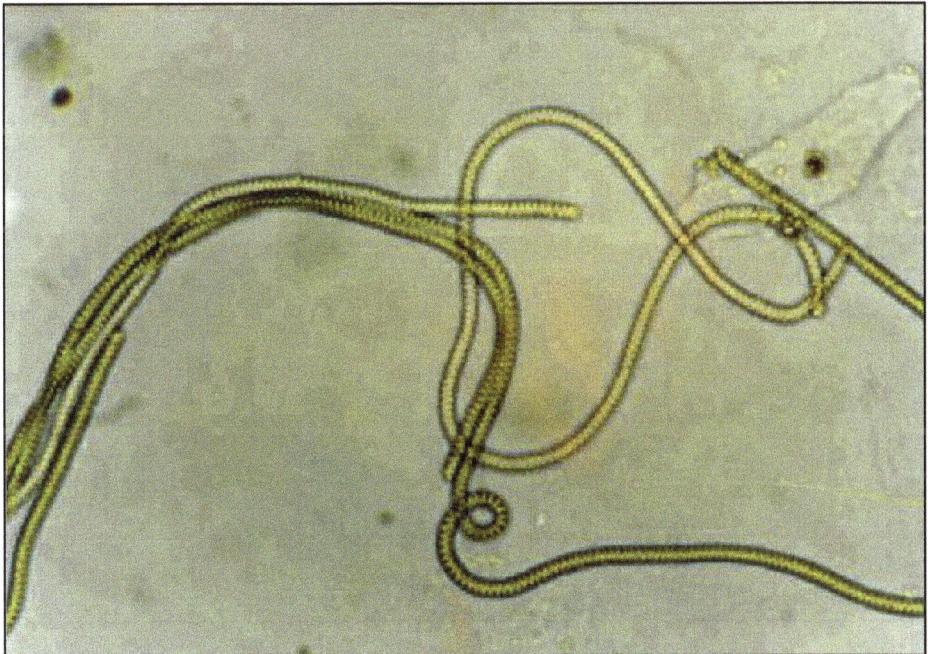

B

Contd...

Figure 15.3–*Contd...*

C

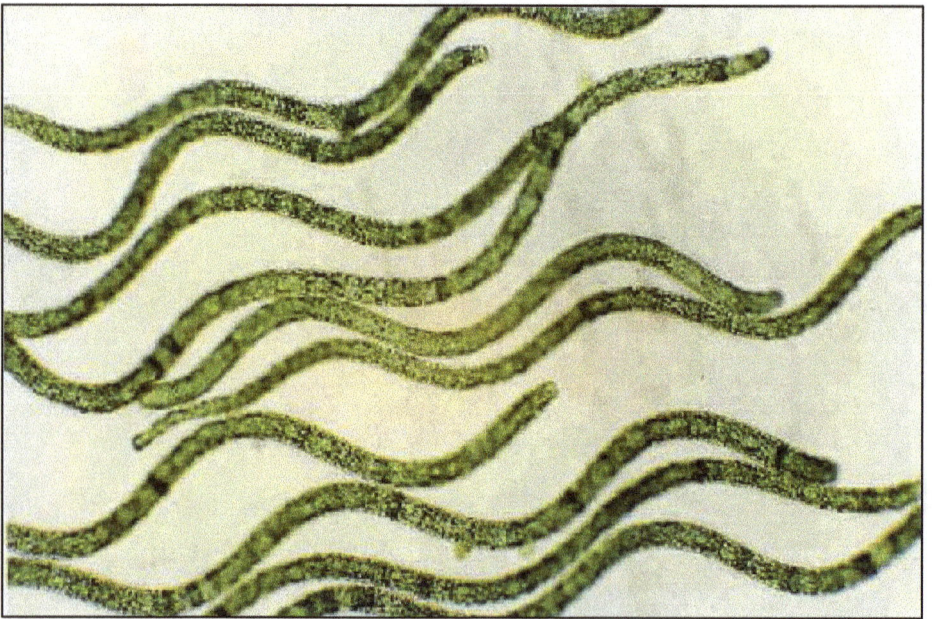

D

Contd...

Figure 15.3–*Contd...*

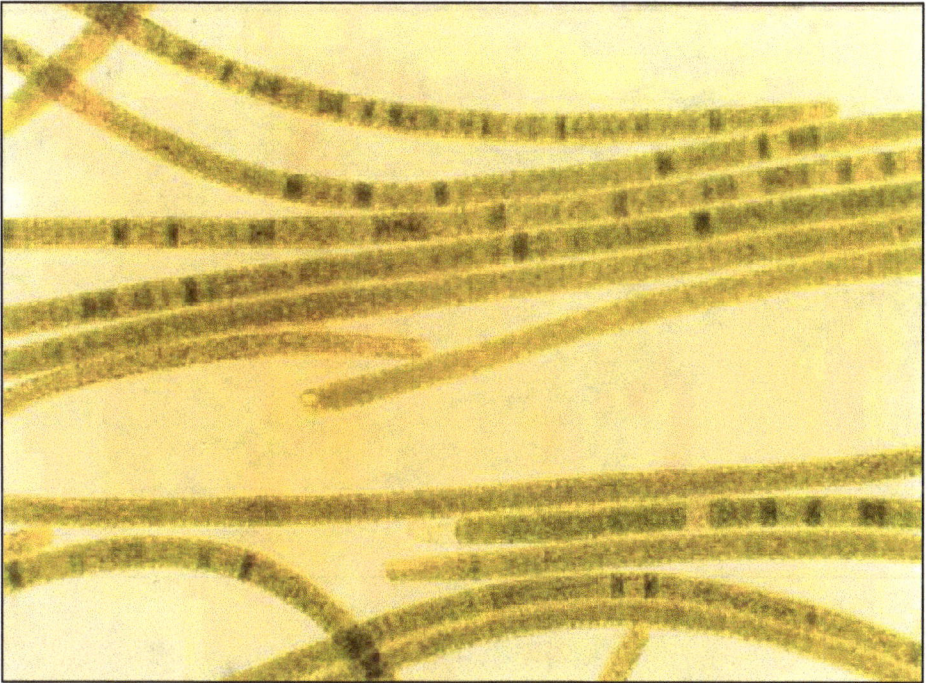

E

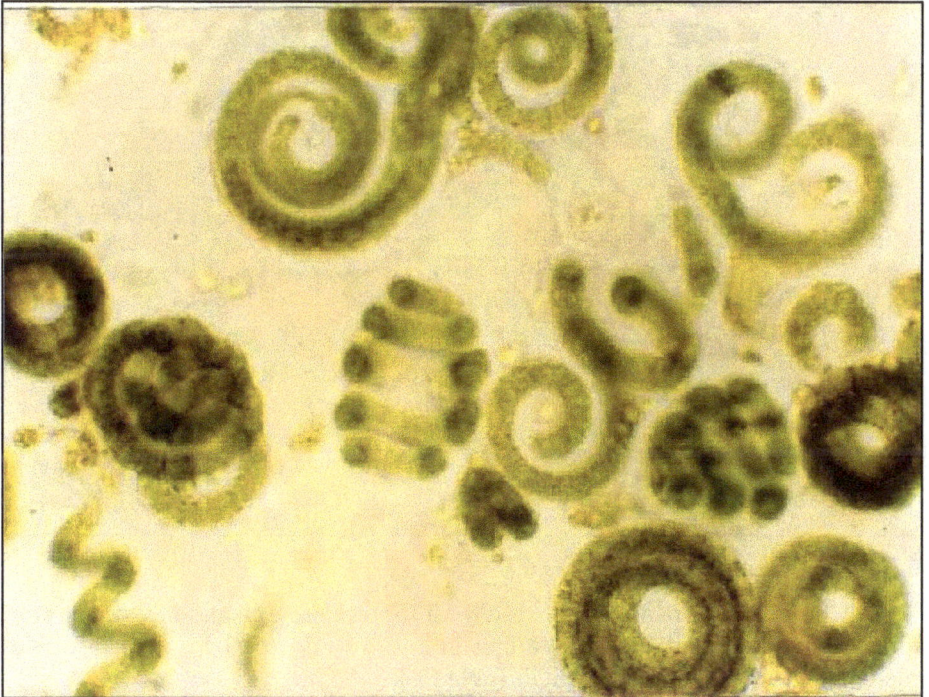

F

Contd...

Figure 15.3–*Contd...*

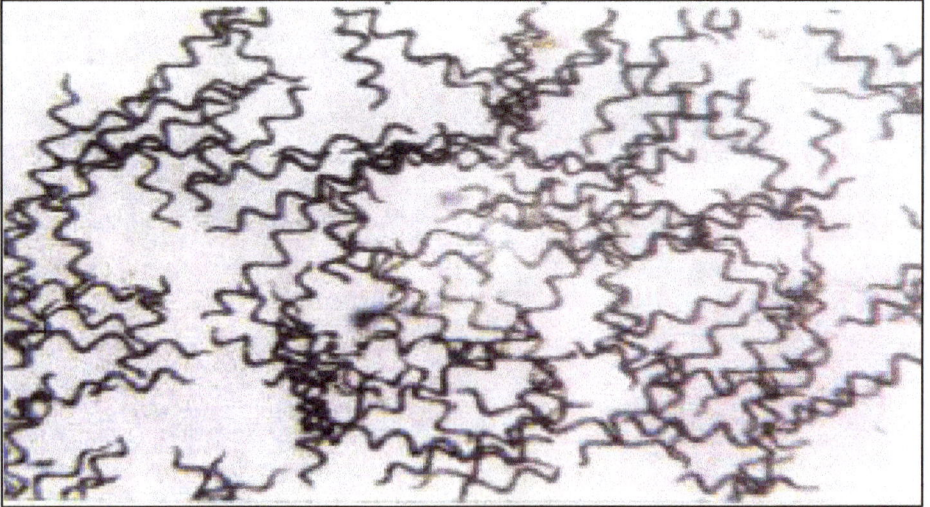

G

H

All these characteristics of lake resulted into an extreme alkaline ecosystem with all different microbial type prevailing in and around the lake. The results of microbiological analysis are given in Table 15.4.

Table 15.4: Microbiological Analysis

Sl.No.	Microorganisms	Sample A	Sample B	Sample C	Sample D
Bacteria					
1.	E. coli	+	+	+	+
2.	S. aureus	+	+	+	+
3.	Bascillus sp.	+	+	+	–
4.	Klebesiella sp.	+	+	+	+
5.	P. aeruginosa	+	+	+	+
6.	Metahnococcus sp.	+	+	+	+
Algae					
1.	Chlorella sp.	+	+	+	–
2.	Clasterium sp.	+	+	+	–
3.	Blue Green Algae (Cyno-bacteria)	+	+	+	–
4.	Spirulina sp.	+	+	+	+
Fungi					
1.	A. niger	+	+	+	–
2.	Fusarium sp.	+	+	+	–

+: Present; –: Absent.

Micro organisms like *Arthospira* and other micro algae are predominant as primary producers are present along with alkaline bacteria and fungi.

The bacteria species are presence related to water borne diseases in higher amount indicating non-potable nature within microscopic view studied and revealed

1. *Spirulina platensis* as blue green spirals are more or less regularly coiled showing constructed cross walls with spiral and trichome breadth 57.12 and 8.92 respectively having spiral distance of 18.75 microns

2. *Spirulina subsala orested* a blue green spiral are regularly coiled or some times loosely coiled showing no constructed cross wall with spiral and trichome breadth 2.15 and 3.22 respectively and spiral distance of 5.35 microns

3. *Spirulina major* a blue green regularly spiraled coiled no constructed cross walls with spiral and trichome breadth 3.57 and 1.43 respectively and spiral distance of 3.57 microns

These isolates possess variation in their morphological characteristics as shown in below (Figure 15.3A-H).

Lonar Lake is a hyper saline environment due to present of higher concentrations of various salts in lake water (Muley and Babar, 1998, Choudhary and Handa, 1978). The characterized salt concentration could be major reason of predominant population of cyno-bacteria such as *Oscillatoria, Synechocystis, Anabaenopsis* and *Spirulina* in the lake. It is confirmed that the cyno-bacteria population is regulated by the concentration of salts and alkalinity.

Conclusion

The present study is summarized in to the following points

1. All the physicochemical parameters of lake water including pH, Chloride (Cl), Total hardness, Total Dissolved Solids, Ca, Mg, Na, K, CO_3, SO_4 etc. are higher than maximum permissible limits set by BIS,1991 for drinking water wheras spring water (Dhar) is potable.

2. The presence of species bacteria related to water borne diseases were also found higher indicating the non potable nature lake water.

Acknowledgement

Author wish to express their sincere gratitude thanks to Dr. P.L. More, Principal, D. S. M. College of Arts, Commerce and Science, Parbhani, Dr. Kshama Khobragade, Head, Dept. of Environmental Science, S.B.E.S. College of Science, Aurangabad and Dr. S.V. Birajdar, Principal, S.B.E.S. College of Science, Aurangabad.

Glossary

Algae–A simple photosynthetic plant that usually lives in moist or aquatic environments. The bodies of algae can be unicellular or multicellular is design.

Alkaline–(1) Having a pH greater than 7. (2) Substance that releases hydroxyl ions (OH–).

Altitude–Vertical distance above sea–level.

Bacteria–Living organisms microscopic in size, which usually consist of a single cell. Most bacteria use organic matter for their food and produce waste products as a result of their life processes.

Basin–A topographic rock structure whose shape is concave downwards.

Biodiversity–The diversity of different species (species diversity), genetic variability among individuals within each species (genetic diversity), and variety of ecosystems (ecosystem diversity). Abbreviation of biological diversity.

Biogeochemical Cycling–Cycling of a single element, compound or chemicals by various abiotic and biotic processes through the various stores found in the biosphere, lithosphere, hydrosphere, and atmosphere.

Biome–Largest recognizable assemblage of animals and plants on the Earth. The distribution of the biomes is controlled mainly by climate.

Chemical–One of the millions of different elements and compounds found naturally and synthesized by humans.

Dissolved Solids–This is referred to chemical substances either organic or inorganic that are dissolved in a waste stream and constitute the residue when a sample is evaporated to dryness.

Ecology–The study of all aspects of How organisms interact with each other and/or their environment.

Ecosystem–Grouping of various organisms interact with each other and their environment

E. coli–Escherichia coli–one of the non pathogenic coliform organisms used to indicate the presence of pathogenic bacteria in water.

Effluent–this is referred to the waste water or other liquid which is raw (Untreated), partially or completely treated and flowing from a reservoir, basin, treatment or treatment plant.

Lake–A body standing water found on the Earth's continental land masses. The water in a lake is normally fresh. Also see eutrophic lake, mesotrophic lake, and logographic lake.

Latitude–Latitude is a north–south measurement of position on the Earth. It is defined by the angle measured from a horizontal plane located at the Earth's center that is perpendicular to the polar axis. A line connecting all places of the same latitude is termed a parallel. Latitude is measured in degrees, minutes, and seconds. Measurements of latitude range from equator (0°) to 90° North and South from this point.

Liquid–A state of matter where molecules have the ability to flow and the surface of this mass displays the property of surface tension.

Longitude–Longitude is a west–east measurement of position on the Earth. It is defined by the angle measured from a vertical plane running through the polar axis and the prime meridian. A line connecting all places of the same longitude is termed a meridian. Longitude is measured in degrees, minutes, and seconds. Measurements of longitude range from prime meridian (0°) to 180° West and East from this point.

Media–This is referred to the material in the trickling filter on which saline accumulates and organisms grow. As settled wastewater trickles over the media, organisms in the saline remove certain types of wastes thereby partially treating the wastewater. This is also referred to the material in a rotating biological contactor (RBC) or in a gravity or pressure filter

Microorganism–Extremely small organism that can only be seen using a microscope.

Pollution–Physical, chemical, or biological change in the characteristics of some component of the atmosphere, hydrosphere, lithosphere, or biosphere that adversely influences the health, survival, or activities of humans or other living organisms.

Pollutant : A substance that has a harmful effect on the health, survival, or activities of humans or other living organism

pH–Scale used to measure the alkalinity or acidity of a substance through the determination of the concentration of hydrogen ions in solution. A pH of 7.0 is neutral. Values below 7.0, to a minimum of 0.0, indicate increasing acidity. Values above 7.0, to a maximum of 14.0, indicate increasing alkalinity.

Phytoplankton–Small photosynthetic organisms, mostly algae and bacteria, found inhabiting aquatic ecosystems.

Reagent–This is referred to a pure chemical substance that is used to make a new products or is used in chemical tests to measure, detect or examine other substances.

Sewage–The water and water–carried solids from homes that flow in sewers to a waste water treatment plant. The preferred term is wastewater.

Sedimentation–This is referred to the process of subsidence and deposition of suspended matter from a wastewater by gravity.

References

1.   APHA (American Public Health Association) (1989). Washington D.C., pp. 1131–1138.

2.   BIS (1991). *Indian Standard Specification for Drinking Water"*, BS 10500.

3.   Choudhary, A.N. and Handa, B.K. (1978). Some aspects of the Geochemistry of the Lonar Lake water. *Indian J. Earth Sci.*, 5: 111–118.

4.   Jain, C.K., Bhatia, K.K.S. and Vijay, T. (1997). Groundwater quality in coastal regions of Andhra Pradesh. *Indian J. Env. Hlth.*, 39(3): 182–192.

5.   Khobragade Kshama (2003). Lonar lake: A great geological monument on the verge of death. *J. Aquatic Biology*, IAAB, Hyderabad, 18(2): 65–68.

6.   Mohd. Musaddiq, Fokmare, A.K. and Rizwan Khan (2001). Microbial diversity and ecology of Lonar Lake, Maharashtra, India. *J. Aqua. Biol.*, IAAB, Hyderabad, 16(2): 1–4.

7.   Muley, R.B. and Md. Babar (1998). Geo-environmental status of Lonar Lake, Maharashtra. In: *Proceeding Vol. of Workshop on 'Quality of Reservoir–1'* at WALMI Aurangabad, pp. 28–33.

8.   Trivedi, R.K. and Goel, P.K. (1986). *Chemical and Biological Methods for Water Pollution Studies*, (Eds.) R.K. Trivedi and P.K. Goel. Environmental Publications, Karad, India, pp. 35–80.

9.   Zaffer, A.R. (1996). Limnology of Hussainsagar Lake, Hyderabad, India. *Phykos*, 5: 126–155.

Part III

Water Resources and
Other Issues

Biodiversity of Aquatic Reseources (2012)   Pages 219-244
Editors: Mamta Rawat & Sumit Dookia
Published by: DAYA PUBLISHING HOUSE, NEW DELHI

Chapter 16

Water Resources of India: A Review

*Pankaj Gupta**

Professor and Head,
Department of Chemistry, Basic Environmental Engineering
and Disaster Management
Alwar Institute of Engineering and Technology
Alwar, Rajasthan

ABSTRACT

Water resources of a country constitute one of its vital assets. India receives annual precipitation of about 4000 km³ (1.17 m × 3.3 million km²) or say *4000 billion cubic meter (4000 km³)* The rainfall in India shows very high spatial and temporal variability and paradox of the situation is that Mousinram near Cherrapunji, which receives the highest rainfall in the world, also suffers from a shortage of water during the non-rainy season, almost every year. The total average annual flow per year for the Indian rivers is estimated as 1953 km³. The total annual replenishable groundwater resources are assessed as 432 km³. The annual utilizable surface water and groundwater resources of India are estimated as 690 km³ and 396 km³ per year, respectively. With rapid growing population and improving living standards the pressure on our water resources is increasing and per capita availability of water resources is reducing day by day. Due to spatial and temporal variability in precipitation the country faces the problem of flood and drought syndrome. Overexploitation of groundwater is leading to reduction of low flows in the rivers, declining of the groundwater resources, and salt water intrusion in aquifers of the coastal areas. Over canal-irrigation in some

* E-mail: pgupta1975@yahoo.com

of the command areas has resulted in water logging and salinity. The quality of surface and groundwater resources is also deteriorating because of increasing pollutant loads from point and non-point sources. The climate change is expected to affect precipitation and water availability. So far, the data collection; processing, storage and dissemination have not received adequate attention. The efforts initiated under the Hydrology Project Phase-I and the development of the Decision Support System proposed under Hydrology Project Phase-II are expected to bridge some of the gaps between the developed advanced technologies of water resources planning, designing and management and their field applications. The paper presents availability and demands of water resources in India as well as describes the various issues and strategies for developing a holistic approach for sustainable development and management of the water resources of the country. It also highlights integration of the blue and green flows and concepts of virtual water transfer for sustainable management of the water resources for meeting the demands of the present, without compromising the needs of future generations.

Keywords: *Rainwater harvesting, Surface water, Water resources, Water requirements, Water transfer.*

Introduction

Irrigated agriculture is dependent on an adequate water supply of usable quality. Water quality concerns have often been neglected because good quality water supplies have been plentiful and readily available. This situation is now changing in many areas. Intensive use of nearly all good quality supplies means that new irrigation projects and old projects seeking new or supplemental supplies must rely on lower quality and less desirable sources. To avoid problems when using these poor quality water supplies, there must be sound planning to ensure that the quality of water available is put to the best use.

The objective of this paper is to help the reader to a better understanding of the effect of water quality upon soil and crops and to assist in selecting suitable alternatives to cope with potential water quality related problems that might reduce production under prevailing conditions of use.

Conceptually, water quality refers to the characteristics of a water supply that will influence its suitability for a specific use, *i.e.,* how well the quality meets the needs of the user. Quality is defined by certain physical, chemical and biological characteristics. Even a personal preference such as taste is a simple evaluation of acceptability. For example, if two drinking waters of equally good quality are available, people may express a preference for one supply rather than the other; the better tasting water becomes the preferred supply. In irrigation water evaluation, emphasis is placed on the chemical and physical characteristics of the water and only rarely is any other factors considered important (APHA,1980 and Dewis and Freitas, 1970).

Water Quality Problems

Water used for irrigation can vary greatly in quality depending upon type and quantity of dissolved salts. Salts are present in irrigation water in relatively small but significant amounts. They originate from dissolution or weathering of the rocks and soil, including dissolution of lime, gypsum and other slowly dissolved soil minerals. These salts are carried with the water to wherever it is used. In the case of irrigation, the salts are applied with the water and remain behind in the soil as water evaporates or is used by the crop.

The suitability of water for irrigation is determined not only by the total amount of salt present but also by the kind of salt. Various soil and cropping problems develop as the total salt content increases, and special management practices may be required to maintain acceptable crop yields. Water quality or suitability for use is judged on the potential severity of problems that can be expected to develop during long-term use.

The problems that result vary both in kind and degree, and are modified by soil, climate and crop, as well as by the skill and knowledge of the water user. As a result, there is no set limit on water quality; rather, its suitability for use is determined by the conditions of use which affect the accumulation of the water constituents and which may restrict crop yield. The soil problems most commonly encountered and used as a basis to evaluate water quality are those related to salinity, water infiltration rate, toxicity and a group of other miscellaneous problems (Halsey, 1974).

Salinity

A salinity problem exists if salt accumulates in the crop root zone to a concentration that causes a loss in yield. In irrigated areas, these salts often originate from a saline, high water table or from salts in the applied water. Yield reductions occur when the salts accumulate in the root zone to such an extent that the crop is no longer able to extract sufficient water from the salty soil solution, resulting in a water stress for a significant period of time. If water uptake is appreciably reduced, the plant slows its rate of growth. The plant symptoms are similar in appearance to those of drought, such as wilting, or a darker, bluish-green colour and sometimes thicker, waxier leaves. Symptoms vary with the growth stage, being more noticeable if the salts affect the plant during the early stages of growth. In some cases, mild salt effects may go entirely unnoticed because of a uniform reduction in growth across an entire field.

Salts that contribute to a salinity problem are water soluble and readily transported by water. A portion of the salts that accumulate from prior irrigations can be moved (leached) below the rooting depth if more irrigation water infiltrates the soil than is used by the crop during the crop season. Leaching is the key to controlling a water quality-related salinity problem. Over a period of time, salt removal by leaching must equal or exceed the salt additions from the applied water to prevent salt building up to a damaging concentration. The amount of leaching required is dependent upon the irrigation water quality and the salinity tolerance of the crop grown (Dewis and Freitas,1970).

Salt content of the root zone varies with depth. It varies from approximately that of the irrigation water near the soil surface to many times that of the applied water at the bottom of the rooting depth. Salt concentration increases with depth due to plants extracting water but leaving salts behind in a greatly reduced volume of soil water. Each subsequent irrigation pushes (leaches) the salts deeper into the root zone where they continue to accumulate until leached. The lower rooting depth salinity will depend upon the leaching that has occurred.

Following irrigation, the most readily available water is in the upper root zone—a low salinity area. As the crop uses water, the upper root zone becomes depleted and the zone of most readily available water changes toward the deeper parts as the time interval between irrigations is extended. These lower depths are usually more salty. The crop does not respond to the extremes of low or high salinity in the rooting depth but integrates water availability and takes water from wherever it is most readily available. Irrigation timing is thus important in maintaining high soil-water availability and reducing the problems caused when the crop must draw a significant portion of its water from the less available, higher salinity soil-water deeper in the root zone. For good crop production, equal importance must be given to maintaining high soil-water availability and to leaching accumulated salts from the rooting depth before the salt concentration exceeds the tolerance of the plant.

For crops irrigated infrequently, as is normal when using surface methods and conventional irrigation management, crop yield is best correlated with the average root zone salinity, but for crops irrigated on a daily, or near daily basis (localized or drip irrigation) crop yields are better correlated with the water-uptake weighted root zone salinity (Rhoades 1982). The differences are not great but may become important in the higher range of salinity. In this paper, discussions are based on crop response to the average root zone salinity.

In irrigated agriculture, many salinity problems are associated with or strongly influenced by a shallow water table (within 2 meters of the surface). Salts accumulate in this water table and frequently become an important additional source of salt that moves upward into the crop root zone. Control of an existing shallow water table is thus essential to salinity control and to successful long-term irrigated agriculture. Higher salinity water requires appreciable extra water for leaching, which adds greatly to a potential water table (drainage) problem and makes long-term irrigated agriculture nearly impossible to achieve without adequate drainage. If drainage is adequate, salinity control becomes simply good management to ensure that the crop is adequately supplied with water at all times and that enough leaching water is applied to control salts within the tolerance of the crop.

Specific uses have different quality needs and one water supply is considered more acceptable (of better quality) if it produces better results or causes fewer problems than an alternative water supply. For example, good quality river water which can be used successfully for irrigation may, because of its sediment load, be unacceptable for municipal use without treatment to remove the sediment. Similarly, snowmelt water of excellent quality for municipal use may be too corrosive for industrial use without treatment to reduce its corrosion potential (Perkins, 1981and Clarke, 1980).

The ideal situation is to have several supplies from which to make a selection, but normally only one supply is available. In this case, the quality of the available supply must be evaluated to see how it fits the intended use. Most of the experience in using water of different qualities has been gained from observations and detailed study of problems that develop following use. The cause and effect relationship between a water constituent and the observed problem then results in an evaluation of quality of degree of acceptability. With sufficient reported experiences and measured responses, certain constituents emerge as indicators of quality-related problems. These characteristics are then organized into guidelines related to suitability for use. Each new set of guidelines builds upon the previous set to improve the predictive capability. Numerous such guidelines have become available covering many types of use (WHO,1973).

Table 16.1: Laboratory Determinations Needed to Evaluate Common Irrigation Water Quality Problems (Rahman, *et al.*, 1979)

Water Parameter	Symbol	Unit[1]	Usual Range in Irrigation Water	
Salinity				
Salt Content				
Electrical Conductivity	EC_w	dS/m	0–3	dS/m
(or)				
Total Dissolved Solids	TDS	mg/l	0–2000	mg/l
Cations and Anions				
Calcium	Ca^{++}	me/l	0–20	me/l
Magnesium	Mg^{++}	me/l	0–5	me/l
Sodium	Na^+	me/l	0–40	me/l
Carbonate	CO_3^-	me/l	0–.1	me/l
Bicarbonate	HCO_3^-	me/l	0–10	me/l
Chloride	Cl^-	me/l	0–30	me/l
Sulphate	SO_4^-	me/l	0–20	me/l
NUTRIENTS[2]				
Nitrate-Nitrogen	$NO_3\text{-}N$	mg/l	0–10	mg/l
Ammonium-Nitrogen	$NH_4\text{-}N$	mg/l	0–5	mg/l
Phosphate-Phosphorus	$PO_4\text{-}P$	mg/l	0–2	mg/l
Potassium	K^+	mg/l	0–2	mg/l
MISCELLANEOUS				
Boron	B	mg/l	0–2	mg/l
Acid/Basicity	pH	1–14	6.0–8.5	
Sodium Adsorption Ratio[3]	SAR	$(me/l)^{1,\,2}$	0–15	

1 dS/m = deciSiemen/metre in S.I. units (equivalent to 1 mmho/cm = 1 millimmho/centi-metre) mg/l = milligram per litre ≅ parts per million (ppm). me/l = milliequivalent per litre (mg/l ÷ equivalent weight = me/l); in SI units, 1 me/l= 1 millimol/litre adjusted for electron charge.

2 NO_3–N means the laboratory will analyse for NO_3 but will report the NO_3 in terms of chemically equivalent nitrogen. Similarly, for NH_4-N, the laboratory will analyze for NH_4 but report in terms of chemically equivalent elemental nitrogen. The total nitrogen available to the plant will be the sum of the equivalent elemental nitrogen. The same reporting method is used for phosphorus.

3 SAR is calculated from the Na, Ca and Mg reported in me/l (*see* Figure 16.1).

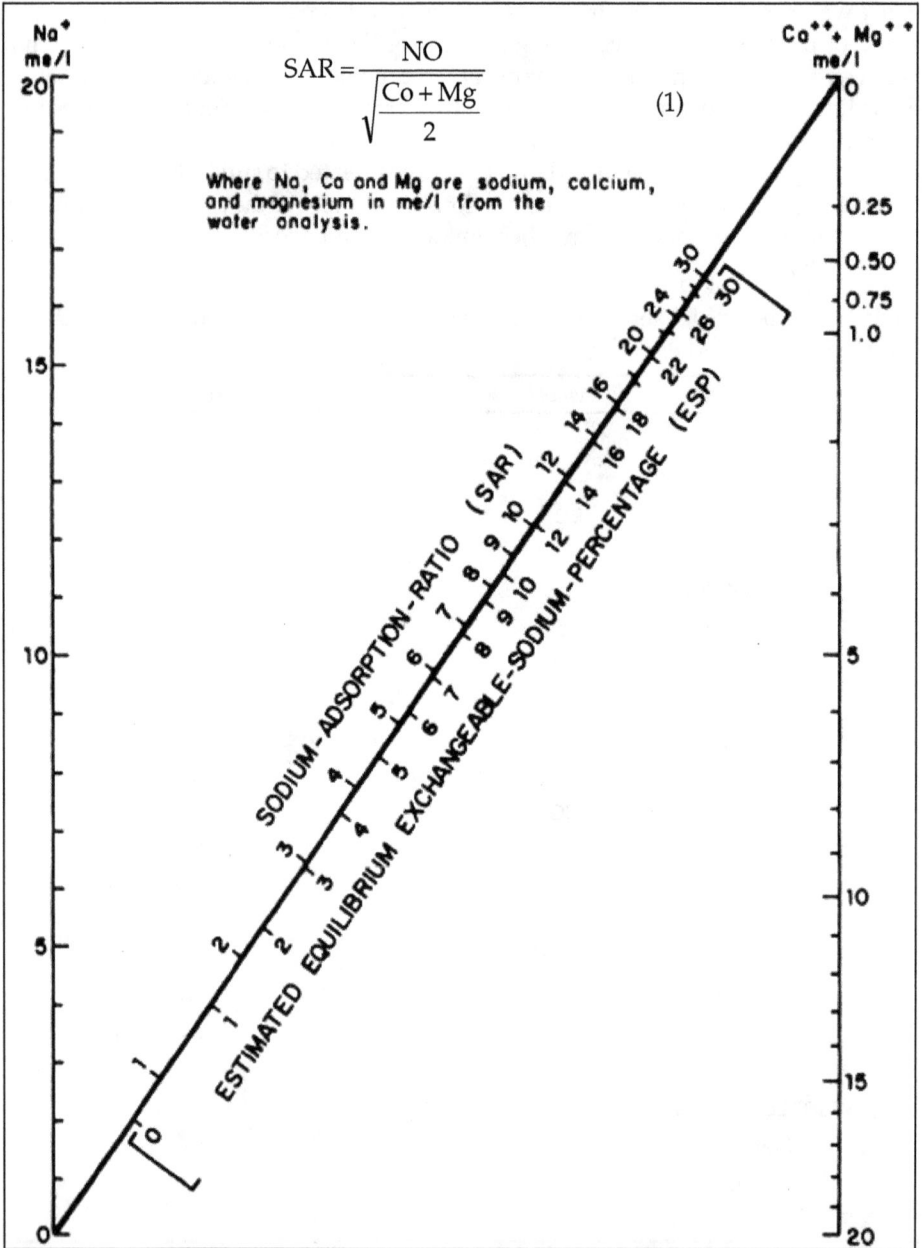

Figure 16.1: SAR Ratio Scale Coordination with ESP (Subramanian, 1979)

Build-up of Soil Salinity

Salts are added to the soil with each irrigation. These salts will reduce crop yield if they accumulate in the rooting depth to damaging concentrations. The crop removes much of the applied water from the soil to meet its evapo-transpiration demand (ET) but leaves most of the salt behind to concentrate in the shrinking volume of soil water. At each irrigation, more salt is added with the applied water. A portion of the added salt must be leached from the root zone before the concentration affects crop yield. Leaching is done by applying sufficient water so that a portion percolates through and below the entire root zone carrying with it a portion of the accumulated salts. The fraction of applied water that passes through the entire rooting depth and percolates below is called the leaching fraction (LF).

$$\text{Leaching fraction (LF)} = \frac{\text{Depth of water leached below the root zone}}{\text{Depth of water applied at the surface}} \qquad (2)$$

After much successive irrigation, the salt accumulation in the soil will approach some equilibrium concentration based on the salinity of the applied water and the leaching fraction. A high leaching fraction (LF = 0.5) results in less salt accumulation than a lower leaching fraction (LF = 0.1). If the water salinity (EC_w) and the leaching fraction (LF) are known or can be estimated, both the salinity of the drainage water that percolates below the rooting depth and the average root zone salinity can be estimated. The salinity of the drainage water can be estimated from the equation:

$$EC_{dw} = \frac{EC_w}{LF} \qquad (3)$$

where,

EC_{dw}: Salinity of the drainage water percolating below the root zone (equal to salinity of soil-water, EC_{sw})

EC_w: Salinity of the applied irrigation water

LF: Leaching fraction

In Example 1, the leaching fraction and water quality are used to predict drainage water quality. The plant, however, is only exposed to this drainage water salinity at the lowest part of the root zone. The salinity in this lower portion of the root zone tends to be higher than in the upper portion due to its much lower leaching fraction. The crop responds, however, to the average root zone soil salinity and not to the extremes of either the upper or lower zones (Mann and Deutscher, 1978).

Example 1: Calculation of Concentration of Deep Percolation from the Bottom of the Root Zone

A crop is irrigated with water of an electrical conductivity (EC_w) of 1 dS/m. The crop is irrigated to achieve a leaching fraction of 0.15 (assumes that 85 per cent of the applied water is used by the crop or evaporates from the soil surface).

Given: EC_w = 1 dS/m

LF = 0.15

Explanation

The concentration of the soil-water percolating below the root zone (EC_{sw}) is equivalent to the concentration of the drainage water (EC_{dw}) accumulating below the root zone. The salinity of the deep percolation from the bottom of the root zone (drainage water) can be estimated by using equation (3):

$$EC_{dw} = EC_{sw} = \frac{EC_w}{LF} \tag{3}$$

$$EC_{dw} = \frac{1}{0.15} = 6.7 \, ds/m$$

The salinity of the soil-water that is percolating from the bottom of the root zone (EC_{dw}) will be approximately 6.7 dS/m.

Equation (3) can also be used to predict average soil-water salinity (EC_{sw}) in the rooting depth if certain assumptions are made regarding water use within the root zone. The guidelines of Table 16.1 assume that 40, 30, 20 and 10 per cent of the water used by the crop comes, respectively, from the upper to lower quarter of the rooting depth. This water use pattern closely fits conditions found under normal irrigation practices. An illustration is given in Example 2 where the above water use pattern is used to estimate average soil-water salinity (EC_{sw}) (Doorenbos and Pruitt, 1977).

Example 2 shows that with a 15 per cent leaching fraction and a 40-30-20-10 water use pattern the average soil-water salinity (EC_{sw}) is approximately 3.2 times more concentrated than the applied irrigation water. At a leaching fraction of 20 per cent, the average EC_{sw} is 2.7 times the salinity of the applied irrigation water (EC_w). The guidelines of Table 1 were developed assuming a 15–20 per cent leaching fraction range which results in average soil-water salinity (EC_{sw}) approximately 3 times that of the applied water. The soil-water salinity (EC_{sw}) is the average root zone salinity to which the plant is exposed. It is difficult to measure. Salinity measurement is normally done on a saturation extract of the soil and referred to as the soil salinity (EC_e). This soil salinity, (EC_e), is approximately equal to one-half of the soil-water salinity (EC_{sw}). As a general rule of thumb, at a 15–20 per cent leaching fraction, salinity of the applied water (EC_w) can be used to predict or estimate soil-water salinity (EC_{sw}) or soil salinity (EC_e) using the following equations:

$$EC_{sw} = 3 \, EC_w \tag{4}$$

$$EC_e = 1.5 \, EC_w \tag{5}$$

$$EC_{sw} = 2 \, EC_e \tag{6}$$

If irrigation practices result in greater or less leaching than the 15–20 per cent LF assumed in the guidelines of Table 16.1, a more correct concentration factor can be calculated using a new estimated average leaching fraction and the procedure illustrated in Example 2. Table 16.2 lists concentration factors for a wide range of leaching fractions (LF = 0.05 to 0.80). The predicted average soil salinity (EC_e) is estimated by multiplying the irrigation water salinity (EC_w) by the appropriate concentration factor for the estimated leaching fraction (see equation (8) in Table 16.2). These predicted average soil salinities reflect changes due to long-term water use and not short term changes that may occur within a season or between irrigations. Figure 16.2 illustrates typical soil salinity profiles that can be identified and are typical of salinity distribution in the crop root zone after several years of irrigation with one water source and closely similar leaching fractions.

Salinity Effects on Crops

The primary objective of irrigation is to provide a crop with adequate and timely amounts of water, thus avoiding yield loss caused by extended periods of water stress during stages of crop growth that are sensitive to water shortages. However, during repeated irrigations, the salts in the irrigation water can accumulate in the soil, reducing water available to the crop and hastening the onset of a water shortage. Understanding how this occurs will help suggest ways to counter the effect and reduce the probability of a loss in yield.

The plant extracts water from the soil by exerting an absorptive force greater than that which holds the water to the soil. If the plant cannot make sufficient internal adjustment and exert enough force, it is not able to extract sufficient water and will suffer water stress. This happens when the soil becomes too dry. Salt in the soil-water increases the force the plant must exert to extract water and this additional force is referred to as the osmotic effect or osmotic potential. For example, if two otherwise identical soils are at the same water content but one is salt-free and the other is salty, the plant can extract and use more water from the salt-free soil than from the salty soil. The reasons are not easily explained. Salts have an affinity for water. If the water contains salt, more energy per unit of water must be expended by the plant to absorb relatively salt-free water from a relatively salty soil-water solution (Prichard, *et al.*, 1983, FAO/UNESCO, 1973).

Example 2: Determination of Average Root Zone Salinity

The average root zone salinity can be calculated using the average of five points in the rooting depth. The following procedure can be used to estimate the average root zone salinity to which the crop responds.

Assumptions

1. Applied water salinity $(EC_w) = 1 \, dS/m$.

2. Crop water demand (ET) = 1000 mm/season.

3. The crop water use pattern is 40-30-20-10. This means the crop will get 40 per cent of its ET demand from the upper quarter of the root zone, 30 per cent from the next

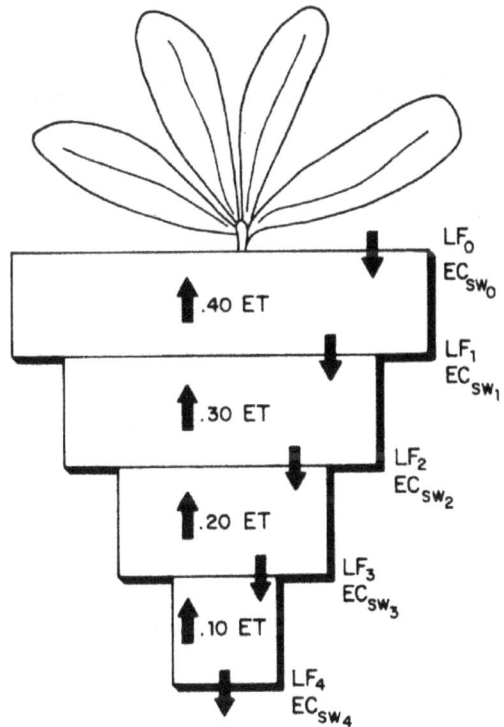

quarter, 20 per cent from the next, and 10 per cent from the lowest quarter. Crop water use will increase the concentration of the soil-water which drains into the next quarter (EC_{sw}) of the root zone.

4. Desired leaching fraction (LF) = 0.15. The leaching fraction of 0.15 means that 15 per cent of the applied irrigation water entering the surface percolates below the root zone and 85 per cent replaces water used by the crop to meet its ET demand and water lost by surface evaporation.

Explanation

1. Five points in the root zone are used to determine the average root zone salinity. These five points are soil-water salinity at (1) the soil surface, (EC_{sw0}); (2) bottom of the upper quarter of the root zone, (EC_{sw1}); (3) bottom of the second quarter depth, (EC_{sw2}); (4) bottom of the third quarter, (EC_{sw3}) and (5) bottom of the fourth quarter or the soil-water draining from the root zone (EC_{sw4}) which is equivalent to the salinity of the drainage water (EC_{dw}).

2. With a LF of 0.15, the applied water (AW) needed to meet both the crop ET and the LF is determined from the following equation:

$$AW = \frac{ET}{1-LF} = 1176 \text{ mm of water} \tag{7}$$

3. Since essentially all the applied water enters and leaches through the soil surface, effectively removing any accumulated salts, the salinity of the soil water at the surface (EC_{sw0}) must be very close to the salinity of the applied water as shown using equation (3) and assuming $LF_0 = 1.0$.

$$EC_{dw_0} = EC_{sw_0} = \frac{EC_w}{LF_0} = \frac{1}{1} = 1 \text{ ds/m}$$

4. The salinity of the soil-water draining from the bottom of each root zone quarter is found by determining the leaching fraction for that quarter using equation (2) and then determining the soil-water salinity using equation (3).

$$LF = \frac{\text{Water leached}}{\text{Water applied}} \qquad EC_{sw} = \frac{EC_w}{LF}$$

For the bottom of the first quarter:

$$LF_1 = \frac{1176 - 0.40(1000)}{1176} = 0.66 \qquad EC_{sw_1} = \frac{EC_w}{LF_1} = 1.5 \text{ ds/m}$$

—at the bottom of the second quarter:

$$LF_2 = \frac{1176 - 0.40(1000) - 0.30(1000)}{1176} = 0.40 \qquad EC_{sw_2} = \frac{EC_w}{LF_2} = 2.5 \text{ ds/m}$$

—at the bottom of the third quarter:

$$LF_3 = \frac{1176 - 0.40(1000) - 0.30(1000) - 0.20(1000)}{1176} = 0.23$$

$$EC_{sw_3} = \frac{EC_w}{LF_3} = 4.3 \, ds/m$$

—at the bottom of the root zone (fourth quarter):

$$LF_4 = \frac{1176 - 0.40(1000) - 0.30(1000) - 0.20(1000) - 0.10(1000)}{1176} = 0.15$$

$$EC_{sw_4} = \frac{EC_w}{LF_4} = 6.7 \, ds/m$$

5. The average soil-water salinity of the root zone is found by taking the average of the five root zone salinities found above:

$$EC_{sw} = \frac{EC_{sw_0} + EC_{sw_1} + EC_{sw_2} + EC_{sw_3} + EC_{sw_4}}{5}$$

$$EC_{sw} = \frac{1.0 + 1.5 + 2.5 + 4.3 + 6.7}{5} = 3.2 \, ds/m$$

6. This calculation shows that the average soil-water salinity of the root zone will be 3.2 times as concentrated as the applied water.

Table 16.2: Concentration Factors (X) for Predicting Soil Salinity $(EC_e)[1]$ from Irrigation Water Salinity (EC_w) and the Leaching Fraction (LF) (Unesco-UNDP, 1970)

Leaching Fraction (LF)	Applied Water Needed (Per cent of ET)	Concentration Factor [2] (X)
0.05	105.3	3.2
0.10	111.1	2.1
0.15	117.6	1.6
0.20	125.0	1.3
0.25	133.3	1.2
0.30	142.9	1.0
0.40	166.7	0.9
0.50	200.0	0.8
0.60	250.0	0.7
0.70	333.3	0.6
0.80	500.0	0.6

1 The equation for predicting the soil salinity expected after several years of irrigation with water of salinity EC_w is:

EC_e (dS/m) = EC_w (dS/m).X   (8)

2 The concentration factor is found by using a crop water use pattern of 40-30-20-10. The procedure is shown in example 2.

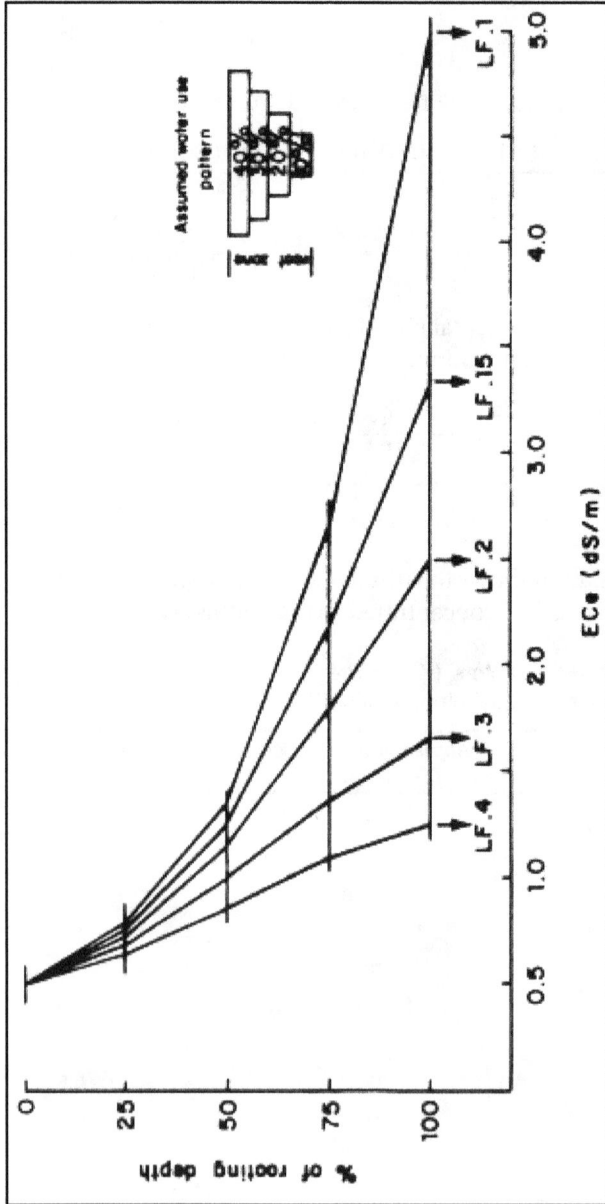

Figure 16.2: Salinity Profile Expected to Develop after Long-term Use of Water of EC_w = 1.0 dS/m at Various Leaching Fractions (LF)

For all practical purposes, the added energy required to absorb water from the salty soil (osmotic potential) is additive to the energy required to absorb water from a salt-free soil (soil-water potential). The cumulative effect is illustrated in Figure 16.3 and results in an important reduction in water available to the crop as salinity increases. Salinity effects are closely analogous to those of drought as both result in water stress and reduced growth. Stunting, leaf damage and necrosis or obvious injury to the plant is only noticeable after prolonged exposure to relatively high salinity.

The previous discussion showed how the concentration of salts in the soil varied with leaching fraction and depth in the root zone and resulted in an increase in concentration as the leaching fraction decreases or with increasing depth in the root zone. As the soil dries, the plant is also exposed to a continually changing water availability in each portion of the rooting depth since the soil-water content (soil-water potential) and soil-water salinity (osmotic potential) are both changing as the plant uses water between irrigations. The plant absorbs water but most of the salt is excluded and left behind in the root zone in a shrinking volume of soil-water. Figure 16.4 shows that following irrigation, the soil salinity is not constant with depth. Following each irrigation, the soil-water content at each depth in the root zone is near the maximum, and the concentration of dissolved salts is near the minimum. Each changes, however, as water is used by the crop between irrigations.

The plant exerts its absorptive force throughout the rooting depth and takes water from wherever most readily available (the least resistance to absorption). Usually this is the upper root zone, the area most frequently replenished by irrigation and rainfall. Since more water passes through this upper root zone, it is more thoroughly leached and the osmotic or salinity effects are much less than at greater depths. Between irrigations, the upper root zone dries more rapidly than the lower because of the proliferation of roots in this zone which extract the readily available soil moisture. As the plant depletes the soil-water, a water extraction pattern develops. The extraction pattern of 40, 30, 20 and 10 per cent for the upper to lower quarters of the root zone is assumed in the guidelines in Table 16.1. This closely fits water extraction patterns under normal irrigation practices and is assumed throughout this paper.

The pattern for water uptake is closely related to the frequency of irrigation. With infrequent irrigations, as assumed for the guidelines in Table 16.1, the typical extraction pattern is 40-30-20-10, but for more frequent irrigations the water uptake pattern is skewed towards greater uptake from the upper root zone and less from the lower and the crop rooting depth tends to be at shallower depths. A typical extraction pattern might be 60-30-7-3. Whatever the frequency, irrigations must be timed to supply adequate water and prevent crop moisture stress between irrigations, especially if soil salinity is also affecting water availability.

Assumptions

1. Salinity in irrigation water x 3 = salinity of soil-water.
2. No removals or additions of salts from the soil-water.
3. Soil-water depletion effects and salinity effects on water availability are additive (EC x.36 = osmotic pressure).

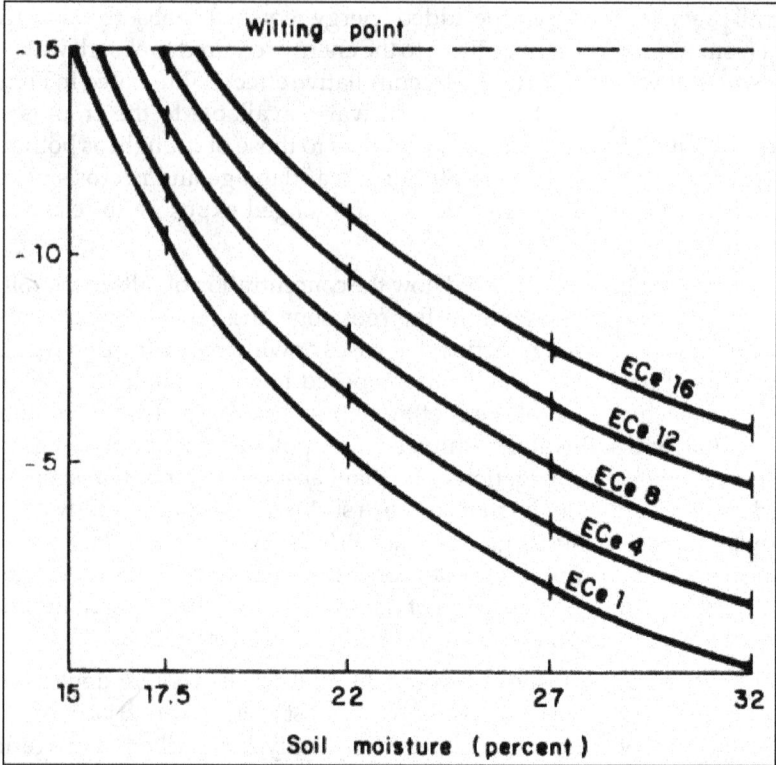

Figure 16.3: Soil Moisture Retention Curves for a Clay-Loam Soil at Varying Degrees of Soil Salinity (EC_e)

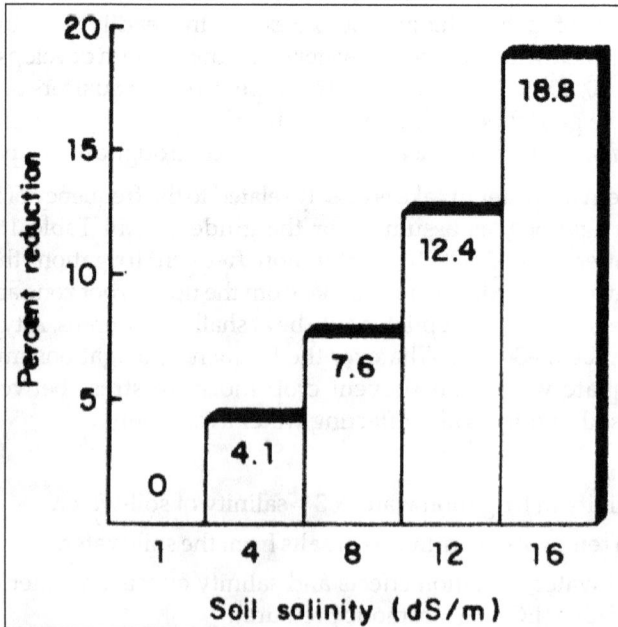

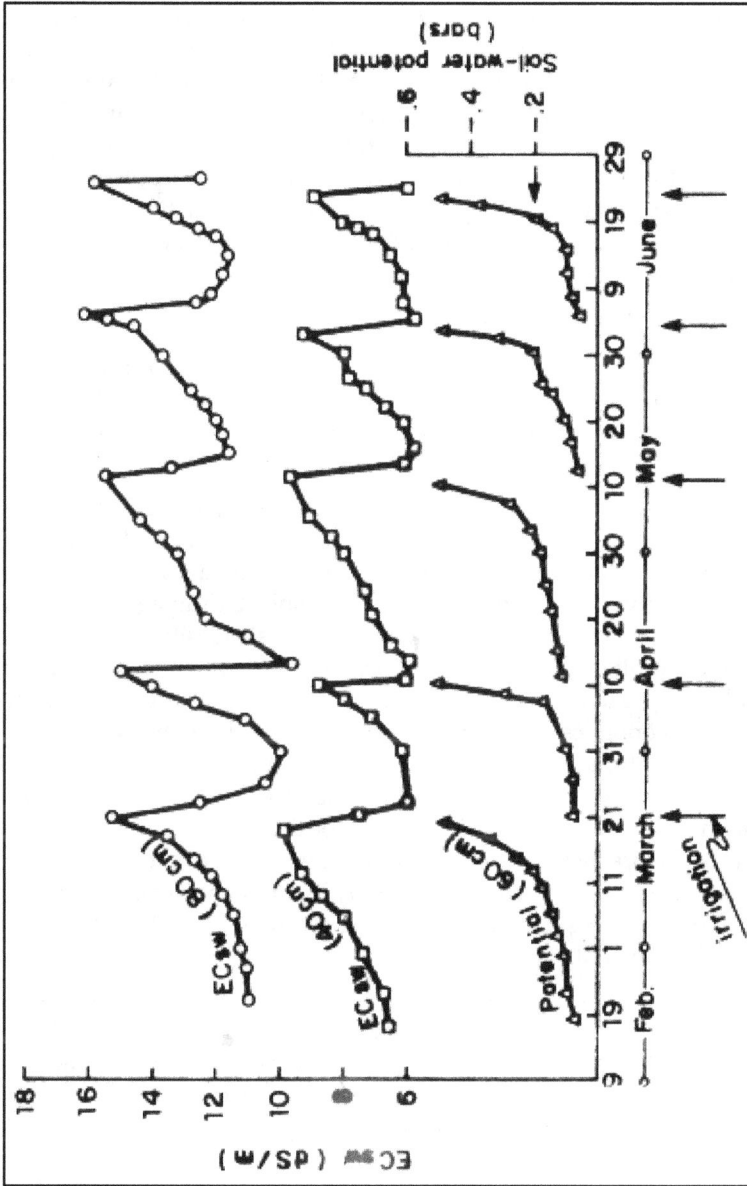

Figure 16.4: Change in Salinity of Soil-Water (EC$_{sw}$) between Irrigations of Alfalfa Due to ET Use of Stored Water (Rhoades, 1972)

4. Available soil-water is difference between per cent soil-water at water holding capacity and at wilting point.

5. Evapotranspiration (ET) by the crop is removing water from the soil.

When the upper rooting depth is well supplied with water, salinity in the lower root zone becomes less important. However, if periods between irrigations are extended and the crop must extract a significant portion of its water from the lower depths, the deeper root zone salinity becomes important particularly if, in the latter stages of a 'dry-down' (soil moisture depletion) period between irrigations, a high crop water demand should occur, such as on a hot, windy day. In this case, absorption and water movement toward the roots may not be fast enough to supply the crop and a severe water stress results. Reduced yields or crop damage can be expected for most crops when there is a shortage of water for a significant period of time (FAO/UNESCO, 1973).

Management of Salinity Problems

The objective of salinity control is to maintain an acceptable crop yield. Several management options are available for salinity control and these will be discussed here as separate options, but in practice a combination may be used to solve the problem.

The preceding section discussed the importance of (1) leaching salts out of the root zone before they build up to levels that might affect yields, and (2) maintaining adequate soil-water availability at all times. Adequate drainage is equally important and long-term salinity control is not otherwise possible. If drainage is adequate, the depth of water required for leaching depends on the salt sensitivity of the crop and the salinity of the applied water. When salinity is high, the depth of leaching water needed may be too great, making it necessary to change to a more salt tolerant crop, provided market economics will allow this. In dealing with a major salinity problem related to water quality, a cropping change is considered a drastic step and will only be taken when less severe options have failed to maintain economic production. Leaching, on the other hand, is a basic step in production even for water of the best quality and must be practiced when necessary to avoid salt accumulation that could ultimately affect production. Leaching can only be done, however, if the drainage below the crop root zone is sufficient to prevent a rise in the water table so that it is not a source of salt by itself (Abrol, *et al.*, 1982).

Drainage, leaching and changes to more salt tolerant crops are used to avoid the impact of long-term salinity build-up but other cultural practices may also be needed to deal with possible short-term or temporary increases in salinity which may be equally detrimental to crop yield. Many cultural practices such as more frequent irrigation, land grading, timing of fertilization and methods of seeding make salinity management easier.

If there is a high level of salinity not resulting from water quality, a soil drainage and reclamation programme may be needed and short-term cropping changes may

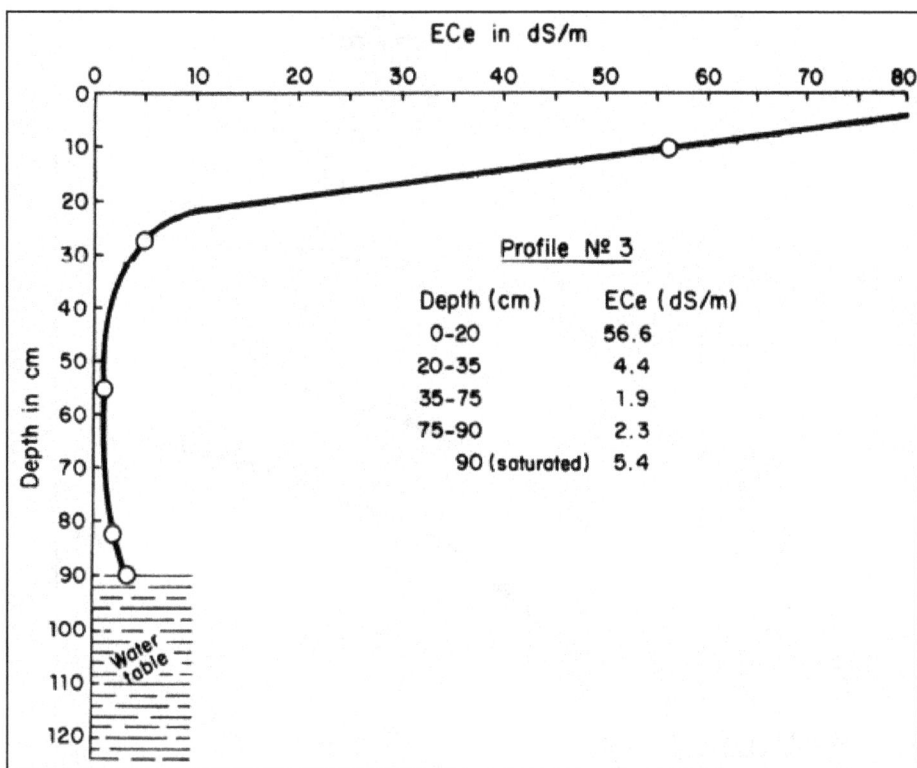

Figure 16.5: Salinity Profile with a High Water Table (Mohamed and Amer, 1972)

also need to be made. After soil reclamation, the permanent cropping pattern will be determined by water quality. In a few instances, an alternative water supply may be available for periodic use or can be blended with a poorer water supply to diminish a quality-related hazard (Nielson and Cannon, 1975). These alternatives, including drainage, leaching, cropping changes and cultural practices, will be discussed in more detail in the following sections.

The salt tolerance data of Table 16.3 are used in the calculation of the leaching requirement. Figure 16.6 can also be used to estimate the leaching requirement if crop tolerance grouping and water salinity are known, as discussed in the previous section. If the exact cropping patterns or rotations are not known for a new area, the leaching requirement must be based on the least tolerant of the crops adapted to the area (Abrol, 1982). In those instances where soil salinity cannot be maintained within acceptable limits of preferred sensitive crops, changing to more tolerant crops will raise the area's production potential. In case of doubt as to the effect of the water salinity on crop production, a pilot study should be undertaken to demonstrate the feasibility for irrigation and the outlook for economic success.

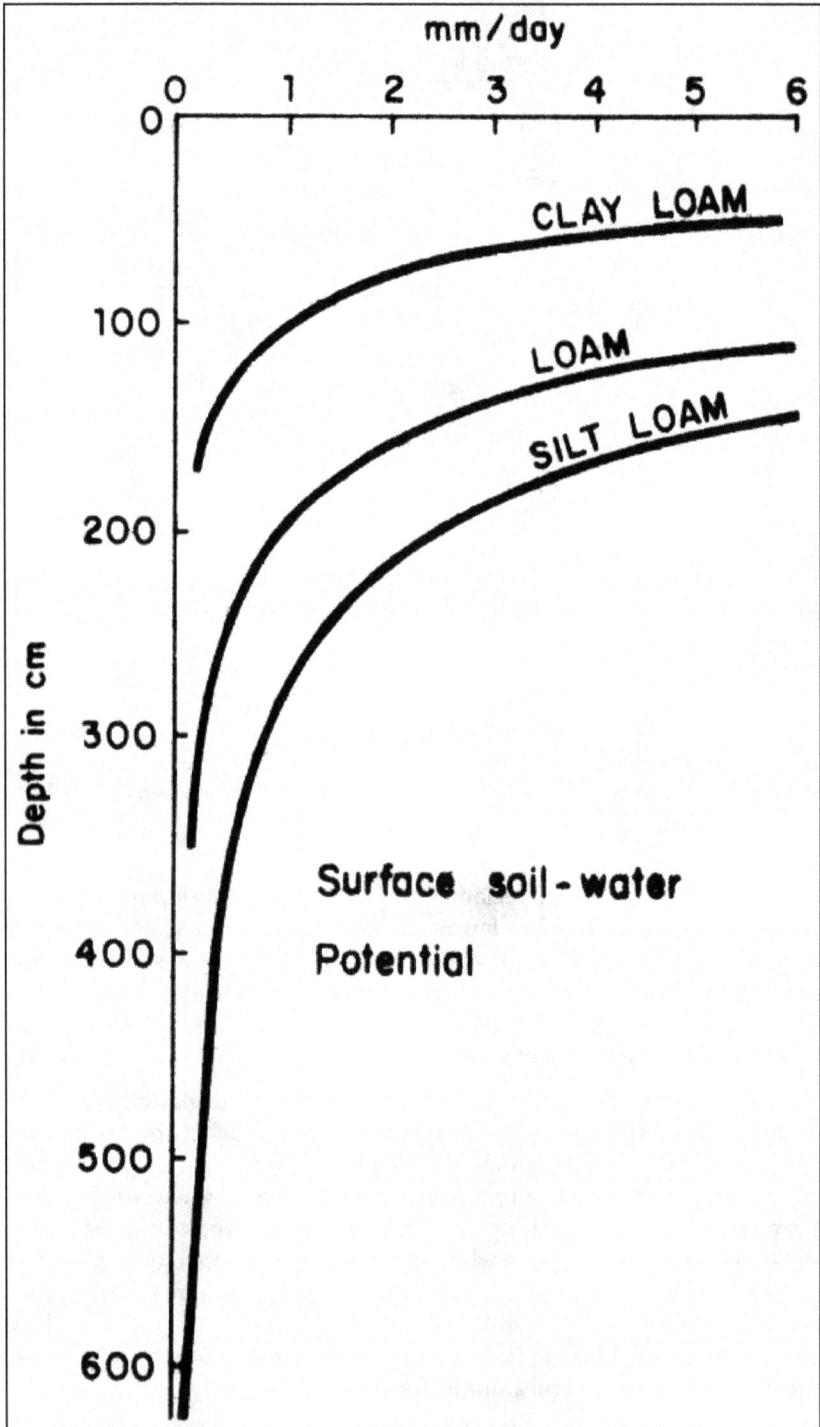

Figure 16.6

Table 16.3: Crop Tolerance and Yield Potential of Selected Crops as Influenced by Irrigation Water Salinity (EC_w) or Soil Salinity (EC_e)[1] Yield Potential[2] (Oster, *et al.*, 1983)

Crops	100%		90%		75%		50%		0% "maximum"[3]	
	EC_e	EC_w	EC_e	EC_w	EC_e	EC_w	EC_e	EC_w	EC_e	EC_w
FIELD CROPS										
Barley (*Hordeum vulgare*)[4]	8.0	5.3	10	6.7	13	8.7	18	12	28	19
Cotton (*Gossypium hirsutum*)	7.7	5.1	9.6	6.4	13	8.4	17	12	27	18
Sugarbeet (*Beta vulgaris*)[5]	7.0	4.7	8.7	5.8	11	7.5	15	10	24	16
Sorghum (*Sorghum bicolor*)	6.8	4.5	7.4	5.0	8.4	5.6	9.9	6.7	13	8.7
Wheat (*Triticum aestivum*)[4],[6]	6.0	4.0	7.4	4.9	9.5	6.3	13	8.7	20	13
Wheat, durum (*Triticum turgidum*)	5.7	3.8	7.6	5.0	10	6.9	15	10	24	16
Soybean (*Glycine max*)	5.0	3.3	5.5	3.7	6.3	4.2	7.5	5.0	10	6.7
Cowpea (*Vigna unguiculata*)	4.9	3.3	5.7	3.8	7.0	4.7	9.1	6.0	13	8.8
Groundnut (Peanut) (*Arachis hypogaea*)	3.2	2.1	3.5	2.4	4.1	2.7	4.9	3.3	6.6	4.4
Rice (paddy) (*Oriza sativa*)	3.0	2.0	3.8	2.6	5.1	3.4	7.2	4.8	11	7.6
Sugarcane (*Saccharum officinarum*)	1.7	1.1	3.4	2.3	5.9	4.0	10	6.8	19	12
Corn (maize) (*Zea mays*)	1.7	1.1	2.5	1.7	3.8	2.5	5.9	3.9	10	6.7
Flax (*Linum usitatissimum*)	1.7	1.1	2.5	1.7	3.8	2.5	5.9	3.9	10	6.7
Broadbean (*Vicia faba*)	1.5	1.1	2.6	1.8	4.2	2.0	6.8	4.5	12	8.0
Bean (*Phaseolus vulgaris*)	1.0	0.7	1.5	1.0	2.3	1.5	3.6	2.4	6.3	4.2
VEGETABLE CROPS										
Squash, zucchini (courgette) (*Cucurbita pepo melopepo*)	4.7	3.1	5.8	3.8	7.4	4.9	10	6.7	15	10
Beet, red (*Beta vulgaris*)[5]	4.0	2.7	5.1	3.4	6.8	4.5	9.6	6.4	15	10
Squash, scallop (*Cucurbita pepo melopepo*)	3.2	2.1	3.8	2.6	4.8	3.2	6.3	4.2	9.4	6.3
Broccoli (*Brassica oleracea botrytis*)	2.8	1.9	3.9	2.6	5.5	3.7	8.2	5.5	14	9.1
Tomato (*Lycopersicon esculentum*)	2.5	1.7	3.5	2.3	5.0	3.4	7.6	5.0	13	8.4
Cucumber (*Cucumis sativus*)	2.5	1.7	3.3	2.2	4.4	2.9	6.3	4.2	10	6.8
Spinach (*Spinacia oleracea*)	2.0	1.3	3.3	2.2	5.3	3.5	8.6	5.7	15	10
Celery (*Apium graveolens*)	1.8	1.2	3.4	2.3	5.8	3.9	9.9	6.6	18	12
Cabbage (*Brassica oleracea capitata*)	1.8	1.2	2.8	1.9	4.4	2.9	7.0	4.6	12	8.1
Potato (*Solanum tuberosum*)	1.7	1.1	2.5	1.7	3.8	2.5	5.9	3.9	10	6.7
Corn, sweet (maize) (*Zea mays*)	1.7	1.1	2.5	1.7	3.8	2.5	5.9	3.9	10	6.7
Sweet potato (*Ipomoea batatas*)	1.5	1.0	2.4	1.6	3.8	2.5	6.0	4.0	11	7.1
Pepper (*Capsicum annuum*)	1.5	1.0	2.2	1.5	3.3	2.2	5.1	3.4	8.6	5.8
Lettuce (*Lactuca sativa*)	1.3	0.9	2.1	1.4	3.2	2.1	5.1	3.4	9.0	6.0
Radish (*Raphanus sativus*)	1.2	0.8	2.0	1.3	3.1	2.1	5.0	3.4	8.9	5.9

Contd...

Table 16.3–Contd...

Crops	100%		90%		75%		50%		0% "maximum"[3]	
	EC_e	EC_w	EC_e	EC_w	EC_e	EC_w	EC_e	EC_w	EC_e	EC_w
Onion (*Allium cepa*)	1.2	0.8	1.8	1.2	2.8	1.8	4.3	2.9	7.4	5.0
Carrot (*Daucus carota*)	1.0	0.7	1.7	1.1	2.8	1.9	4.6	3.0	8.1	5.4
Bean (*Phaseolus vulgaris*)	1.0	0.7	1.5	1.0	2.3	1.5	3.6	2.4	6.3	4.2
Turnip (*Brassica rapa*)	0.9	0.6	2.0	1.3	3.7	2.5	6.5	4.3	12	8.0
Wheatgrass, tall (*Agropyron elongatum*)	7.5	5.0	9.9	6.6	13	9.0	19	13	31	21
Wheatgrass, fairway crested (*Agropyron cristatum*)	7.5	5.0	9.0	6.0	11	7.4	15	9.8	22	15
Bermuda grass (*Cynodon dactylon*)[7]	6.9	4.6	8.5	5.6	11	7.2	15	9.8	23	15
Barley (forage) (*Hordeum vulgare*)[4]	6.0	4.0	7.4	4.9	9.5	6.4	13	8.7	20	13
Ryegrass, perennial (*Lolium perenne*)	5.6	3.7	6.9	4.6	8.9	5.9	12	8.1	19	13
Trefoil, narrowleaf birdsfoot[8] (*Lotus corniculatus tenuifolium*)	5.0	3.3	6.0	4.0	7.5	5.0	10	6.7	15	10
Harding grass (*Phalaris tuberosa*)	4.6	3.1	5.9	3.9	7.9	5.3	11	7.4	18	12
Fescue, tall (*Festuca elatior*)	3.9	2.6	5.5	3.6	7.8	5.2	12	7.8	20	13
Wheatgrass, standard crested (*Agropyron sibiricum*)	3.5	2.3	6.0	4.0	9.8	6.5	16	11	28	19
Vetch, common (*Vicia angustifolia*)	3.0	2.0	3.9	2.6	5.3	3.5	7.6	5.0	12	8.1
Sudan grass (*Sorghum sudanense*)	2.8	1.9	5.1	3.4	8.6	5.7	14	9.6	26	17
Wildrye, beardless (*Elymus triticoides*)	2.7	1.8	4.4	2.9	6.9	4.6	11	7.4	19	13
Cowpea (forage) (*Vigna unguiculata*)	2.5	1.7	3.4	2.3	4.8	3.2	7.1	4.8	12	7.8
Trefoil, big (*Lotus uliginosus*)	2.3	1.5	2.8	1.9	3.6	2.4	4.9	3.3	7.6	5.0
Sesbania (*Sesbania exaltata*)	2.3	1.5	3.7	2.5	5.9	3.9	9.4	6.3	17	11
Sphaerophysa (*Sphaerophysa salsula*)	2.2	1.5	3.6	2.4	5.8	3.8	9.3	6.2	16	11
Alfalfa (*Medicago sativa*)	2.0	1.3	3.4	2.2	5.4	3.6	8.8	5.9	16	10
Lovegrass (*Eragrostis* sp.)[9]	2.0	1.3	3.2	2.1	5.0	3.3	8.0	5.3	14	9.3
Corn (forage) (maize) (*Zea mays*)	1.8	1.2	3.2	2.1	5.2	3.5	8.6	5.7	15	10
Clover, berseem (*Trifolium alexandrinum*)	1.5	1.0	3.2	2.2	5.9	3.9	10	6.8	19	13
Orchard grass (*Dactylis glomerata*)	1.5	1.0	3.1	2.1	5.5	3.7	9.6	6.4	18	12
Foxtail, meadow (*Alopecurus pratensis*)	1.5	1.0	2.5	1.7	4.1	2.7	6.7	4.5	12	7.9
Clover, red (*Trifolium pratense*)	1.5	1.0	2.3	1.6	3.6	2.4	5.7	3.8	9.8	6.6
Clover, alsike (*Trifolium hybridum*)	1.5	1.0	2.3	1.6	3.6	2.4	5.7	3.8	9.8	6.6
Clover, ladino (*Trifolium repens*)	1.5	1.0	2.3	1.6	3.6	2.4	5.7	3.8	9.8	6.6

Contd...

Table 16.3–Contd...

| Crops | 100% | | 90% | | 75% | | 50% | | 0% | |
| | | | | | | | | | "maximum"[3] | |
	EC_e	EC_w	EC_e	EC_w	EC_e	EC_w	EC_e	EC_w	EC_e	EC_w
Clover, strawberry (*Trifolium fragiferum*)	1.5	1.0	2.3	1.6	3.6	2.4	5.7	3.8	9.8	6.6
FRUIT CROPS[10]										
Date palm (*Phoenix dactylifera*)	4.0	2.7	6.8	4.5	11	7.3	18	12	32	21
Grapefruit (*Citrus paradisi*)[11]	1.8	1.2	2.4	1.6	3.4	2.2	4.9	3.3	8.0	5.4
Orange (*Citrus sinensis*)	1.7	1.1	2.3	1.6	3.3	2.2	4.8	3.2	8.0	5.3
Peach (*Prunus persica*)	1.7	1.1	2.2	1.5	2.9	1.9	4.1	2.7	6.5	4.3
Apricot (*Prunus armeniaca*)[11]	1.6	1.1	2.0	1.3	2.6	1.8	3.7	2.5	5.8	3.8
Grape (*Vitus sp.*)[11]	1.5	1.0	2.5	1.7	4.1	2.7	6.7	4.5	12	7.9
Almond (*Prunus dulcis*)[11]	1.5	1.0	2.0	1.4	2.8	1.9	4.1	2.8	6.8	4.5
Plum, prune (*Prunus domestica*)[11]	1.5	1.0	2.1	1.4	2.9	1.9	4.3	2.9	7.1	4.7
Blackberry (*Rubus sp.*)	1.5	1.0	2.0	1.3	2.6	1.8	3.8	2.5	6.0	4.0
Boysenberry (*Rubus ursinus*)	1.5	1.0	2.0	1.3	2.6	1.8	3.8	2.5	6.0	4.0
Strawberry (*Fragaria sp.*)	1.0	0.7	1.3	0.9	1.8	1.2	2.5	1.7	4	2.7

1 Adapted from Maas and Hoffman (1977) and Maas (1984), these data should only serve as a guide to relative tolerances among crops. Absolute tolerances vary depending upon climate, soil conditions and cultural practices. In gypsiferous soils, plants will tolerate about 2 dS/m higher soil salinity (EC_e) than indicated but the water salinity (EC_w) will remain the same as shown in this table.

2 ECe means average root zone salinity as measured by electrical conductivity of the saturation extract of the soil, reported in deciSiemens per metre (dS/m) at 25°C. ECw means electrical conductivity of the irrigation water in deciSiemens per metre (dS/m). The relationship between soil salinity and water salinity ($EC_e = 1.5\ EC_w$) assumes a 15–20 per cent leaching fraction and a 40-30-20-10 per cent water use pattern for the upper to lower quarters of the root zone. These assumptions were used in developing the guidelines in Table 16.1.

3 The zero yield potential or maximum EC_e indicates the theoretical soil salinity (EC_e) at which crop growth ceases.

4 Barley and wheat are less tolerant during germination and seeding stage; EC_e should not exceed 4–5 dS/m in the upper soil during this period.

5 Beets are more sensitive during germination; EC_e should not exceed 3 dS/m in the seeding area for garden beets and sugar beets.

6 Semi-dwarf, short cultivars may be less tolerant.

7 Tolerance given is an average of several varieties; Suwannee and Coastal Bermuda grass are about 20 per cent more tolerant, while Common and Greenfield Bermuda grass are about 20 per cent less tolerant.

8 Broadleaf Birdsfoot Trefoil seems less tolerant than Narrowleaf Birdsfoot Trefoil.

9 Tolerance given is an average for Boer, Wilman, Sand and Weeping Lovegrass; Lehman Lovegrass seems about 50 per cent more tolerant.

10 These data are applicable when rootstocks are used that do not accumulate Na+ and Cl-rapidly or when these ions do not predominate in the soil. If either ions do, refer to the toxicity discussion in Section 4.

11 Tolerance evaluation is based on tree growth and not on yield.

Table 16.4: Recommnded Maximum Concentrations of Trace Elements in Irrigation Water[1]

Element	Recommended Maximum Concentration[2] (mg/l)	Remarks
Al (aluminium)	5.0	Can cause non-productivity in acid soils (pH < 5.5), but more alkaline soils at pH > 7.0 will precipitate the ion and eliminate any toxicity.
As (arsenic)	0.10	Toxicity to plants varies widely, ranging from 12 mg/l for Sudan grass to less than 0.05 mg/l for rice.
Be (beryllium)	0.10	Toxicity to plants varies widely, ranging from 5 mg/l for kale to 0.5 mg/l for bush beans.
Cd (cadmium)	0.01	Toxic to beans, beets and turnips at concentrations as low as 0.1 mg/l in nutrient solutions. Conservative limits recommended due to its potential for accumulation in plants and soils to concentrations that may be harmful to humans.
Co (cobalt)	0.05	Toxic to tomato plants at 0.1 mg/l in nutrient solution. Tends to be inactivated by neutral and alkaline soils.
Cr (chromium)	0.10	Not generally recognized as an essential growth element. Conservative limits recommended due to lack of knowledge on its toxicity to plants.
Cu (copper)	0.20	Toxic to a number of plants at 0.1 to 1.0 mg/l in nutrient solutions.
F (fluoride)	1.0	Inactivated by neutral and alkaline soils.
Fe (iron)	5.0	Not toxic to plants in aerated soils, but can contribute to soil acidification and loss of availability of essential phosphorus and molybdenum. Overhead sprinkling may result in unsightly deposits on plants, equipment and buildings.
Li (lithium)	2.5	Tolerated by most crops up to 5 mg/l; mobile in soil. Toxic to citrus at low concentrations (<0.075 mg/l). Acts similarly to boron.
Mn (manganese)	0.20	Toxic to a number of crops at a few-tenths to a few mg/l, but usually only in acid soils.
Mo (molybdenum)	0.01	Not toxic to plants at normal concentrations in soil and water. Can be toxic to livestock if forage is grown in soils with high concentrations of available molybdenum.
Ni (nickel)	0.20	Toxic to a number of plants at 0.5 mg/l to 1.0 mg/l; reduced toxicity at neutral or alkaline pH.
Pb (lead)	5.0	Can inhibit plant cell growth at very high concentrations.
Se (selenium)	0.02	Toxic to plants at concentrations as low as 0.025 mg/l and toxic to livestock if forage is grown in soils with relatively high levels of added selenium. An essential element to animals but in very low concentrations.
Sn (tin)	—	—

Contd...

Table 16.4–Contd...

Element	Recommended Maximum Concentration[2] (mg/l)	Remarks
Ti (titanium)	—	Effectively excluded by plants; specific tolerance unknown.
W (tungsten)	—	—
V (vanadium)	0.10	Toxic to many plants at relatively low concentrations.
Zn (zinc)	2.0	Toxic to many plants at widely varying concentrations; reduced toxicity at pH > 6.0 and in fine textured or organic soils.

Source: Loehr R.C. (ed)1977)

1 Adapted from National Academy of Sciences (1972) and Pratt (1972).

2 The maximum concentration is based on a water application rate which is consistent with good irrigation practices (10,000 m^3 per hectare per year). If the water application rate greatly exceeds this, the maximum concentrations should be adjusted downward accordingly. No adjustment should be made for application rates less than 10 000 m^3 per hectare per year. The values given are for water used on a continuous basis at one site.

References

1. Abrol, I.P. (1982). Technology of chemical, physical and biological amelioration of deteriorated soils. In: *Presented at Panel of Experts on Amelioration and Development of Deteriorated Soils* in Egypt, 2–6 May, Cairo.

2. Abrol, I.P., Bhumbla, D.R. and Bhanqqua, G.P. (1972). Some observations on the effect of groundwater table on soil salinization. In: *Proc. Internat. Symp. on New Developments in the Field of Salt Affected Soils*, 4–9 December, Ministry of Agriculture, ARE, Cairo, p. 41–48.

3. Ahmed, B., Kemper, W.D., Haider, G. and Niazi, M.A. (1979). Use of gypsum stones to lower the sodium adsorption ratio of irrigation water. *Soil Science. Soc. Amer. J.*, 43: 698–702.

4. Amer, S.A. (1983). The study and improvement of soil productivity under plastic tunnels and open field conditions. Technical Report No. G: NECP/BAH/501/KUW, FAO, Rome.

5. APHA (American Public Health Association) (1980). *Standard Methods for the Examination of Water and Wastewater*, 15th Edition. APHA–AWWA–WPCF, Washington DC., 1000 p.

6. Ayers, R.S. andWestcot, D.W. (1976). *Water Quality for Agriculture*. FAO Irrigation and Drainage Paper 29, FAO, Rome, 97 p.

7. Back, W. and Hanshaw, B. (1970). Comparison of chemical hydrogeology of carbonate peninsulas of Florida and Yucatan. *J. Hydrol.*, 10: 330–368.

8. Bingham, F.T., Mahler, R.J. and Sposito, G. (1979). Effects of irrigation water composition on exchangeable sodium status of a field soil. *Soil Science*, 127(4): 248–252.

9.   Bouwer, H. (1969). Salt balance, irrigation efficiency and drainage design. *Amer. Soc. civil engineers (ASCE), Proc. 95(IRI)*: 153–170.

10. Chapman, H.D. and Pratt, P.F. (1961). *Methods of Analysis for Soils, Plants and Waters*. University of California, Division of Agricultural Science.

11. Clarke, F.E. (1980). *Corrosion and Encrustation in Water Wells: A Field Guide for Assessment, Prediction and Control*. FAO Irrigation and Drainage Paper 34. FAO, Rome, 95 p.

12. Crites, R.W. (1974). *Irrigation with Wastewater in Bakersfield, California*. Report by Metcalf and Eddy Inc., Palo Alto, California.

13. Dewan, H.C. *et al.* (1978). *Irrigation Water Quality in the Yemen Arab Republic*. Report by Soil and Water Research Section, Central Agricultural Research Organization, UNDP/FAO/YEM 73/010, 9 p.

14. Dewis, J. and Freitas, F. (1970). *Physical and Chemical Methods of Soil and Water Analysis*. FAO Soils Bulletin 10, FAO, Rome, 275 p.

15. Doorenbos, J. and Kassam, A.H. (1979). *Yield Response to Water*. FAO Irrigation and Drainage Paper 33, FAO, Rome, 193 p.

16. Doorenbos, J. andPruitt, W.O. (1977). *Guidelines for Predicting Crop Water Requirements (Revised)*. FAO Irrigation and Drainage Paper 24, FAO, Rome, 143 p.

17. Dutt, G.R., Pennington, D.A. and Turner, F.Jr.(1984). Irrigation as a solution to salinity problems of river basins. In: *Salinity in Water Courses and Reservoirs*, (Ed.) R.H. French. Ann Arbor Science, pp. 465–472.

18. EPA(1979). *Process Design Manual for Land Treatment of Wastewater*. US Environmental Protection Agency.

19. FAO/Unesco (1973). *Irrigation, Drainage and Salinity: An International Sourcebook*. Paris, Unesco/Hutchinson (Publishers), London, 510 p.

20. Feachem, R., McGarry, M. and Mara, D. (Eds.) (1977). *Water, Wastes and Health in Hot Climates*. Wiley, Chichester, UK.

21. Halsey, D. (1974). *Growth and Salt Accumulation with Desert Drip Irrigation*. California Citograph, December, p. 47–48.

22. Hibbs, C.M. and Thilsted, J.P. (1983). Toxicosis in cattle from contaminated well water. *Veterinary and Human Toxicology*, 25(4).

23. Hughes, H.E. and Hanan, J.J. (1978). Effect of salinity in water supplies on greenhouse rose production. *J. Amer. Hortic. Sci.*, 103(5): 694–699.

24. Issar, A., Bein, A. and Mitchaeli, A. (1972). On ancient water on the upper Nubian sandstone aquifer in Central Sinai and Southern Israel. *J. Hydrol.*, 17: 353–374.

25. Lawrence, A.R. and Dharmaguna-Wardena, H.A., (1983). *Vertical* recharge to a confined limestone aquifer in Northwest Sri Lanka. *J. Hydrol.*, 63: 287–297.

26. Loehr, R.C. (Ed.) (1977). *Land as Waste Management Alternative*. Ann Arbor Science, Michigan.

27. Mann, A.W. and Deutscher, R.L. (1978). Hydrogeochemistry of a calcrete-containing aquifer near Lakeway, Western Australia. *J. Hydrol.*, 38: 357–377.

28. Mather, T.H. (1984). *Environmental Management for Vector Control in Rice Fields.* FAO Irrigation and Drainage Paper No. 41. FAO, Rome, 152 p.

29. National Research Council of Canada (1974). Land for waste management. In: *Proc. of an International Conference*, Ottawa.

30. Nielson, R.F. and Cannon, O.S. (1975). Sprinkling with salty well water can cause problems. *Utah Science*, Agricultural Experiment Station, 36(2): 61–63.

31. Oster, J.D. and Rhoades, J.D. (1983). Irrigation with saline water. In: *Soil and Water Newsletter*. University of California Cooperative Extension, Fall 1983, 56: 1–3.

32. Perkins, P.H. (1981). The corrosion resistance of concrete-sanitary engineering structures. *Concrete International*, April, p. 75–81.

33. Pratt, P.F. (1972). *Quality Criteria for Trace Elements in Irrigation Waters.* California Agricultural Experiment Station, 46 p.

34. Prichard, T.L. *et al.* (1983). Relationships of irrigation water salinity and soil-water salinity. *California Agriculture*, July–August, 37(7): 11–15.

35. Pulawski, B. and Obro, H. (1976). Groundwater study of a volcanic area near Bandung, Java, Indonesia. *J. Hydrol.*, 28: 53–72.

36. Rahman, W.A. and Rowell, D.L. (1979). The influence of magnesium in saline and sodic soils: A specific effect or a problem of cation exchange? *J. Soil Science*, 30: 535–546.

37. Sanha, B.K. and Singh, N.T.(1976). Salt distribution around roots of wheat under different transpiration rates. *Plant and Soil*, 44: 141–147.

38. Soil Improvement Committee (1975). *Western Fertilizer Handbook.* California Fertilizer Association, Sacramento, California. 250 p.

39. Suarez, D.L. (1982). Graphical calculation of ion concentrations in calcium carbonate and/or gypsum soil solutions. *J. Environmental Quality*, 11: 302–308.

40. Subramanian, V. (1979). Chemical and suspended-sediment characteristics of rivers of India. *J. Hydrol.*, 44: 37–55.

41. Tanji, K. (1976). *Irrigation Tailwater Management.* Water Science Report No. 4011 (1975–76 Annual Report), UC Davis, April.

42. Taylor, W.H. (1977). *Concrete Technology and Practice*, 4th Edition. McGraw-Hill, Sydney, 450 p.

43. Tipton and Kalmbach Inc. (1972). *Feasibility of the Kufra Agricultural Project.* 1971 Programme Development, 19 p.

44. Tuolumne (1980). Regional water district. Operations manual for the district. Tuolumne regional water District, Sonora, California.

45. Ulrich, A. and Mostafa, M.A.E. (1976). Calcium nutrition of the sugarbeet. *Communications in Soil Science and Plant Nutrition*, 7(5): 483–495.

46. Unesco-UNDP (1970). *Research and Training on Irrigation with Saline Water.* Technical Report of UNDP Project, Tunisia, 5: 256 p.

47. United States Bureau of Reclamation (1975). *Concrete Manual,* 8th Edition, Washington DC, 350 p.

48. Vermeiren, L. and Jobling, G.A. (1980). *Localized Irrigation: Design, Installation, Operation and Evaluation.* FAO Irrigation and Drainage Paper 36, FAO, Rome.

49. WHO (1973). *Reuse of Effluents: Methods of Wastewater Treatment and Health Safeguards.* Report of a WHO meeting of experts. Tech. Rep. No. 517, WHO, Geneva.

50. Worthington, E.B. (Ed.) (1977). *Arid Land Irrigation in Developing Countries: Environmental Problems and Effects.* Pergamon Press, Oxford, 463 p.

Biodiversity of Aquatic Reseources (2012)
Editors: **Mamta Rawat & Sumit Dookia**
Published by: **DAYA PUBLISHING HOUSE, NEW DELHI**

Pages **245–256**

Chapter 17

Water Stress and Economic Considerations

M.A. Khan* and Hemant Mangal

Department of Geography,
Govt. Lohia P.G. College, Churu – 331 001, Rajasthan

Water resources are sources of water that are useful or potentially useful to humans. Uses of water include agricultural sector, industrial sector, household, recreational and environmental activities. Virtually all of these human activities require freshwater for their day to day uses. About 97 per cent of water on the Earth is salty, leaving only 3 per cent as freshwater of which slightly over two third is frozen in glaciers and polar ice caps. The remaining unfrozen freshwater is mainly found as groundwater, with only a small fraction present above ground or in the air.

Freshwater is a renewable resource, yet the world's supply of clean, freshwater is steadily decreasing. Water demand already exceeds supply in many parts of the world and as the world population continues to rise, so too does the water demand. Awareness of the global importance of preserving water for ecosystem services has only recently emerged as, during the 20th century, more than half the world's wetlands have been lost along with their valuable environmental services. Biodiversity-rich freshwater ecosystems are currently declining faster than marine or land ecosystems.

* Corresponding Author.

Distribution of Earth's Water

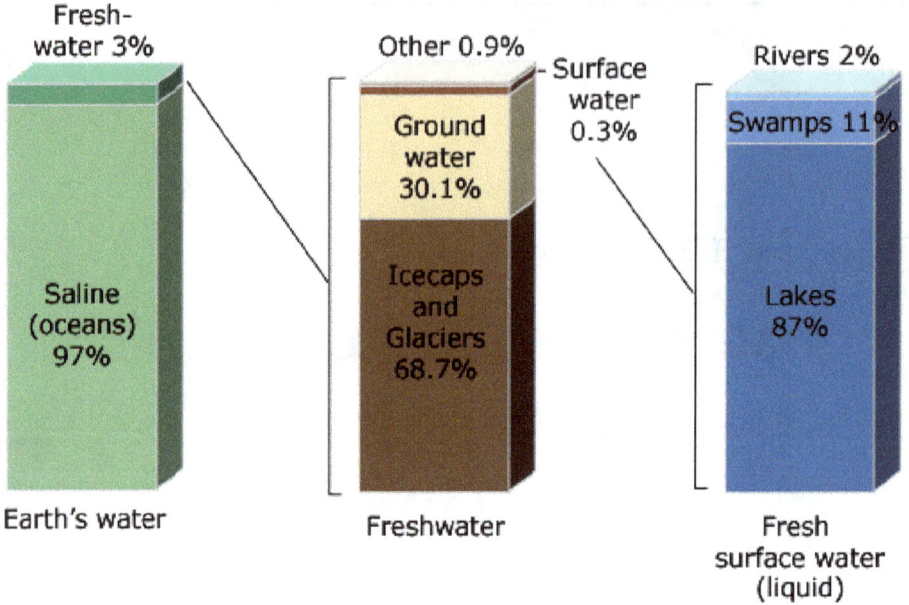

Fresh-
water 3%

Other 0.9%

Surface
water
0.3%

Rivers 2%

Ground
water
30.1%

Swamps 11%

Saline
(oceans)
97%

Icecaps
and
Glaciers
68.7%

Lakes
87%

Earth's water

Freshwater

Fresh
surface water
(liquid)

Sources of Freshwater

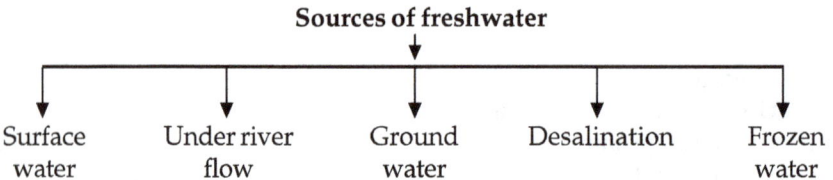

Sources of freshwater

| Surface water | Under river flow | Ground water | Desalination | Frozen water |

Surface Water

Surface water is naturally replenished by precipitation and naturally lost through discharge to the oceans, evaporation, and sub-surface seepage. Although the only natural input to any surface water system is precipitation within its watershed, the total quantity of water in that system at any given time is also dependent on many other factors. These factors include storage capacity in lakes, wetlands and artificial reservoirs, the permeability of the soil beneath these storage bodies, the runoff characteristics of the land in the watershed, the timing of the precipitation and local evaporation rates. All of these factors also affect the proportions of water lost.

Human activities can have a large and sometimes devastating impact on these factors. Humans often increase storage capacity by constructing reservoirs and decrease it by draining wetlands. Humans often increase runoff quantities and velocities by paving areas and channelising stream flow. The total quantity of water available at any given time is an important consideration. Humans can also cause surface water to be "lost" (*i.e.* become unusable) through pollution. Brazil is the country estimated to have the largest supply of freshwater in the world, followed by Russia and Canada.

Under River Flow

Throughout the course of the river, the total volume of water transported downstream will often be a combination of the visible free water flow together with a substantial contribution flowing through sub-surface rocks and gravels that underlie the river and its floodplain called the "hyporheic zone". For many rivers in large valleys, this unseen component of flow may greatly exceed the visible flow. The hyporheic zone often forms a dynamic interface between surface water and true ground-water receiving water from the groundwater when aquifers are fully charged and contributing water to ground-water when groundwaters are depleted. This is especially significant in karsts areas where pot-holes and underground rivers are common.

Groundwater

Sub-surface water, or groundwater, is freshwater located in the pore space of soil and rocks. Sometimes it is useful to make a distinction between sub-surface water that is closely associated with surface water and deep sub-surface water in an aquifer and sometimes called "fossil water". Sub-surface water can be thought of in the same terms as surface water: inputs, outputs and storage. The critical difference is that due to its slow rate of turnover, sub-surface water storage is generally much larger compared to inputs than it is for surface water. This difference makes it easy for humans to use sub-surface water unsustainably for a long time without severe consequences. Nevertheless, over the long term the average rate of seepage above a sub-surface water source is the upper bound for average consumption of water from that source. The natural input to sub-surface water is seepage from surface water. The natural outputs from sub-surface water are springs and seepage to the oceans.

If the surface water source is also subject to substantial evaporation, a sub-surface water source may become saline. This situation can occur naturally under endorheic bodies of water, or artificially under irrigated farmland. In coastal areas, human use of a sub-surface water source may cause the direction of seepage to ocean to reverse which can also cause soil salinization. Humans can increase the input to a sub-surface water source by building reservoirs or detention ponds.

Desalination

Desalination is an artificial process by which saline water (generally sea water) is converted to freshwater. The most common desalination processes are distillation and reverse osmosis. Desalination is currently expensive compared to most alternative sources of water, and only a very small fraction of total human use is satisfied by desalination. It is only economically practical for high-valued uses (such as household and industrial uses) in arid areas. The most extensive use is in the Persian Gulf.

Frozen water: Several schemes have been proposed to make use of icebergs as a water source, however to date this has only been done for novelty purposes. Glacier runoff is considered to be surface water. The Himalayas contain some of the most extensive and rough high altitude areas on Earth as well as the greatest area of glaciers and permafrost outside of the poles. Ten of Asia's largest rivers flow from there and more than a billion people's livelihoods depends on them. To complicate matters, temperatures are rising more rapidly here than the global average. In Nepal the temperature has risen with 0.6 degree over the last decade, whereas the global warming has been around 0.7 degree over the last hundred years.

Uses of Freshwater

Uses of freshwater can be categorized as consumptive and non-consumptive (sometimes also called "renewable"). A use of water is consumptive if that water is not immediately available for another use. Losses to sub-surface seepage and evaporation are considered consumptive, as is water incorporated into a product (such as farm produce). Water that can be treated and returned as surface water, such as sewage, is generally considered non-consumptive if that water can be put to additional use.

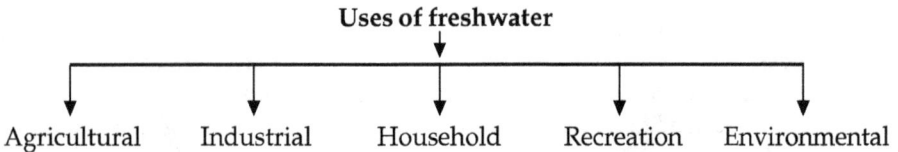

Uses of freshwater

| Agricultural | Industrial | Household | Recreation | Environmental |

Agricultural

It is estimated that 69 per cent of worldwide water use is for irrigation, with 15-35 per cent of irrigation withdrawals being unsustainable. In some areas of the world irrigation is necessary to grow any crop at all, in other areas it permits more profitable crops to be grown or enhances crop yield. Various irrigation methods involve different trade-offs between crop yield, water consumption and capital cost of equipment and structures. Irrigation methods such as furrow and overhead sprinkler irrigation are usually less expensive but are also typically less efficient, because much of the water evaporates, runs off or drains below the root zone. Other irrigation methods considered to be more efficient include drip or trickle irrigation, surge irrigation, and some types of sprinkler systems where the sprinklers are operated near ground level. These types of systems, while more expensive, usually offer greater potential to minimize runoff, drainage and evaporation. Any system that is improperly managed can be wasteful; all methods have the potential for high efficiencies under suitable conditions, appropriate irrigation timing and management. One issue that is often insufficiently considered is salinization of sub-surface water.

Aquaculture is a small but growing agricultural use of water. Freshwater commercial fisheries may also be considered as agricultural uses of water, but have generally been assigned a lower priority than irrigation. As global populations grow, and as demand for food increases in a world with a fixed water supply, there are efforts underway to learn how to produce more food with less water, through improvements in irrigation methods and technologies, agricultural water management, crop types, and water monitoring.

Industrial

It is estimated that 15 per cent of worldwide water use is industrial. Major industrial users include power plants, which use water for cooling or as a power source (*i.e.* hydroelectric plants), ore and oil refineries, which use water in chemical processes, and manufacturing plants, which use water as a solvent. The portion of industrial water usage that is consumptive varies widely, but as a whole is lower than agricultural use. Water is used in power generation. Hydro-electricity is electricity obtained from hydropower. Hydroelectric power comes from water driving a water turbine connected to a generator. Hydroelectricity is a low-cost, non-polluting, renewable energy source. The energy is supplied by the sun. Heat from the sun evaporates water, which condenses as rain in higher altitudes, from where it flows down.

Water is also used in many industrial processes and machines, such as the steam turbine and heat exchanger, in addition to its use as a chemical solvent. Discharge of untreated water from industrial uses is pollution. Pollution includes discharged solutes (chemical pollution) and discharged coolant water (thermal pollution). Industry requires pure water for many applications and utilizes a variety of purification techniques both in water supply and discharge.

Household

It is estimated that 15 per cent of worldwide water use is for household purposes. These include drinking water, bathing, cooking, sanitation, and gardening. Basic household water requirements have been estimated by Peter Gleick at around 50 litres per person per day, excluding water for gardens. Drinking water is water that is of sufficiently high quality so that it can be consumed or used without risk of immediate or long term harm. Such water is commonly called potable water. In most developed countries, the water supplied to households, commerce and industry is all of drinking water standard even though only a very small proportion is actually consumed or used in food preparation.

Recreation

Recreational water use is usually a very small but growing percentage of total water use. Recreational water use is mostly tied to reservoirs. If a reservoir is kept fuller than it would otherwise be for recreation, then the water retained could be categorized as recreational usage. Release of water from a few reservoirs is also timed to enhance whitewater boating, which also could be considered a recreational usage. Other examples are anglers, water skiers, nature enthusiasts and swimmers. Recreational usage is usually non-consumptive. Golf courses are often targeted as using excessive amounts of water, especially in drier regions. It is, however, unclear whether recreational irrigation (which would include private gardens) has a noticeable effect on water resources. This is largely due to the unavailability of reliable data. However, the actual statistical effect of this reassignment is close to zero.

Additionally, recreational usage may reduce the availability of water for other users at specific times and places. For example, water retained in a reservoir to allow boating in the late summer is not available to farmers during the spring planting season. Water released for whitewater rafting may not be available for hydroelectric generation during the time of peak electrical demand. Environmental: Explicit environmental water use is also a very small but growing percentage of total water use. Environmental water usage includes artificial wetlands, artificial lakes intended to create wildlife habitat, fish ladders, and water releases from reservoirs timed to help fish spawn. Like recreational usage, environmental usage is non-consumptive but may reduce the availability of water for other users at specific times and places. For example, water release from a reservoir to help fish spawn may not be available to farms upstream.

Water Stress

It is a direct threat to man on its existence and world peace.

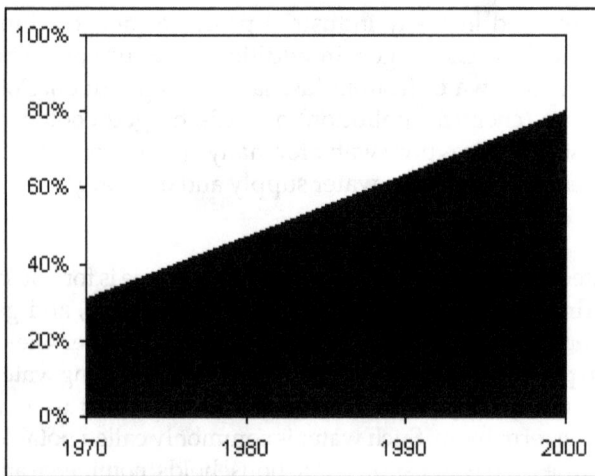

Figure 17.2: Best Estimate of the Share of People in Developing Countries with Access to Drinking Water (1970–2000)

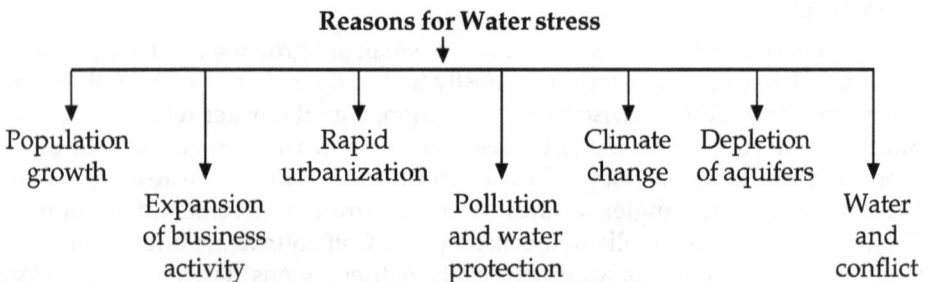

Reasons for Water stress

Population growth		Rapid urbanization		Climate change	Depletion of aquifers	
	Expansion of business activity		Pollution and water protection			Water and conflict

The concept of water stress is relatively simple, according to the World Business Council for Sustainable Development (WBCSD), it applies to situations where there is not enough water for all uses, whether agricultural, industrial or domestic. Defining thresholds for stress in terms of available water per capita is more complex, however, entailing assumptions about water use and its efficiency. Nevertheless, it has been proposed that when annual per capita renewable freshwater availability is less than 1,700 cubic meters, countries begin to experience periodic or regular water stress. Below 1,000 cubic meters, water scarcity begins to hamper economic development and human health and well-being.

Population Growth

In 2000, the world population was 6.2 billion. The UN estimates that by 2050 there will be an additional 3.5 billion people with most of the growth in developing countries that already suffer water stress. Thus, water demand will increase unless there are corresponding increases in water conservation and recycling of this vital resource.

Expansion of Business Activity

Business activity ranging from industrialization to services such as tourism and entertainment continues to expand rapidly. This expansion requires increased water services including both supply and sanitation, which can lead to more pressure on water resources and natural ecosystems.

Rapid Urbanization

The trend towards urbanization is accelerating. Small private wells and septic tanks that work well in low-density communities are not feasible within high-density urban areas. Urbanization requires significant investment in water infrastructure in order to deliver water to individuals and to process the concentrations of wastewater–both from individuals and from business. These polluted and contaminated waters must be treated or they pose unacceptable public health risks. In 60 per cent of European cities with more than 100,000 people, groundwater is being used at a faster rate than it can be replenished. Even if some water remains available, it costs more and more to capture it.

Climate Change

Climate change could have significant impacts on water resources around the world because of the close connections between the climate and hydrologic cycle. Rising temperatures will increase evaporation and lead to increases in precipitation, though there will be regional variations in rainfall. Overall, the global supply of freshwater will increase. Both droughts and floods may become more frequent in different regions at different times, and dramatic changes in snowfall and snowmelt are expected in mountainous areas. Higher temperatures will also affect water quality in ways that are not well understood. Possible impacts include increased eutrophication. Climate change could also mean an increase in demand for farm irrigation, garden sprinklers, and perhaps even swimming pools.

Depletion of Aquifers

Due to the expanding human population, competition for water is growing such that many of the world's major aquifers are becoming depleted. This is due both for direct human consumption as well as agricultural irrigation by groundwater. Millions of pumps of all sizes are currently extracting groundwater throughout the world.

Irrigation in dry areas such as northern China and India is supplied by groundwater, and is being extracted at an unsustainable rate. Cities that have experienced aquifer drops between 10 to 50 meters include Mexico City, Bangkok, Manila, Beijing, Madras and Shanghai.

Pollution and Water Protection

Water pollution is one of the main concerns of the world today. The governments of many countries have striven to find solutions to reduce this problem. Many pollutants threaten water supplies, but the most widespread, especially in underdeveloped countries, is the discharge of raw sewage into natural waters; this method of sewage disposal is the most common method in underdeveloped countries, but also is prevalent in quasi-developed countries such as China, India and Iran. Sewage, sludge, garbage, and even toxic pollutants are all dumped into the water. Even if sewage is treated, problems still arise. Treated sewage forms sludge, which may be placed in landfills, spread out on land, incinerated or dumped at Sea. In addition to sewage, nonpoint source pollution such as agricultural runoff is a significant source of pollution in some parts of the world, along with urban storm water runoff and chemical wastes dumped by industries and governments.

Water and Conflict

The only known example of an actual inter-state conflict over water took place between 2500 and 2350 BC between the Sumerian states of Lagash and Umma. Yet, despite the lack of evidence of international wars being fought over water alone, water has been the source of various conflicts throughout history. When water scarcity causes political tensions to arise, this is referred to as water stress. Water stress has led most often to conflicts at local and regional levels. This can be seen in the water conflicts between different states *i.e.* Kaveri river water conflict (Tamil Nadu-Karnataka), Indira Gandhi Nahar Priyojna water issue (Rajasthan-Punjab), Narmada dam issue (Madhya Pradesh-Gujarat) and India's conflict over water with Pakistan, China and Bangladesh. Using a purely quantitative methodology, Thomas Homer-Dixon successfully correlated water scarcity and scarcity of available arable lands to an increased chance of violent conflict.

Water stress can also exacerbate conflicts and political tensions which are not directly caused by water. Gradual reductions over time in the quality and/or quantity of freshwater can add to the instability of a region by depleting the health of a population, obstructing economic development, and exacerbating larger conflicts.

Conflicts and tensions over water are most likely to arise within national borders, in the downstream areas of distressed river basins. Areas such as the lower regions of China's Yellow River or the Chao Phraya River in Thailand, for example, have already been experiencing water stress for several years. Additionally, certain arid countries which rely heavily on water for irrigation, such as China, India, Iran, and Pakistan, are particularly at risk of water-related conflicts. Political tensions, civil protest, and violence may also occur in reaction to water privatization. The Bolivian Water Wars of 2000 are a case in point.

World Water Supply and Distribution

Food and water are two basic human needs. However, global coverage figures from 2002 indicate that, of every 10 people about 5 have a connection to a piped water supply at home (in their dwelling, plot or yard); 3 make use of some other sort of improved water supply, such as a protected well or public standpipe and 2 are still having no accessibility to piped water distribution system. In addition, 4 out of every 10 people live without improved sanitation.

At Earth Summit 2002 governments approved a Plan of Action to:

☆ Halve by 2015 the proportion of people unable to reach or afford safe drinking water. The Global Water Supply and Sanitation Assessment 2000 Report (GWSSAR) defines "Reasonable access" to water as at least 20 liters/person/day from a source within one kilometer of the user's home.

☆ Halve the proportion of people without access to basic sanitation. The GWSSR defines "Basic sanitation" as private or shared but not public disposal systems that separate waste from human contact.

In 2025, water shortages will be more prevalent among poorer countries where resources are limited and population growth is rapid, such as the Middle East, Africa, and parts of Asia. By 2025, large urban and peri-urban areas will require new infrastructure to provide safe water and adequate sanitation. This suggests growing conflicts with agricultural water users, who currently consume the majority of the water used by humans.

Generally speaking the more developed countries of North America, Europe and Russia will not see a serious threat to water supply by the year 2025; not only because of their relative wealth, but more importantly their populations will be better aligned with available water resources. North Africa, the Middle East, South Africa and northern China will face very severe water shortages due to physical scarcity and a condition of overpopulation relative to their carrying capacity with respect to water supply. Most of South America, Sub-Saharan Africa, Southern China and India will face water supply shortages by 2025; for these latter regions the causes of scarcity will be economic constraints to developing safe drinking water, as well as excessive population growth.

1.6 billion People have gained access to a safe water source since 1990. The proportion of people in developing countries with access to safe water is calculated to have improved from 30 per cent in 1970 to 71 per cent in 1990, 79 per cent in 2000 and 84 per cent in 2004. This trend is projected to continue.

Economic Considerations

Water supply and sanitation require a huge amount of capital investment in infrastructure such as pipe networks, pumping stations and water treatment works. It is estimated that Organisation for Economic Co-operation and Development (OECD) nations need to invest at least USD 200 billion/year to replace aging water infrastructure to guarantee supply, reduce leakage rates and protect water quality.

International attention has focused upon the needs of the developing countries. To meet the Millennium Development Goals targets of halving the proportion of the population lacking access to safe drinking water and basic sanitation by 2015, current annual investment on the order of USD 10 to USD 15 billion would need to be roughly doubled. This does not include investments required for the maintenance of existing infrastructure.

Once infrastructure is in place, operating water supply and sanitation systems entails significant ongoing costs to cover personnel, energy, chemicals, maintenance and other expenses. The sources of money to meet these capital and operational costs are essentially either user fees, public funds or some combination of the two. But this is where the economics of water management start to become extremely complex as they intersect with social and broader economic policy. Such policy questions are beyond the scope of this article, which has concentrated on basic information about water availability and water use. They are, nevertheless, highly relevant to understanding how critical water issues will affect business and industry in terms of both risks and opportunities.

Business Response

The World Business Council for Sustainable Development (WBCSD) in its H_2O Scenarios engaged in a scenario building process to:

☆ Clarify and enhance understanding by business of the key issues and drivers of change related to water.

☆ Promote mutual understanding between the business community and non-business stakeholders on water management issues.

☆ Support effective business action as part of the solution to sustainable water management.

Conclusion

Human activities can have a large and sometimes devastating impact on water. The total quantity of water available at any given time is an important consideration. Some human water users have an intermittent need for water.

Sustainable water management is a global challenge, since every single person's food, everyday usable goods, and energy are all produced with freshwater use. Freshwater is a sustainable, renewable resource, yet through poor management, water can be polluted or lost to inaccessible reservoirs. Over a third of the world's population is suffering critical water stress and the demand for freshwater is growing every year.

Freshwater comes from many different sources, including groundwater systems, freshwater lakes and rivers, and Arctic regions' ice and snow cover. Yet, freshwater only makes up 2.5 per cent of the total water volume on Earth, and of that total freshwater volume, only 1 per cent of freshwater is a usable supply for human consumption and production.

It concludes that:

☆ Economic activities cannot survive in a society that thirsts.

☆ One does not have to be in the water business to have a water crisis.

☆ Business is part of the solution, and its potential is driven by its engagement.

☆ Growing water issues and complexity will drive up costs.

☆ Precarious Situation of Rajasthan.

The entire world has become increasingly aware of shortage of freshwater, imminent in some countries and regions. These include India, with 16 per cent of the humanity but less than 3 per cent of global freshwater resources. The poor water availability is exacerbated by its uneven spread over regions and time of the year. Rajasthan is very much at a disadvantage even in the Indian context. This large but possibly the driest state suffers from a disproportionately poor availability of water when compared to its potential large users, people, animals and agriculture:

Rajasthan Water Resources and Potential Users

State Parameter	Share of Nation (in per cent)
Area	10.40
Population	5.40
Livestock	18.70
Cultivable area	13.88
Surface water	1.16
Groundwater	1.70

The situation has worsened over time due to a rapid increase in use-related parameters. The population growth rate of the state is among the highest in the country. Demand for water from hitherto insubstantial uses, such as industry, tourism and recreation, as well as sanitation and environmental purposes, has been growing apace. The supply, however, has remained unchanged. The primary source is the scanty and uncertain precipitation, confined to just two months of the year. Nearly a third of the state is arid and another 30 per cent semi-arid, which implies that nearly two-thirds of the state suffers from recurrent water scarcity.

References

1. UN World Water Development Report.

2. International Water Resources Association.

3. Institute for Water Resources–USACE.

4. "Threats to water resources" by the Environment Agency.

5. Ancient Irrigation from the University of California, Geology Department.

6. Future Sources of Freshwater.

7. World Water Supply and Demand: 1995 to 2025 from the International Water Management Institute.

8. Porous cities, new directions in urban water usage.

9.   Water and the Future of Life on Earth.

10. Water and Cities: Acting on the Vision.

11. FAO Water Portal Food and Agriculture Organization of the United Nations.

12. Gleick, P.H. (Ed.) (1993). *Water in Crisis: A Guide to the World's Freshwater Resources.* Oxford University Press, Oxford, UK.

13. Hails, C., Loh, J. and Goldfinger, S. (2006). *Living Planet Report 2006.* World Wide Fund, Gland, Switzerland.

14. Hoekstra, A.Y. (Ed.) (2003). *Virtual Water Trade: Proceedings of the International Expert Meeting on Virtual Water Trade.* Delft, The Netherlands, 12–13 December 2002, Value of Water Research Report Series No.12, UNESCO–IHE, Delft, The Netherlands.

15. Hoekstra, A.Y. (2006). *The Global Dimension of Water Governance: Nine Reasons for Global Arrangements in Order to Cope with Local Water Problems.* Value of Water Research Report Series No.20, UNESCO–IHE,Delft, the Netherlands.

16. Hoekstra, A.Y. and Chapagain, A.K. (2008). *Globalization of Water: Sharing the Planet's Freshwater Resources.* Blackwell Publishing, Oxford, UK.

17. Morgan, J.P. (2008). *Watching Water: A Guide to Evaluating Corporate Risks in a Thirsty World.* New York, USA.

18. Kampman, D.A., Hoekstra, A.Y. and Krol, M.S. (2008). *The Water Footprint of India.* Value of Water Research Report Series No.32, UNESCO–IHE, Delft, The Netherlands.

19. Lenzen, M., Murray, J., Sack, F. and Wiedmann, T. (2007). Shared producer and consumer responsibility: Theory and practice. *Ecological Economics,* 61(1): 27–42.

20. Postel, S.L., Daily, G.C. and Ehrlich, P.R. (1996). Human appropriation of renewable freshwater. *Science,* 271: 785–788.

21. Rees, W.E. and Wackernagel, M. (1994). Ecological footprints and appropriated carrying capacity: Measuring the natural capital requirements of the human economy, In: Jansson, A.M., Hammer, M., Folke, and Smakhtin, V., Revenga, C., and Döll, P. (2004) Taking into account environmental water requirements in global–scale water resources assessments Comprehensive Assessment Research Report 2, IWMI, Colombo, Sri Lanka.

22. TCPA (2008). *Sustainable Water Management: Eco-towns Water Cycle Worksheet.* Town and Country Planning Association, London, UK.

23. The Independent (2008). Forget carbon: You should be checking your water footprint. *The Independent,* April 21st.

24. UNESCO (2006). *Water, a Shared Responsibility: The United Nations World Water Development Report 2.* UNESCO Publishing, Paris, France/Berghahn Books, Oxford, UK.

25. Verkerk, M.P., Hoekstra, A.Y. and Gerbens-Leenes, P.W. (2008). *Global Water Governance: Conceptual Design of Global Institutional Arrangements.* Value of Water Research Report Series No.26, UNESCO–IHE, Delft, The Netherlands.

Biodiversity of Aquatic Reseources (2012)
Editors: **Mamta Rawat & Sumit Dookia**
Published by: **DAYA PUBLISHING HOUSE, NEW DELHI**

Pages **257-269**

Chapter 18

Regrowth and Survival of Bacterial Pathogens on the Surfaces of Household Containers Used for the Storage of Drinking Water in Rural Supply of Western Rajasthan

Mamta Rawat

Ecology and Rural Development Society,
1-A-43, Kudi Housing Board, Jodhpur – 342 005, Rajasthan

ABSTRACT

The present study investigated quality of different type of water used for drinking water in few rural villages of Jodhpur district of western Rajasthan. To investigate the regrowth and survival of indicator microorganisms on the surface of household containers during the storage of drinking water. This period of 48 hour was chosen as the study period because results from the questionnaire indicated that the largest percentage of households stores their water for that length of time. The experiment was performed to test drinking water as it is collected and stored by rural communities. No disinfection of household containers was done during the study period. Indicator and emerging pathogenic bacteria were measured during the study period. Attached indicator microorganisms consisted of indicator bacteria (TC, FC and FS) and pathogenic bacteria (*Salmonella, Shigella, Vibrio* and pathogenic *E.coli* 0157:H7), although the yield (average count) for the pathogenic bacteria was lower than that of indicator bacteria. However, the lowest yield of indicator microorganisms was noted for presumptive *E. coli*. Whereas the occurrence and survival of TC and FS was noted on the surface of household containers during the entire period of the experimental study, FC and rest pathogenic bacteria occurred from time to time. The regrowth of indicator microorganisms occurred 48 h to all types of test waters. This length of time mostly resulted in the regrowth of TC (with an increase in bacterial counts) while the persistence of other indicator organism groups on the surface of the slides was apparent. A comparison between earthen and metallic containers showed that more TC (average count) regrew on earthen than on metallic containers.This study revealed that both types of household containers supported the growth and survival of indicator microorganisms due to the bad quality of the intake water before storage. The storage of drinking water for 48 h mainly resulted in the regrowth of TC. Nevertheless, the persistence of other indicator microorganisms was observed on the surface of household containers.

Keyword: Regrowth, Survival, Pathogen bacteria, Household earthen containers, Drinking water.

Introduction

Water-related diseases continue to be one of the major health problems globally. An estimated 4 billion cases of diarrhoea annually represented 5.7 per cent of the global disease burden in the year 2000 (WHO 2002). Most of these deaths are due to inadequate potable water supplies, poor hygiene practices and insufficient sanitation infrastructures (WHO, 2002). The importance of water in the transmission of diseases in humans remains the main concern of the WHO (1976). Also it was claimed that 25 per cent of all the worlds' hospital beds were occupied by the people with disease caused by polluted water or diseases related to water. Diarrhoeal diseases account for 4.3 per cent of the total global disease burden (62.5 million DALYs). An estimated 88 per cent of this burden is attributable to unsafe drinking water supply, inadequate sanitation, and poor hygiene. These risk factors are second, after malnutrition, in contributing to the global burden of disease. (Pruss *et al.*, 2002).

Due to the long distance between the drinking water sources and villages in the arid regions of Jodhpur district of Thar Desert, household containers are used for the storage of drinking water. Also it is well known that microorganisms attach on the surface walls of such containers during storage and develop at the expense of the low concentration of carbon in water. Attached bacteria can detach from the surface walls and lead to continuous contamination of the water phase. A large variety of different heterotrophic bacteria from pathogenic to opportunistic bacteria such as *Legionella* species, *Pseudomonas, Mycobacter, Campylobacter, Klebsiella, Aeromonas, Helicobacter pylori, Salmonella* and *E.coli* and Somatic coliphages and F-RNA coliphages have been isolated from biofilm (Engel *et al.*, 1980, Wadowsky *et al.*, 1982; Burke *et al.*, 1984; Armon *et al.*, 1997; Mackay *et al.*, 1998 and Momba *et al.*, 2002).There is also a problem of bacterial regrowth which is the increase of bacteria contained in the water without subsequent additional contamination and introduction of metabolizable substances (Schoenen,1986). Since water storage is a must in these arid areas, it is also essential to assess the impact of this method of storage on water quality. The study thus aims to determine out whether indicator and emerging bacteria re grow on storing containers and survive during the length of time rural communities store their drinking water.

Material and Methods

Study Area and Drinking Water Sources

The study was undertaken in 15 villages of three blocks of Jodhpur district of Rajasthan, a part of Thar Desert. The area had limited natural water resources and an extreme climate, with temperatures ranging from 4 °C in January to 50 °C in June with an average annual rainfall of 156 mm. However treated water is supplied to these villages which are stored in *Groundwater level resources* (GLR). This water is further collected by residents and stored in clay or earthen and metallic pitcher vessels within the household. The inhabitants also use rainwater harvesting (RWH), sources for all water uses. Another drinking water source was wells, which was not very good, enriched with cations and anions, hence unsuitable for drinking. All studied villages in the area had similar types of drinking water sources.

Data Collection, Sampling of Test Waters and Inoculation of Metal Slides

Proper sterilized glassware was used to collect sample water from sources and further from containers within houses. The research includes microbiological measures of contamination and chemical aspects of water quality. The 30 households were visited thrice in month for a year and water samples were taken from the clay and metallic drinking water pitcher found inside the household and from the original water sources. For capturing of bacteria from storage containers metal coupons or slides were suspended in the studied containers for a maximum of 48 hours. During the study period, all containers were always covered with their lids. The study deals not only with indicator bacteria (total coliforms, faecal coliforms and faecal streptococcus) but also with emerging pathogens like *Salmonella*, *Shigella*, pathogenic *E.coli* 0157:H7, *Vibrio* species.

Detachment and Calculation of Bacteria from Metal Slides

The slides were aseptically removed from the household containers after 24 and 48 hours and further transferred into sterile Borosil glassware containing 80 ml saline solutions. The attached microorganisms were released using vortex mixer.

The following equation was used to calculate the number of attached bacteria:

$cfu/cm^2 = ND/$ surface area of slides, where N is the number of bacteria and D is the dilution factor.

Detection of Microorganisms

Indicator Bacteria

Ten ml of water samples were filtered through a sterile membrane filter and further it was placed on the McConkey agar, EMB agar and Pfizer Selective

Enterococcus agar for enumeration of Total coliform (TC), Feacal Coliform (FC) (*E.coli*) and Feacal Streptococcus (FS) (*Streptococcus feacalis*) respectively (APHA,1998).

Emerging Pathogens

The same traditional procedure of membrane filter was used.

For enrichment of *Salmonella*, Selenite F broth and tetra thionate broth for were used and further streaked on Xylose-lysine-deoxycholate (XLD) agar and Hi chrome *Salmonella* agar plates were incubated for growth. Small red colonies with black center on XLD agar and mauve colored colonies on Hichrome agar depicted *Salmonella* colonies. Further biochemical confirmation was carried out with the aid of ready prepared biochemical kits. The red-headed pink colonies without black centers isolated from XLD agar were considered as *Shigella* and transferred to nutrient agar, and incubated at $37 \pm 2°C$ overnight. Further a series of biochemical and serological confirmation test were carried out for complete verification of *Shigella* species. Alkaline peptone water was used for enrichment of *Vibrio* and TCBS agar for its respective yellow colored colonies. Again further biochemical tests conducted for verification. For *E.coli* 0157:H7, about 100ml of sample and 50 ml of triple strength lauryl tryptose broth was incubated at 37°C for 24 hours. It was then streaked on sorbitol McConkey (SMAC) agar containing potassium tellurite.The sorbitol negative strains (SMAC) were tested for enzyme β-d-glucuronisade using MUG(4-methylumbelliferyl-β-d glucuronide) as substrate.A colony of the test isolate was applied to MUG impregnated paper, moistened with a drop of saline and incubated at 37 for 20 min. The absence of enzyme was indicated by the lack of fluorescence when the filter paper was examined under U.V. light. The MUG negative colonies were further serotype using 0157 and H7 antiserum (APHA, 1998).

Field-Based Observational Studies

The study covered few rural communities of Bilara, Pipar and Osian blocks covering fifteen villages (Table 18.1). For the study, households in each randomly selected village were selected representing the different drinking water sources used in the village. The presence of a piped water connection, water storage and toilet facility was registered for each household. A questionnaire related to the types of household containers and the length time for storage was administered to households of each villages. Around 108 water samples were taken for the present study. The data from this interview revealed that 63 per cent and 37 per cent of households use earthen and metallic pots respectively. It was also observed that average storage duration was 48 hours. The experimental study was carried over a period of 12 months from January to December 2007.

The drinking water sources used included the open/closed RWH tank, the well of somewhat less importance and piped supply in GLR's each used by these household. All studied thirty households used pitchers made of traditional earthen clay and metal pitcher. During the study period water was collected from the pitchers and at the same time a sample from the source from which the water originated. For the source samples to reflect the actual source contamination at the time of water collection, the water samples were collected in the morning–just after most households had collected water. A total of 108 water samples were taken during the study period.

Table 18.1: Sampling Stations along GPS Reading

Sl.No.	Sampling Station	GPS		
		Latitude	Longitude	Altitude
	Block: Osian			
1.	Surpura	26 31 28.43	73 18 16.20	838 ft
2.	Birai	26 32 12.09	73 22 12.12	821 ft
3.	Bucheti	26 31 38.64	73 17 59.27	847 ft
4.	Bodvi	26 29 57.43	73 21 37.67	861 ft
5.	Sevki khurd	26 28 52.12	73 22 36.34	864 ft
	Block: Bhopalgarh			
6.	Nandiya prabavati	26 30 12.72	73 16 43.55	820 ft
7.	Burkiya	26 27 09.76	73 16 28.61	789 ft
8.	Devtada	26 28 21.12	73 25 23.52	920 ft
9.	Godaras	26 28 09.87	73 23 22.36	912 ft
10.	Ratkuriya	26 28 10.12	73 23 28.36	902 ft
	Block: Bilara			
11.	Barna	26 12 34.12	73 40 20.01	907 ft
12.	Pichyak	26 12 58.43	73 41 19.08	914 ft
13.	Nehri	26 12 43.21	73 41 20.08	901 ft
14.	Jhurli	26 12 47.11	73 41 16.71	900 ft
15.	Bhavi	26 12 46.12	73 41 17.12	902 ft
16.	Kaparda	26 08 55.04	73 44 02.57	954 ft

Results and Discussion

The objective of this study was to determine whether a buildup and regrowth of indicator organisms on the surface of household containers would occur upon the length of time (48 hours) used by most of the households for the storage of their drinking water. Not only the sources but also the presence of biofilm in drinking water should not be overlooked. Biofilm can harbor and protect the very pathogens of concern (Doolittle *et al.*, 1996, Quingnon *et al.*, 1997) and effort to manage and control their growth should be made. Also a concern when addressing the delivery of safe drinking water is biofilm. Biofilm involve microbial cells attaching to pipe surfaces and multiplying to form a film/slime layer on the pipe that can harbor and protect bacteria from disinfectants. Work has been done to show that biofilm provide protection from the nutrient poor conditions maintained in drinking water. Szewzk *et al.* (1994) showed that *E. coli* was capable of scavenging nutrients from excretions of the biofilm associated water bacteria. The results revealed that in both earthen and metal containers-stored water, the adhesion of indicator microorganisms was detected after an exposure time of 24 hrs. Earthen based materials supported more attachment of indicator microorganisms than the metal based material (Table 18.2).

Table 18.2: Counts of Indicator and Pathogenic Concentration on the Surface of the Household Containers after the Storage of Groundwater, RWH Water and Supplied Municipal Water

Type and Duration of Storage	TC	FC	FS	Pseudomonas	Salmonella	Shigella	Vibrio	E. coli O157:H7
(a) Range of Counts								
RWH water (cfu/100ml)	$3.69\text{-}902 \times 10^3$	80-150	15-195	82-210	72-104	53-80	10-45	15-29
Earthen for 24 hrs (cfu/cm²)	68-102	<1-10	<1-19	7-102	<1-9	<1-5	<1-3	<1
Earthen for 48 hrs (cfu/cm²)	85-415	<1-16	<1-25	15-198	1-22	<1-6	<1-6	<1
Metallic for 24 hrs (cfu/cm²)	18-95	<1-8	<1-13	1-53	<1-8	<1-3	<1-1	<1
Metallic for 48 hrs (cfu/cm²)	35-254	<1-11	<1-21	4-87	1-17	<1-6	<1-2	<1
Groundwater (cfu/100ml)	$5.99\text{-}150 \times 10^3$	22-75	32-95	51-93	66-95	56-120	–	5-22
Earthen for 24 hrs (cfu/cm²)	79-281	12-25	8-17	9-77	<1-15	<1-16	–	<1
Earthen for 48 hrs (cfu/cm²)	116-510	18-38	13-26	13-89	<1-22	<1-21	–	<1
Metallic for 24 hrs (cfu/cm²)	69-127	1-10	<1-7	1-18	<1-5	<1-16	–	<1
Metallic for 48 hrs (cfu/cm²)	95-342	7-18	<1-19	4-42	<1-6	<1-22	–	<1
Piped supply (cfu/100ml)	$47\text{-}80 \times 10^3$	<1-50	<1-32	35-90	12-27	15-40	–	–
Earthen for 24 hrs (cfu/cm²)	20-112	<1-9	<1-13	7-38	<1-4	<1-7	–	–
Earthen for 48 hrs (cfu/cm²)	31-283	<1-13	<1-16	12-66	<1-6	<1-11	–	–
Metallic for 24 hrs (cfu/cm²)	5-85	<1-4	<1-5	1-21	<1	<1-3	–	–
Metallic for 48 hrs (cfu/cm²)	8-104	<1-9	<1-6	1-45	<1-4	<1-5	–	–

Contd...

Table 18.2—Contd...

(b) Average counts

Type and Duration of Storage	TC	FC	FS	Pseudomonas	Salmonella	Shigella	Vibrio	E. coli 0157:H7
RWH water (cfu/100ml)	189×10^3	132	144	172	80	63	15	20
Earthen for 24 hrs (cfu/cm²)	85	<8	<15	53	<5	<3	2	<1
Earthen for 48 hrs (cfu/cm²)	270	<12	<19	67	11	<4	3	<1
Metallic for 24 hrs (cfu/cm²)	60	<4	<8	31	<3	<2	1	<1
Metallic for 48 hrs (cfu/cm²)	148	<9	<13	53	9	<4	2	<1
Groundwater (cfu/100ml)	188×10^3	48	54	76	42	80	–	17
Earthen for 24 hrs (cfu/cm²)	140	17	14	52	<7	<9	–	<1
Earthen for 48 hrs (cfu/cm²)	290	23	19	63	<13	<13	–	<1
Metallic for 24 hrs (cfu/cm²)	90	6	<4	11	<2	<7	–	<1
Metallic for 48 hrs (cfu/cm²)	210	13	<12	230	<3	<9	–	<1
Piped supply (cfu/100ml)	69×10^3	<21	<22	49	<1	31	–	–
Earthen for 24 hrs (cfu/cm²)	60	<4	<9	22	<3	<4	–	–
Earthen for 48 hrs (cfu/cm²)	180	<9	<12	35	<5	<7	–	–
Metallic for 24 hrs (cfu/cm²)	42	<2	<3	18	<1	<2	–	–
Metallic for 48 hrs (cfu/cm²)	88	<6	<5	26	<2	<3	–	–

Although there are earlier studies citing a direct relationship between the storage materials and the quality of water, in the present context, the formation of biofilms could also be related to many factors prevailing in the intake of water, *e.g.* temperature, turbidity and microbial occurrence (Tables 18.2 and 18.3). Biofilm formation is usually encouraged on the surface of a material if that material is able to supply the require nutrients for the bacterial growth (Momba *et al.*, 2000). This could clearly explain the survival and regrowth of indicator organisms on earthen and metal containers during the storage of drinking water. The adhesion of indicator microorganisms on the surface of household containers could also be supported by the level of temperature detected in drinking water during storage (Table 18.3). The ability of bacteria to grow and survive over a wide range of temperature has been demonstrated. Previous studies have shown that temperature above 5°C (Lund and Ormerod, 1995) and 15°C (Le Chevallier *et al.*, 1996 and Power and Nagy, 1999) have been associated with biofilm formation in drinking water. Research studies have linked between high turbidity and growth of microorganisms (Le Chevallier, 1990). Turbidity is in fact serves as a source of water borne pathogens, viruses and protozoa. Consequently the occurrence and persistence of microorganisms on the surface of earthen and metal could be related to the level of turbidity in test water (Table 18.3). The most important factor involved in the attachment of bacteria was the occurrence and survival of the bacteria in the intake water (Table 18.2). It is well known that coliforms comprise a heterogeneous group which includes bacteria from the genera *Escherichia, Citrobacter, Klebsiella,* and *Enterobacter* (APHA, 1998). Although during present study these genera were not analysed, they could be released into the container-stored waters and deteriorate the quality of water more and more. Therefore, their presence in drinking water could have a negative impact on consumer health since they could cause disease like gastritis, dysentery, etc. Total Coliform, was constantly present in initial test water from different water sources districts along with piped supply of Jodhpur during the entire study period. In looking at emerging pathogen concerns, Szewzyk *et al.* (2000) assert that the modern routine monitoring of drinking water is not practical but also that detection of the classical indicators fails to indicate contamination by a larger number of emerging pathogens. Among pathogenic bacteria *Salmonella* and *Shigella* were present in all three types of water sources. However, Vibrio was only recorded in RWH water sources at low concentration and showed its presence time to time. And pathogenic *E.coli* 0157:H7, showed its erratic presence in lower count (average) only in the RWH and groundwater sources (Table 18.1).

In this study, the regrowth of bacteria occurred 48 hrs after the exposure of slides to all types of test waters. This regrowth was observed more with TC than with other indicator organisms groups. A comparison between earthen and metallic containers showed that more TC grew on earthen than on metal containers. Sharan *et al.* (2010), observed that the copper based water disinfection methods are more efficient under environment conditions, compared to others. Similar observations were observed experimentally by Sudha *et al.* (2009) under laboratory conditions. Power and Nagy (1999) in their studies of drinking water distribution system of Sydney found that regrowth was present within the system and that certain parameters points correlated with the presence of high bacterial numbers.

Table 18.3: Physico-chemical Values (Average) of Rural Areas of Jodhpur District, from Three Different Sources

Parameters	Water Sources					
	RWH Source		Groundwater		Piped Supply	
	Stored in Earthen Vessel	Stored in Metallic Vessel	Stored in Earthen Vessel	Stored in Metallic Vessel	Stored in Earthen Vessel	Stored in Metallic Vessel
pH	7.56±0.22	7.82±0.15	8.15±0.85	8.20±0.42	7.12±0.38	7.18±0.31
Turbidity (NTU)	12.45±3.77	14.72±2.48	24.86±6.67	26.18±5.89	8.44±0.70	9.08±0.88
DO (mg/lt)	7.33±2.01	7.08±1.22	6.09±1.03	5.55±321	4.56±0.82	5.27±0.80
Alkalinity (mg/lt)	123±29.89	135±13.93	534±143	587±122	146±81	140±38
Hardness (mg/lt)	220±76.70	180±12.42	470±99	530±83	155±42	85±19
Ca (mg/lt)	128±13.88	143±18.6	227±38	318±45	94±13	50±21
Mg (mg/lt)	83±9.29	90±12.21	108±46.22	188±1.06	61±21	27±13
Nitrate (µg/lt)	1.02±0.42	0.99±0.28	2.56±0.08	2.73±0.33	0.46±0.08	0.57±0.08
Silicate (mg/lt)	0.22±0.09	0.43±0.08	0.76±0.33	0.82±0.28	0.28±0.12	0.37±0.04
Phosphate (µg/lt)	0.77±0.06	0.90±0.33	1.06±0.61	1.83±0.23	0.29±.0.04	0.67±0.01

The regrowth and survival of indicators microorganisms on the surface of household containers consisted of all bacteria although the yield for the *Vibrio* and pathogenic *E.coli* 0157:H7 was lower than that of *Salmonella* and *Shigella* among pathogenic bacteria (Table 18.2). The lowest yield of indicators microorganisms was then noted for presumptive *E.coli*. Whereas the occurrence and survival of TC was always found during storage on the surface of household containers, other microorganisms occurred from time to time (Table 18.2).

The storage of water for hours or even days allows the possibility of faecal contamination of otherwise good quality drinking water inside the household. Children may, in particular, cause contamination when they put their faecally contaminated hands or utensils into the household water container. In the studied villages, water storage is a major necessity both for those who are connected to a non-continuous water supply system and those who depend on drinking water supply system and those who depend on drinking water sources located outside the household perimeter. Under hot climatic conditions even households with a continuous water supply often store water in traditional clay pitcher, because of the cooling effect caused by evaporation through the porous clay. It has frequently been observed that the microbiological quality of water in vessels in the home is lower than that at the source, suggesting that contamination is widespread during collection, transport, storage and drawing of water (Van Zijl 1966; Lindskog and Lindskog 1988). This contamination may lessen the health benefits of water source improvements.

Understanding of bacterial life cycles, including alternating shifts between planktonic (not attached) and surface–attached stages, is necessary for understanding of the persistence survival and growth of pathogenic microorganisms in the drinking water systems. Flushing a water system will only remove planktonic organisms and reorganizations of the free water phase by organisms from biofilm may occur within a few hours (Dukan *et al.*, 1996). Cell detachment from biofilms has been shown to happen as a result of physical disturbances (*i.e.* showing stress, abrasion and collision of particles) (Chang *et al.*, 1991, Chang and Rittman, 1988, Peyton and Characklis, 1992) or alternatively by the ability of the bacteria to activity detach (Marshall, 1991; Marshall and Blainey, 1991 and Szewzyk and Schink, 1987).

Conclusion

Majority of household in present research used earthen containers for water storage. The findings thus revealed that both types of containers supported the regrowth and survival of bacteria during the length of time. The storage of drinking water mainly resulted in the regrowth of TC. Nevertheless the persistence of other indicator and pathogenic bacteria was observed on the surface of household containers, which could further deteriorate the quality of the water in the containers. A suitable education and information programme on water safety and personal hygiene of household containers should be given in rural communities in order to minimize contamination of potable water stored in containers. In this regard, it is important to keep rural communities informed about the quality of drinking water and teach them to operate simple water treatment facilities before and during the storage.

Acknowledgements

The first author is thankful to Department of Science and Technology for providing funds to carry out this research work under SERC-OYS Scheme.

References

1. American Public Health Association (1998). *Standard Methods for the Examination of Water and Wastewater*, 20th Edition, American Public Health Association, Washington, D.C.

2. Armon, R., Strosvetzky, J., Arbel, T. and Green, M. (1997). Survival of *Legionella pneumophila* and *Salmonella typhimurium* in biofilm systems. *Water Sci. Technol.*, 35(11/12): 293–300.

3. Burke, V., Robinson, J., Gracey, M., Peterson, D. and Partridge, K. (1984). Isolation of *Aeromonas hydrophila* from a metropolitan water supply: Seasonal correlation with clinical isolates. *Appl. Environ Microbiol.*, 48: 361–366.

4. Chang, H.T. and Rittmann, B.E. (1988). Comparative study of biofilm shears loss on different adsorptive media. *J. Water Pollut. Control Fed.*, 60: 362–368.

5. Chang, H.T., Rittmann, B.E., Amar, D., Heim, R., Ehlinger, O. and Lesty, Y. (1991). Biofilm detachment mechanisms in a liquid-fluidized bed. *Biotechnol. Bioeng.*, 38: 499–506.

6. Doolittle, M.M., Cooney, J.J. and Caldwell, D.E. (1996). Tracing the interaction of bacteriophage with bacterial biofilms using fluorescent and chromogenic probes. *J. Ind. Microbiol.*, 16: 331–341.

7. Dukan, S., Levi, Y., Piriou, P., Guyon F. and Villon P. (1996). Dynamic modelling of bacterial growth in drinking–water networks. *Water Res.* 30(9): 1991–2002.

8. Engel, H.W.B., Berwald, L.G. and Havelaar, A.H. (1980). The occurrence of *Mycobacterium kansasii* in tap water. *Tubercle*, 61: 21–26.

9. Le Chevallier (1990). Coliform regrowth in drinking water: a review. *J. AWWA*, 82(11): 74.

10. Le Chevallier, M.W., Welch, N.J. and Smith, D.B. (1996). Full scale studies of factors related to coliform regrowth in drinking water. *Appl. Environ. Microbiol.*, 62(7): 2201–2211.

11. Lindskog, R.U. and Lindskog, P.A. (1988). Bacteriological contamination of water in rural areas: an intervention study from Malawi. *Journal of Tropical Medicine and Hygiene*, 91: 1–7.

12. Lund, V. and Ormerod, K. (1995). The influence of distribution processes on biofilm formation in water distribution systems. *Water Res.*, 29(4): 1013–1021.

13. Mackay, W.G., Gribbon, L.T., Barer, M.R. and Reid, D.C. (1998). Biofilms in drinking water systems: A possible reservoir for *Helicobacter pylori*. *Water Sci. Technol.*, 38(12): 181–185.

14. Marshall, K.C. (1988). Adhesion and growth of bacteria at surfaces in oligotrophic habitats. *Can. J. Microbiol.*, 34: 503–506.

15. Marshall, K.C. and Blainey, B.L. (1991). Role of bacterial adhesion in biofilm formation and biocorrosion. In: *Biofouling and Biocorrosion in Industrial Water Systems*, (Ed.) H. C. Flemming. Springer-Verlag, Berlin, 29–46 pp.

16. Momba, M.N.B. and Kaleni, P. (2002). Regrowth and survival of indicator microorganisms on the surfaces of household containers used for the storage of drinking water in rural communities of South Africa. *Water Research*, 36: 3023–3028.

17. Momba, M. N. B. and Mnqumevu, B. V. (2000). Detection of feacal coliform bacteria and heterotrophic plate count bacteria attached to household containers during storage of drinking water in rural communities. *WISA Annual Conference*, Sun City, Africa.

18. Peyton, B.M. and Characklis, W.G. (1992). Kinetics of biofilm detachment. *Water Sci. Technol.*, 26: 1995–1998.

19. Power, K.N. and Nagy (1999). Relationships between bacterial regrowth and some physical and chemical parameters within Sydney's drinking water. *Water Res.*, 33(3): 741–750.

20. Pruss A., Kay D., Fewtrell L. and Bartram J. (2002). Estimating the burden of disease from water, sanitation, and hygiene at a global level. *Environ. Health Perspect.*, 110(5): 537–42.

21. Quignon, F., Kiene, L., Levi, Y., Sardin, M. and Schwartzbrod, L. (1997). Virus behaviour within a distribution system. *Water Sci. Technol.*, 35: 311–318.

22. Schoenen, D. (1986). Neure unterchugen zur widerverkeiming des trinkwassere zentrablott for bakteriologieund. *Hygiene*, 83: 70–78.

23. Sharan, R., Chhibber, S., Attri, S. and Robert H. Reed, R.H. (2010). Inactivation and injury of *Escherichia coli* in a copper water storage vessel: effects of temperature and pH. *Antonie van Leeuwenhoek*, 97(1): 91–97.

24. Sudha, V., Singh, K., Prasad, S. and Venkatasubrama, P. (2009). Killing of enteric bacteria in drinking water by a copper device for use in the home: Laboratory evidence. *Transactions of the Royal Society of Tropical Medicine and Hygiene*, 103(8): 819–822.

25. Szewzyk, U. and Schink, B. (1987). Surface colonization by and life cycle of *Pelobacter acidigallici* studied in a continuous-flow microchamber. *J. Gen. Microbiol.*, 134: 183–190.

26. Szewzyk, U., Manz, W., Amann, R., Schleifer, K. H. and Stenström, T. A. (1994). Growth and *in situ* detection of a pathogenic *Escherichia coli* in biofilms of a heterotrophic water-bacterium by use of 16S-and 23S-rRNA-directed fluorescent oligonucleotide probes. *FEMS Microbiol. Ecol.*, 13: 169–176.

27. Szwezyk, U., Szewzyk, R., Manz, W. and Schleifer, K.H. (2000). Microbiological safety of drinking water. *Annu. Rev. Microbiol.*, 54: 81–127.

28. Van Zijl, W.J. (1966). Studies on diarrhoeal diseases in seven countries by the WHO diarrhoeal diseases advisory team. *Bulletin of the World Health Organisation* 35: 249–261.

29. Wadowsky, R.M., Yee, R.B., Mazmar, L., Wing, E.J. and Dowling, N.J. (1982). Hot water systems as a source of *Legionella pneumophila* in hospitals and non-hospital plumbing fixtures. *Appl. Environ. Microbiol.*, 43: 1104–1110.

30. World Health Organization (1976). Water needs in relation to health. *Water Int.*, (14): 7–8.

31. World Health Organization (2002). *World Health Report 2002: Reducing Risks, Promoting Healthy Life.* Geneva: WHO. (http: //www.who.int/whr/2002/, accessed November 2009).

Part IV

Miscellaneous Aspects

Biodiversity of Aquatic Reseources (2012)
Editors: *Mamta Rawat & Sumit Dookia*
Published by: DAYA PUBLISHING HOUSE, NEW DELHI

Pages **273–282**

Chapter 19

Study of Soil Enzyme, Carbon, Nitrogen Content and C/N Ratio during Vermicomposting of Eri and Tasar Culture Waste

Mamata Pandey[1]*, T.V. Rao[1] and S.K. Satapathy[2]

[1]*School of Life science, Sambalpur University,
Jyotivihar – 768 019, Orissa*
[2]*RTRS, Baripada, Mayurbhanj, Orissa*

ABSTRACT

Vermicomposting is a process of mineralization and mobilization of nutrients from organic waste using earthworm. This process release carbon and nitrogen and transferred the soil in to quality compost. Along with this the process of decomposition some crucial enzymes like dehydrogenase are released by the microbial population in the compost. The use of waste from sericulture *of Philosamia ricini* (Eri) and *Antheraea mylitta* (Tasar) in the vermicomposting has been studied and the quality of the compost evaluated. In this evaluation carbon, nitrogen content and dehydrogenase activity were assessed. During the 60 day of composting the soil was analyzed on various days *i.e.* 0, 15, 30, 45 and 60 and the

* Corresponding Author E-mail: pndy_mmt@rediffmail.com

parameter of decomposition studied. The carbon and carbon nitrogen ratio were decreased where as nitrogen and dehyodrogenase were increased during decomposition. In this view the waste from sericulture gives better quality compost.

Keywords: Eri and Tasar waste, Vermicomposting, Dehydrogenase, C/N ratio.

Introduction

Rearing silk is the second largest cottage industry in the world. Besides mulberry other species producing silk are Eri (*Philosamia ricini*), Tasar (*Antheraea mylitta*) and Muga (*Antheraea assama*). Non–mulberry silk culture is an ethnic practice in Orissa, especially in tribal dominated western Orissa, which provides livelihood to thousands of people. The sericulture generates large amount of waste containing organic matter like larval excreta, leaf litter, dead larvae, moth and cocoons (Das *et al.*, 2003). Presently the waste from this industry has limited utility like in poultry, fish feed. But it can also be used in the production of good quality compost. Waste material from mulberry culture has been used for decomposition by earthworms (vermicomposting) to produce better type of organic manure (Kale 1995; Gunathilagaraj and Ravignanam 1996; Kar 2004). Vermicomposting is a biotechnological process in which earthworms enhance the waste conversion and produce compost (Deka *et al.*, 2003). Along with the earthworms microbes augment the compost formation by releasing nutrients from the organic matter. While the process is dependent on the density and diversity of microbes (Suthar, 2007), it is also determined by the type and amount of organic compound in the waste matter. Mostly the contribution of microbes is correlated with the amount of exo-enzymes they release during decomposition. These enzymes participate in the cycling of mineral elements and regulate availability of nutrients to plants (Taylor *et al.*, 2002). According to several workers various organic waste materials can be used for vermicomposting (Elvira *et al.*, 1998, Garg and Kaushik, 2005, Garg *et al.*, 2006 and Aira *et al.*, 2006). In the present investigation we have assayed dehydrogenase an extra cellular enzyme catalyzing oxidation reduction reaction (Nannipieri *et al.*, 1979) along with C, N, and C/N ratio in the compost.

Material and Methods

Experimental Setup

Bulk surface soil (1-10cm) collected from the meadows in the Sambalpur University was mixed thoroughly for experiment. The soil was acidic pH–6.4, yellow in color (Literati) with clay (62 per cent), silt (21 per cent), and sand (17 per cent). A locally available species of epigeic earthworm *Perionyx excavatus* was used for vermicomposting. After collection, the worms were cultured in a tank with feed mix of cow dung and soil. Only adult earthworms were used for decomposition.

Sericulture Waste

Eggs of *Philosamia ricini* (kindly supplied by the Sericulture Department of Government of Orissa) were reared in temperature controlled Eri culture Laboratory of Life Sciences, Sambalpur University. After hatching the larvae were fed with fresh tender castor leaves thrice a day. The wastes produced (larval excreta, leaf litter, dead larvae, moth and cocoons) were collected and air dried. The rearing wastes of Tasar culture were collected from Basic Seed Multiplication and Training Center (BSMTC), Central Silk Board, Kerai, Sundargarh (Orissa). The Vermicomposting beds were prepared by using cement tanks (30 cm length x 45 cm height x 25 cm breadth). At the bottom of tank 8 kg of stone chips and 8 kg of coarse sand were layered. Mixture of cow dung and sericulture wastes (1:1) was layered on the top of it. The experiment consisted of 8 tanks as replicates.

Experimental Design

The experiment consisted of 2 treatments, which were replicated twice. The control contained only cow dung (CD) as substrate. This experiment is designed as follows:

Soil + CD + *P. excavatus*   –C (Control)

Soil + CD + Tasar waste + *P. excavatus*   –T (Tasar)

Soil + CD + Eri waste + *P. excavatus*   –E (Eri)

About 50 adult earthworms were introduced to each tank and 30-40 per cent soil moisture was maintained in them. The experiment was observed for 60 days with samples taken for analysis at intervals of 0, 15, 30, 45 and 60 days.

Sampling of Soil

The soil samples were taken randomly by using a cylindrical soil sampler of size 5 cm in height and area 3.14 cm^2. The samplers were inserted in tank and samples were collected randomly. Then the soil sample was packed in a plastic zipper bag and placed in deep-fridge at (0 degree temp) to slowdown the rate of microbial growth. Results of 3 independent experiments on different occasions were used for calculation.

Organic Carbon

Percentage of organic carbon was determined by Walkley and Black's titration method (Jackson 1973). 1gm air dried and sieved soil was taken in a 500ml conical flask. 10ml of 1N $K_2Cr_2O_7$ and 20ml of concentrated H_2SO_4 was added to it, mixed thoroughly and allowed to stand for 30 minutes to complete the reaction. The reaction mixture was diluted with 200ml water and 10ml H_3PO_4. 10ml of sodium fluoride solution and 2ml of diphenylamine indicator was added to it. The whole solution was titrated with 1N $(NH_4)_2 Fe (SO_4)_2 6H_2O$ to give a brilliant green colour. A blank without soil is run simultaneously.

Organic carbon in soil was calculated by the following formula.

$$\text{g per cent of organic carbon.} = V_1\text{-}V_2 \text{ X } 003 \text{ X } 100/\text{wt of soil}$$

where,

V_1: ml $FeSO_4$ solution for blank

V_2: ml $FeSO_4$ solution for soil sample.

Total Nitrogen (Autokjeltech Method)

Soil samples of 1g soil, 12ml concentrated H_2SO_4 and 2 catalyst tablets were taken in sample tubes and digested for about 30 min. It was then kept for cooling. Digested sample was diluted with 60 ml of distilled water. The diluted sample was taken in the distillation unit along with 60 ml of 40 per cent NaOH. And the vapors were passed into conical flask. To it 25 ml of 4 per cent boric acid, 2 drops of bromocresol green and 1 drop of methyl red was added and kept for 5 minutes. Then it was titrated with 1N HCl to get a light pink color.

From the above Nitrogen is estimated by the following formula

$$\frac{(T - B) \times N \times 14.007 \times 100}{wt - of\ sample}$$

where,

        T: Volume at sample consumed by HCl.

        B: Volume at blank consumed by HCl.

        N: Strength of HCl.

Dehydrogenase Assay

Dehydrogenase activity was assayed following the methods of Casida *et al.* (1964) by the reduction of 2, 3, 5–tri phenyl tetrazolium chloride (TTC). 2g of soil sample was treated with 0.1g of $CaCO_3$ followed by 2ml of 1 per cent (w/v) tri phenyl tetrazolium chloride (TTC) and incubated for 24 hr. at 37°C. The tri phenyl formazan (TPF) formed from the reaction mixture was extracted with methanol and measured at 485 nm in a spectrophotometer. Dehydrogenase activity was expressed in microgram formazan g soil/hr.

Result and Discussion

Vermicomposting is a process of decomposition, mineralization and separation of the nutrients from organic matter by earthworms (Logakanthi *et al.*, 2000). Any decomposable organic matter can be used in this process along with soil. Usually cow dung is the main organic matter for vermicomposting analyses. (Brar *et al.*, 1999, Gunathilagaraj and Ravignanam, 1996). Earlier reports have shown that the composting potential of any waste can be assessed by adding to the cow dung control. (Umamaheswari and Vijayalakshmi 2003, Suthar and Singh, 2008). In our investigation the waste matter of Eri and Tasar was mixed in a fixed proportion to the cow dung control for analysis of the nutrient content and soil dehydrogenase activity.

Carbon content of the compost during 60 days decomposition in control and Seri waste supplemented soil are depicted in Figure 19.1(a). Initially the soil carbon content was 75 per cent which declined to 57 per cent and 46 per cent in Eri and Tasar waste after 60 days showing a decline of 18 and 29 per cent respectively. Suthar (2007) reported similar decline in carbon content during decomposition of farmyard waste. Carbon is the most abundant of all nutrients because it is utilized by the organisms for their growth (Warman and Termeer, 1996, Zachariah and Chhonkar,

Figure 19.1: (a) Organic Carbon (b) Total Nitrogen (c) Dehydrogenase Activity of Air Dried Soil on Various Days of Vermicomposting with Amendments

(C) Control no amendment, (E) Eri waste supplement, (T) Tasar waste supplement.

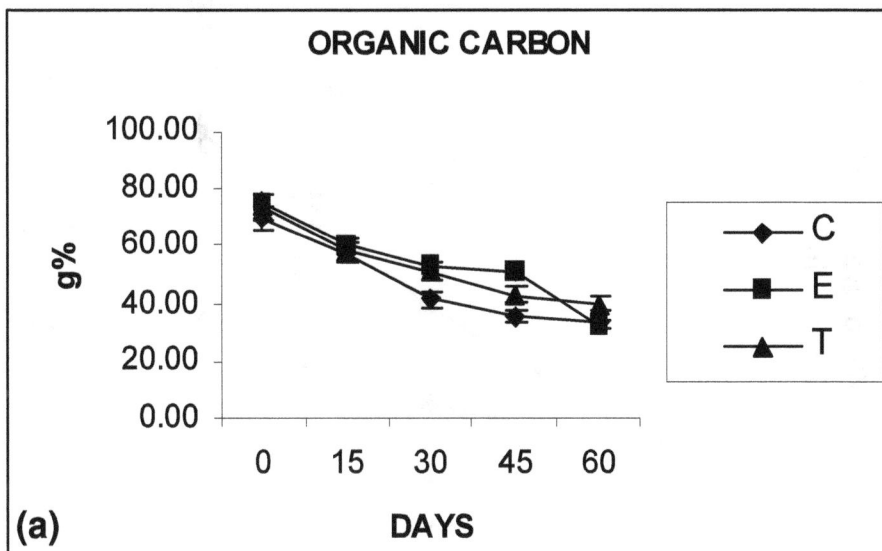

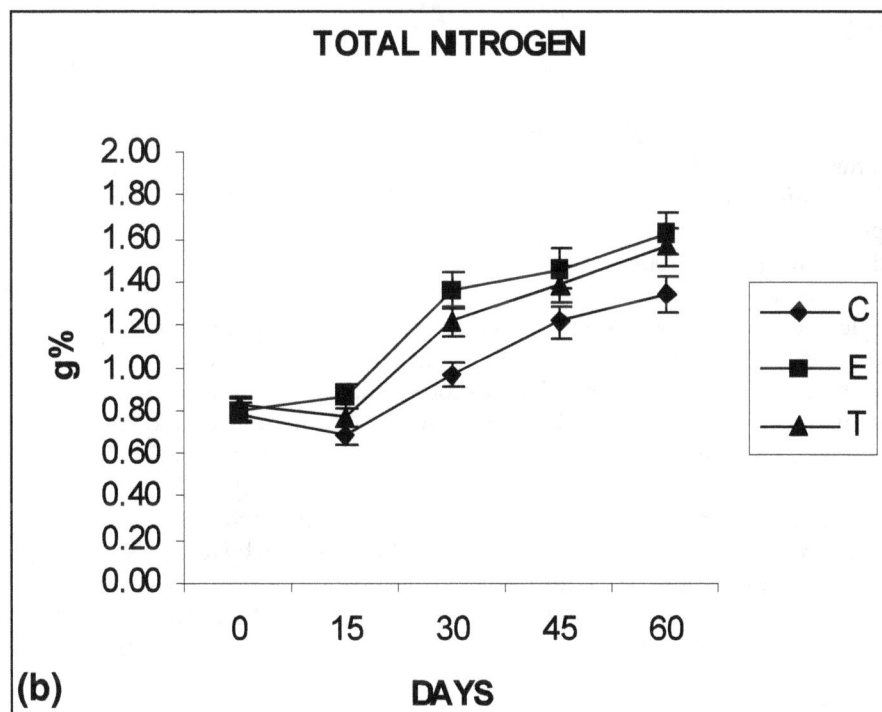

Contd...

Figure 19.1–Contd...

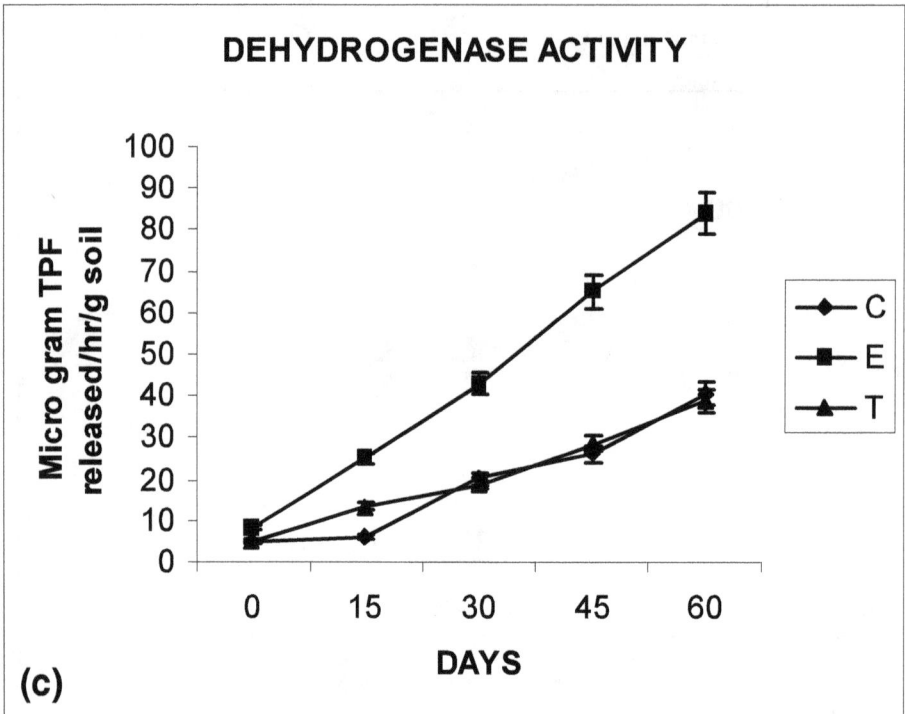

DEHYDROGENASE ACTIVITY

(c)

2004). It is released to the atmosphere in the form of carbon dioxide by the decomposer organisms. Hence the process of decomposition can be assessed by measuring the carbon content on various stages of composting. The other nutrient which is equally important in the compost is nitrogen (Brady 1990), several workers (Deka *et al.*, 2003, and Sangwan *et al.*, 2008) have reported that nitrogen content increased during decomposition of various organic wastes. Our results in Eri and Tasar waste supplements soil nitrogen content increased by about 98 per cent and 99 per cent respectively in 60 day (Figure 19.1b). While carbon is the major energy source, nitrogen determines the microbial population in a composting system. During decomposition amount of carbon is reduced while that of nitrogen is increased, and consequently decreases C/N ratio (Hamouda *et al.*, 1998). In the present investigation loss in C/N ratio (Table 19.1) during 60 days of decomposition can be attributed to the relative loss of carbon in the form of carbon dioxide associated with the accumulation of nitrogen by various sources. Ideal compost is characterized by the C/N ratio of 25:1 (Brady 1990) our results of Eri and Tasar waste decomposition can be compared to the earlier reports of Logakanthi *et al.*, 2000, Garg and Kaushik 2005. In our investigation we have observed the C/N ratio with Eri supplements 20:1 and in case of Tasar it is 26:1 indicating the quality of the compost.

Dehydrogenase has a greater roll to regulate nutrient in soil. It is directly influenced by the presence of microbes (Garcia *et al.*, 1997). In our investigation dehydrogenase activity increased steadily during 60 day of decomposition (Figure

Figure 19.2: Correlation Plots Between Carbon Content and Dehydrogenase Activity of (a) Control (b) Eri (c) Tasar of Air Dried Soil on Various Days of Vermicomposting

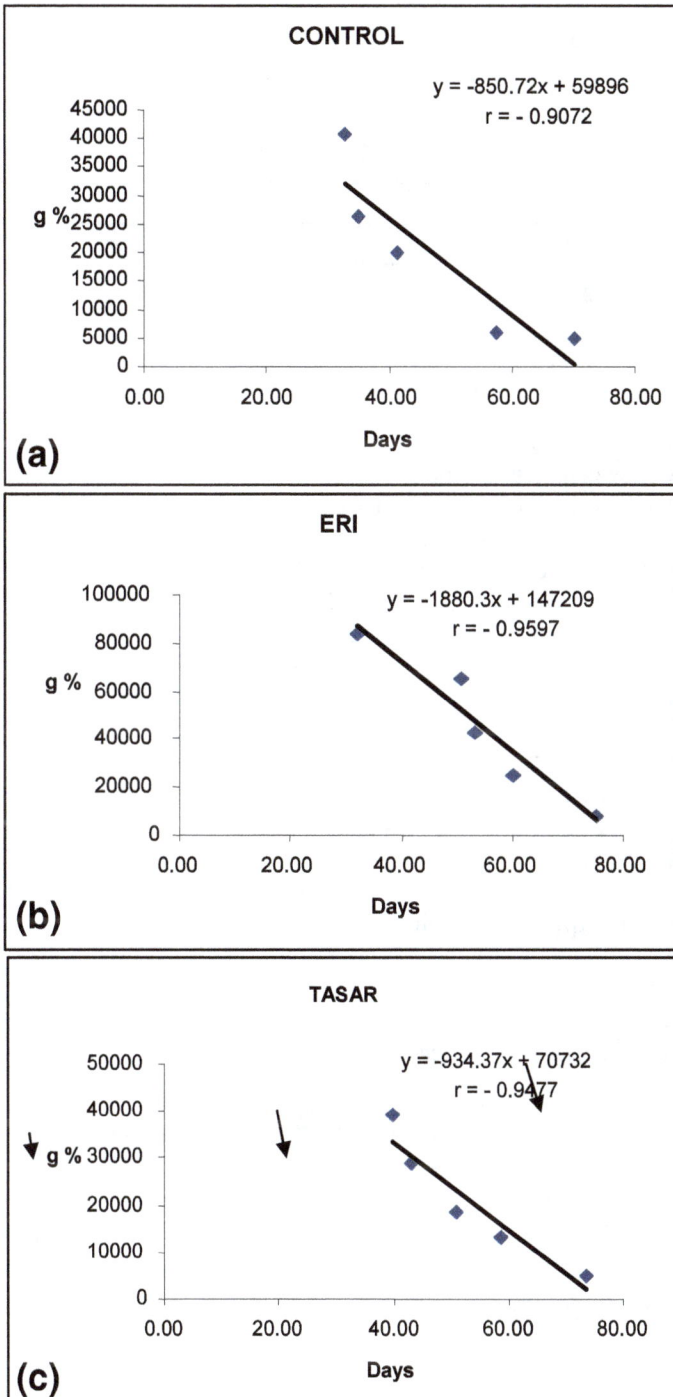

19.1c) by about 8 to 10 times in control, Eri and Tasar supplements. However in case of Eri supplemented compost the increase is maximal of about 10.5 times. Eri waste contains a large proportion of leaf residues which might be the reason for enhanced microbial population and consequently the dehydrogenase activity. Similar explanation was given by Sriramachandrasekharan *et al.* (1997) working with green manure.

Table 19.1: Carbon and Nitrogen Ratio of Soil on Different Days of Composting

Test	Days				
	0	15	30	45	60
Control	88.75	84.55	42.87	28.94	24.56
Eri	93.70	69.48	39.36	34.88	20.22
Tasar	89.61	76.42	42.01	31.00	25.53

Carbon content in a decomposing matter represents the source of building blocks to synthesize new cells. On the other hand the new cells will release the carbon in form of carbon dioxide decreasing the total carbon content whereas dehydrogenase activity is considered to be the measure of increasing microbial population in the compost (Beyer *et al.*, 1993). In our results this proposition was verified by the correlation analysis between carbon and dehydrogenase activity on different days of composting (Figures 19.2a, b, c). The coefficients ('r') between them were,–0.9072,–0.9597,–0.9477 in control, Eri and Tasar respectively showing significant negative correlation between them.

Conclusion

Silk worm culture being a supportive vocation for small farmers of India, it can be made more productive than the present condition by utilizing waste generated out of them. From above results it is evident that by vermicomposting of Eri and Tasar waste produces better manure than other similar organic waste materials can be obtained and thus supplement their meager income.

References

1. Aira, M., Monroy, F. and Dominguez, J. (2006). C to N ratio strongly affects population structure of *Eisenia foetida* in vermicomposting system. *Eur. J. of Soil Biol.*, 42: 127–131.

2. Brar, B.S., Dhillon, N.S. and Vig, A.C. (1999). Integrated use of farm yard manure, biogas slurry and inorganic phosphate in P nutrition of wheat crop. *J. Ind. Soc. Soil Sci.*, 47: 264–268.

3. Beyer, L., Wachendorf, C., Elsner, D.C. and Knabe, R. (1993). Suitability of dehydrogenase activity assay as index of soil biology activity. *Biol. Fertil. Soil.*, 16: 52–56.

4. Brady, N.C. (1990). *The Nature and Properties of Soils*, 10th Edition, MacMillan, New York, 34–58 pp.

5. Casida, L.E.Jr., Klein, D.A. and Santoro, T. (1964). Soil dehydrogenase activity. *Soil Sci.*, 98(6): 371–376.

6. Das, P.K., Bhogesha, K., Rajanna, L. and Dandin, S.B. (2003). Vermiculture as a means of Seri cultural waste management technology an eco-friendly approach. In: *Proc. Nat. Symp. Bioresources, Biotech. and Bioenterprise*, p. 398–402.

7. Deka, P.K., Paul, S.K. and Borah, A. (2003). Vermitechonology for economic use and control of water hyacinth. *Enviromedia*, 22(3): 385–387.

8. Elvira, C.L., Sampedro, L., Benitez, E. and Nogales, R. (1998). Vermicomposting of sludge from paper mill and dairy industry with *Eisenia foetida* a pilot scale study. *Bioresource Tech.*, 63: 205–211.

9. Garg, V.K. and Kaushik, P. (2005). Vermistabilization of textile mill sludge spiked with poultry droppings by an epigeic earthworm *Eisenia foetida*. *Bioresource Tech.*, 96: 1063–1071.

10. Garg, V.K., Kaushik, P. and Dilbaghi, N. (2006). Vermiconservation of waste water sludge from textile mill mixed with anaerobic ally digested biogas plant slurry employing *Eisenia foetida*. *Eco. and Env. Safety*, 65(3): 412–419.

11. Garcia, C., Hernandez, T. and Costa, F. (1997). Potential use of dehydrogenase activity as an index of microbial activity in degraded soils. *Communications in Soil Science and Plant Analysis*, 28: 123–134.

12. Gunathilagaraj, K. and Ravignanam, T. (1996). Vermicomposting of sericultural waste. *Madras Agric. J.*, 83(7): 455–457.

13. Hamouda, M.F., Abu Qdais, H.A. and Newham, J. (1998). Evaluation of municipal solid waste composting kinetics. *Res. Cons. and Recycling*, 23: 209–223.

14. Jackson, M.L. (1973). *Soil Chemical Analysis*. Prentice Hall of India Pvt Ltd., New Delhi, 498 pp.

15. Kar, J., Guru, B.C. and Nayak, B.K. (2004). Reproductive and commercial productivity of the cultivated Eri silk moth. *Samia ricini*, Donovan (1798) in Orissa. *Bull. Ind. Acd. Seri.*, 8(1): 87–92.

16. Kale, R.D. (1995). Vermicomposting has a bright scope. *Ind. Silk*, 34: 6–9.

17. Logakanthi, S., Rajesh, J. and Vijayalaxmi, G.S. (2000). Biodung-vermicomposting: A novel method for green waste recycling. *Ind. J. Env. Prot.*, 20(11): 850–854.

18. Nannipieri, P., Pedrizzini, P.F., Arcara, P.G., Pidvanellis, C. (1979i). Changes in the exoenzyme activity and biomass during soil microbial growth. *Soil Sci.*127: 26–34.

19. Sangwan, P., Kaushik, C.P. and Garg, V.K. (2008). Feasibility of utilization of horse dung spiked filter cake in vermicomposter using exotic earthworm *Eisenia foetida*. *Bioresource Tech.*, 99: 2442–2448.

20. Sriramachandrasekharan, M.V., Ramanathan, G. and Ravichandran, M. (1997). Effect of different organic manures on enzyme activities in a flooded rice soil. *Oryza*, 34: 39–42.

21. Suthar, S. (2007). Vermicomposting potential of *Perionyx sansibaricus* (Perrier) in different waste materials. *Bioresource Tech.*, 98: 1231–1237.

22. Suthar, S. and Singh, S. (2008). Vermicomposting of domestic waste by using two epigeic (*Perionyx excavatus* and *Perionyx sansibaricus*). *Int. J. Environ. Sci. Tech.*, 5(1): 99–106.

23. Taylor, J.P., Wilson, B., Mills, M.S. and Burns, R.G. (2002). Comparison of microbial number and enzyme activity in surface soil using various techniques. *Soil Biol. and Biochemistry*, 34: 387–401.

24. Umamaheswari, S. and Vijayalakshmi, G.S. (2003). Vermicomposting of paper mill sludge using an African earthworm species *Eudrilus eugeniae* (Kinberg) with a note on its physico-chemical features. *Enviromedia*, 22(3): 339–341.

25. Warman, P.R. and Termeer, W.C. (1996). Composting and evolution of race track manure, grass clippings and sewage sludge. *Bioresource Tech.*, 55: 95–101.

26. Zachariah, A.S. and Chhonkar, P.K. (2004). Biochemical properties of compost as influenced by earthworms and feed martial. *J. Ind. Soc. Soil Sci.*, 52(2): 155–159.

Biodiversity of Aquatic Reseources (2012)
Editors: *Mamta Rawat & Sumit Dookia*
Published by: DAYA PUBLISHING HOUSE, NEW DELHI

Pages **283-288**

Chapter 20

Analysis of Sugar Industry Molasses, Uses and its Economical Importance

Poonam Nomulwar and P.M. Patil

*Department of Zoology and Fishery Science,
N.E.S. Science College, Nanded, Maharashtra*

ABSTRACT

Main by product of sugar industry is molasses. This molasses is utilized in distillery for the production of the alcohol. The composition of molasses based distillery industry effluent 'spent wash' is totally depending on the quality of molasses. The wastewater from distillery is a strong waste called as spentwash. The analysis on molasses and spentwash has been conducted. The color and odor was dark brown and jaggery in smell. The pH is very acidic. The COD and BOD values are also very high for molasses and spentwash. The study has been conducted on economical importance of the various by–products of the molasses.

Keywords: Molasses, By-products, Economical importance, Spentwash.

Introduction

India is the largest manufacturer of cane sugar in the world. Molasses is a major byproduct of the sugar industry, which is again a raw material for the manufacture of industrial alcohol, ethanol. Molasses is the by-product of sugar refining that contains all the nutrients from the raw sugar cane plant. It is richest in sources of sugars, carbon, B-vitamins, iron, calcium, sodium and potassium.

Molasses which is also known as black strap molasses or treacle is a dark brown viscous liquid obtained as a by-product in processing Cane Sugar. It contains nearly 45 per cent uncrystallized, fermentable Sugar and some Sucrose. It is a valued by-product of foreign liquor, as a table syrup and food flavourant, Ethyl Alcohol, Acetic Acid, Citric Acid, Glycerin and Yeast. It is also used as food for farm animals and in the manufacture of several processed tobaccos.

Molasses or treacle is a thick syrup by-product from the processing of the sugarcane or sugar beet into sugar. (In some parts of the US, molasses also refers to sorghum syrup.) The word molasses comes from the Portuguese word "*melao*", which comes from mel, the Portuguese word for "honey". The quality of molasses depends on the maturity of the sugar cane or beet, the amount of sugar extracted, and the method of extraction. Because of its unusual properties, molasses has several uses beyond that of a straightforward food additive. It can be used as a chelating agent to remove rust, as the base material for fermentation into rum, as the carbon source for in situ remediation of chlorinated hydrocarbons, and it can even is mixed with sand to make mortar for brickwork.

The components of molasses include

1. Major components (water, sugar,non-sugars)
2. Minor components (Trace elements, vitamins, growth substance)

Materials and Methods

The effluent was collected from sugar and distillery industry, a mill situated in Basamat Nagar, Hingoli district Maharashtra state, India. The collected samples were brought to the laboratory and stored at 4°C. Effluents were analyzed by standard methods for the examinations of water and wastewater (APHA, AWWA, 1995).

Results and Discussion

Characteristics of Cane Sugar Molasses

Characteristics	Average Value
Colour	Dark Brown
Odour	Jaggery smell
pH	3.2 to 3.5
COD	1,24,000–1,24,500
BOD(5-days, 20°C,mg/l)	30,000–30,100
Total Solids	1,49,500
Chloride	1,500–1,600
Sulphate	1070–1250
Phosphate	175–185
Nitrate	750
Magnesium	1010

In India, sugar is refined using sulphur dioxide; hence, the molasses and ethanol produced have high sulphur content. The heavy metal content of Indian molasses is also higher compared to that of beetroot-based molasses from other countries. According to Sahu(1999), the concentration of iron and zinc in Indian molasses is 410 micrograms per gram ($\mu g/g$) and 477 $\mu g/g$ respectively, which is 10 to 30 times higher than that of Cuban or Brazilian molasses.

Thousands of tones of molasses are produced annually as waste by–product. It is acidic in nature having good reserve of calcium and carbohydrates. Dhar and Mukherjee (1936) have reported dramatic and complete reclamation of alkali soils within six months with an application of 60 to 90 tones of molasses per hector. The native acidity as well as acid produced as a consequence of decomposition of carbohydrates fraction of molasses neutralises the soil alkali. Calcium contained in it gets into soil solution and replaces exchangeable sodium from the exchange complex and improves the physical condition of soil. The gums and polysaccharides produced during microbial decomposition of molasses improve the physical properties of soil. The nutrient reserves of molasses add to soil fertility. As such molasses poses serious difficulty in handling and transportation which needs to be sorted out by mixing it with other amendments(Somani and Totawat, 1993).

Molasses produced from the small scale Khandsari industry are of a different quality with low chemical but high sugar content going up to 65 per cent. The subsequent process of treatment of the effluent is influenced by the presence of organic and inorganic inhibitors in the molasses. These are contributed by the sugar cane as well as the process of manufacture of sugar.

The composition of cane molasses based distillery industry effluent 'spent wash' is totally depending on the quality of molasses. So it is conceivable to use effluent particularly the spent wash directly for irrigation purposes because of its high nutrient content. However, it is highly acidic for direct disposal and apart from damaging the crop it could be harmful to soil (Bhasin, 1995). According to the Mali (2002), the characteristic of spent wash depend on the quality of molasses in the production of ethanol. The pH of spentwash is 4.2 which is very acidic in nature. The total solids, chloride, potassium 2,40,000 mg/l, 14,500 mg/l, 12,500mg/l respectively. The COD and BOD values very high for spent wash. The colour of spent wash was dark brown in color and odour was found to be jaggery smell.

Parthsarthy *et al.* (1967), reported the total solids, chlorides of the cane sugar based distillery industry effluent spent wash was 90,000–1,10,000 mg/l, 10,150–11,300 mg/l respectively. Trivedy (1995), noted the COD of the spentwash as 60,000 to 1,50,000 mg/l and sulphate as 4,200 mg/l. Parthsarthy *et al.* (1997), studied the total nitrogen from the spent wash 1200–1500 mg/l.

Any study of the treatment processes for the effluent must first take into account the quality of the molasses. Composition of molasses varies widely depending upon the processes employed in sugar manufacture. The sugar industry in India utilizes chemical treatment for efficient extraction. Molasses are often stored for long period in open pits and may even get diluted with the rain water. Beet–root molasses are entirely different from sugarcane molasses

Economical Importance

Sugar mills produce around 8 million tonnes (mt) of molasses, which is projected to increase to 10.5 mt over the next ten years. About 14 per cent of the molasses produced today is exported, without any value addition. On the other hand, the country imports molasses-derived chemicals.

Ethyl Alcohol

It is probably the best known fermentation product produced from sugar by yeast using low cost waste carbohydrate source, like molasses. Before the fermentation of a scientific understanding of fermentation of fruits to alcohol was well established for centuries.

Oxalic Acid

Used in textile and detergent, plastic, pharmaceutical and film industries, oxalic acid is a "sunset" chemical in the developed countries. In India, oxalic acid is produced from sugar, by a dozen-odd companies, most of them Mumbai-based. T he total installed capacity is 30,000 tons and the production is of the order of 22,000 tons, of which about 9,000 tons are exported. To produce one ton of oxalic acid, 2.5 tons of molasses would be required. The global demand is estimated at 250,000 tons, which is projected to grow at three per cent every year.

Citric Acid

Unlike oxalic acid, the technology for producing citric acid from molasses is established and available. This chemical is mainly used as a preservative in food processing, pharmaceutical and cosmetic industries. It is also used to manufacture sodium citrate, used in the production of detergents. Seven tons of molasses would be required to produce one ton of citric acid. The global demand for this chemical is estimated at 703,200 tons in 2001, which is projected to increase to 1,040,900 tons by 2011. Correspondingly, the domestic demand is set to rise from 12,000 tons now to 38,475 tons by 2011. Here again, a huge export opportunity exists, as India, being a major producer of molasses, can produce this chemical at competitive rates.

Lactic Acid

Described as "the chemical of the future" because of its use in the manufacture of polylactic acid, it is reckoned that polylactic acid could substantially replace plastic bags, since it is bio-degradable. Five tonnes of molasses would be needed to produce one ton of lactic acid. Molasses-based products such as lysiene, a nutrient that the human body needs but does not produce, glutamic acid, itaconic acid, gloconic acid, ephedrine hydrochloride and yeast.

Glutamic Acid

It is used as flavour enhances in the form of monosodium glutamate and is produced by culture of caryne bacterium glutamicum and *Bervitbacterium falcum*.

Glutanic Acid

It can be produced by chemical or electrolytic oxidation of a sugar solution or by fermentation of glucose with the help of special strains of *Sperquillus*.

Intaconic Acid

It is an unsaturated dibaic acid. It can be commercially produced by the microbial fermentation using special strains of *Spergillus terrius*. This is used as a raw material for the manufacture of alkyed, resin, variety of thermoplastics, lubricants and additives, soil conditioners and ion–exchange resins.

Alcohol

Molasses are fermented in distilleries to produce alcohol commercially. India has been producing about 1.7 billion litre of liquor utilizing 75-80 per cent molasses produced in the country. Though the free production of liquor using molasses is restricted for social reasons, its application in the preparation of alcohol based chemicals is also economically viable.

India has the largest chemical industry in the world using sugar cane molasses to produce Acetaldehyde, Acetic Acid, Poly Vinyl Chloride (PVC) and Mono Ethylene Glycol (MEG) is being produced by molasses only. Elsewhere these are produced using Petrochemicals. Ethanol is planned to be a substitute for fossil fuels.

Brown Sugar

It is used in home and food industry to develop the rich molasses type flavor in cookies, candies and similar products. It consists of sugar crystals coated in a molasses syrup with natural flavor and color.

Eco-Molasses

Eco-Molasses is the by-product of sugar refining that contains all the nutrients from the raw sugar cane plant. It is richest in sources of sugars, carbon, B-vitamins, iron, calcium, sodium and potassium. With continued applications, Eco-Molasses encourages a soil environment that helps reduce thatch. Eco-Molasses may be added to foliar sprays to enhance adhesion to leaf surfaces.

Major Benefits of Applying Eco-Molasses

1. Improves Fertilizer Efficacy
2. Increase Leaf Color
3. Increase Microbial Activity
4. Increase Plant Health

The gist of the study is that the chemicals derived from molasses could be produced cost effectively and be profitably exported.

References

1. APHA and AWWA (1995). *Standard Methods for the Examination of Water and Wastewater*, 20th Edition, Washington USA.

2. Bhasin, S.D. (1995). Distillery industry in India: Indian effluent scenario. *Indian National Academy of Engg.*, pp. 303–320.

3. Dhar and Mukherjee (1936). *J. Indian Chem. Soc.*, 12: 436.

4. Mali, D.S. (2002). Aerobic composting of spentwash. *J. Ecotox. Environ. Monit.*, 12(2): 101–104.

5. Parthasarthy, T.D. Thangadurai and N. Basheer Ahmed (1967). Anaerobic stabilization of distillery waste from cane sugar industry. *Ind. J. Env. Hlth.*, 9: 110–117.

6. Sahu, K.S. (1999). *Down to Earth* Vol. 7 Issue: 1999 0131: Jan 1999.

7. Somani, L.L. and Totawat, K.L. (1993). *Management of Salt Affected Soils and Wastes.* Agrotech Publication, Udaipur, p. 37–39.

8. Trivedy, R.K. (1995). Encyclopedia of environmental pollution and control *Industry and Water Pollution*, p. 57–59.

Index

H

www.ingramcontent.com/pod-product-compliance
Lightning Source LLC
Chambersburg PA
CBHW060247230326
41458CB00094B/1488